Fundamentals of C# Programming

for Information Systems

George C. Philip
The University of Wisconsin Oshkosh

Contributions by Dr. Jakob Iversen
The University of Wisconsin Oshkosh

Copyright © 2019 Prospect Press, Inc. All rights reserved.

No part of this publication may be reproduced, stored in a retrieval system or transmitted in any form or by any means, electronic, mechanical, photocopying, recording, scanning or otherwise, except as permitted under Sections 107 or 108 of the 1976 United States Copyright Act, without either the prior written permission of the Publisher, or authorization through payment of the appropriate per-copy fee to the Copyright Clearance Center, Inc. 222 Rosewood Drive, Danvers, MA 01923, website www.copyright.com. Requests to the Publisher for permission should be addressed to the Permissions Department, Prospect Press, 47 Prospect Parkway, Burlington, VT 05401 or e-mail to Beth.Golub@ProspectPressVT.com.

Founded in 2014, Prospect Press serves the academic discipline of Information Systems by publishing innovative textbooks across the curriculum including introductory, emerging, and upper level courses. Prospect Press offers reasonable prices by selling directly to students. Prospect Press provides tight relationships between authors, publisher, and adopters that many larger publishers are unable to offer. Based in Burlington, Vermont, Prospect Press distributes titles worldwide. We welcome new authors to send proposals or inquiries to Beth.golub@prospectpressvt.com.

Editor: Beth Lang Golub
Production Management: Rachel Paul
Cover Illustration and Design: Annie Clark

Ebook ISBN: **978-1-943153-55-8**
Available from Redshelf.com and VitalSource.com

Black & White Printed Paperback ISBN: **978-1-943153-56-5**
Available from Redshelf.com

Full-Color Printed Paperback ISBN: **978-1-943153-57-2**
Available from Redshelf.com

For more information, visit https://prospectpressvt.com/titles/c-sharp-programming/

To Mary, my wife of more than 40 years,
for her invaluable love, patience and support.

To my children, Samantha and Joshua,
for the fulfillment they have brought to my life, and for their love.

To my nephews and nieces,
Finney, Vava, Philipkuttan, Julie, Georgie, Wesley, Jerrymon,
Kochumon, Kochumol, Faith, Titty, Tony, Sarah, Sharon and Molsey,
who give me faith in the next generation.

Brief Table of Contents

Chapter 1: Introduction to Visual Studio and Programming 1

Chapter 2: Data Types 53

Chapter 3: Decision Structures and Validation 82

Chapter 4: Iteration Structure: Loops 118

Chapter 5: Methods 140

Chapter 6: Graphical User Interface Controls 174

Chapter 7: Arrays 205

Chapter 8: Sequential Files and Arrays 226

Chapter 9: Collections 276

Chapter 10: Graphical User Interface: Additional Controls 298

Chapter 11: Multi-form Applications and Menus 312

Chapter 12: Databases 335

Chapter 13: Object-Oriented Programming 414

Chapter 14: Inheritance 450

Chapter 15: Introduction of Web Applications Development 491

Chapter 16: Web Development: Multipage and Database Applications 523

Appendix A: Binary Files 559

Appendix B: Creating Databases 562

Appendix C: Answers to Review Questions 565

Appendix D: HTML Elements 584

Index 586

Expanded Table of Contents

Chapter 1: Introduction to Visual Studio and Programming ... 1
   1.1 Introduction to Programming ... 1
   1.2 Introduction to Visual Studio ... 5
   Tutorial 1: Creating a Form: Ice Cream Cost ... 9
   1.3 Visual Studio Environment ... 11
   1.4 Adding Controls to a Form ... 16
   1.5 The Code Editor Window ... 20
   1.6 Introductory Programming Concepts ... 25
   1.7 Data Types ... 26
   1.8 Working with Variables ... 28
   Tutorial 2: Developing the Code ... 32
   1.9 Doing Calculations ... 35
   1.10 Formatting and Displaying Output in a Label ... 36
   1.11 Additional Concepts: Constants, Scope of Variables and ListBox ... 41
   Tutorial 3: Ice Cream Cost with Sales Tax ... 41
   1.12 Exception Handling: Catching Errors ... 49
   Exercises ... 51

Chapter 2: Data Types ... 53
   2.1 Working with Different Numeric Data Types ... 53
   Tutorial 1: Ice Cream Total with Data Types ... 55
   2.2 Casting ... 56
   2.3 Working with Strings and Characters ... 64
   Tutorial 2: String Operations ... 65
   2.4 Working with Dates and Times ... 70
   Tutorial 3: DateTime Operations ... 70
   Exercises ... 80

Chapter 3: Decision Structures and Validation ... 82
   3.1 Flowchart for Decision Structure ... 82
   3.2 if else Statement ... 83
   Tutorial 1: Ice Cream Total with Discount ... 84
   3.3 if else if Statement ... 92
   3.4 Nested if Statements ... 94
   Tutorial 2: Compute Grade ... 95

3.5 The switch Statement 98
Tutorial 3: switch Statement 100
3.6 Data Validation 101
Tutorial 4: Late Validation 103
Tutorial 5: Immediate Validation Using Validating Event 106
Exercises 113
Rentals Assignment 1 115
Help Desk Assignment 1 116

Chapter 4: Iteration Structure: Loops 118

4.1 The while Loop 119
4.2 Console Applications 120
Tutorial 1: while Loop to Verify Password 121
4.3 Incrementing Variables in a Loop 122
Tutorial 2: while Loop: Financial Planning Application 125
4.4 The do-while Loop 130
4.5 The for Loop 131
Tutorial 3: for Loop: Financial Planning Application 134
4.6 The break and continue Statements 137
Exercises 139

Chapter 5: Methods 140

5.1 Introduction to Methods 140
Tutorial 1: Methods: Compute Heart Rate 143
5.2 Passing Values to Methods 144
5.3 Passing Arguments by Value 149
5.4 Methods That Return a Value 152
5.5 Top-Down Design 154
5.6 Passing Arguments by Reference 162
5.7 Enumerations: Limiting Parameter Values 168
Tutorial 2: Enumeration: Enumeration Type Parameter 170
Exercises 172

Chapter 6: Graphical User Interface Controls 174

6.1 Working with ScrollBars 175
Tutorial 1: Theater Tickets Sale Application 175
6.2 Working with CheckBoxes 178
6.3 Working with RadioButtons 180
6.4 Validating Input 182

6.5 Getting User Input from Controls 183

6.6 Working with ComboBoxes 186

6.7 Working with ListBoxes 196

6.8 Windows Presentation Foundation (WPF) 201

Exercises 203

Chapter 7: Arrays 205

7.1 Introduction to One-Dimensional Arrays 205

Tutorial 1: Working with 1-D Arrays: Test Scores Application 208

7.2 Assigning a Value to an Element of an Array 211

7.3 Accessing Elements of an Array Using the Index 212

7.4 Accessing Elements of an Array Using the foreach Loop 215

7.5 Copying an Array 217

7.6 Looking Up Values in an Array 221

Tutorial 2: Lookup Units Sold Using 1-D Arrays 222

Exercises 224

Chapter 8: Sequential Files and Arrays 226

8.1 Introduction to Text Files 226

Tutorial 1: Reading Text Files 229

8.2 Splitting a Row: Split Method 231

8.3 Reading Data from Files into Arrays 234

8.4 Additional Methods of Arrays 241

Tutorial 2: Looking Up Phone Numbers Using Arrays 242

Tutorial 3: Using a ComboBox to Select the Search Name 246

8.5 Writing to Text Files 248

Tutorial 4: Writing Scores to Text File 248

8.6 SaveFileDialog Control 255

8.7 Passing Arrays to Methods 259

Tutorial 5: Passing Arrays to Compute Average 259

8.8 Two-Dimensional (2-D) Arrays 263

Tutorial 6: 2-D Array to Find the Sales for a Product 267

Exercises 271

Rentals Assignment 2 272

Help Desk Assignment 2 274

Chapter 9: Collections 276

9.1 Introduction to Collections 276

9.2 List<T> Collection 278

Tutorial 1: Looking Up Sales Data Using a List...279
9.3 Dictionary<TKey, TValue> Collection 285
Tutorial 2: Dictionary with Product# as Key and Price as Value 287
Tutorial 3: Dictionary with Product# as Key and Sales Array as Value..............................293
Exercises 297

Chapter 10: Graphical User Interface: Additional Controls............................298

10.1 ListView: Features 298
10.2 Adding Items to ListView...299
Tutorial 1: Selecting a Show from a ListView 302
10.3 Accessing Data from a ListView 302
10.4 Forms with Tab Pages...308
Tutorial 2: Multipage Form with TabControl 308
Exercises 310

Chapter 11: Multi-form Applications and Menus 312

11.1 Introduction...312
11.2 Accessing an Existing Form from Another..313
Tutorial 1: Multi-form Financial Planning Application...315
11.3 Sharing a Method between Multiple Forms...318
Tutorial 2: Sharing the LoadDictionary Method between Forms 319
11.4 Closing a Parent Form 325
11.5 Menus...328
11.6 ToolStrips...332
Exercises 334

Chapter 12: Databases..335

12.1 Introduction...336
12.2 Creating Database Objects..336
Tutorial 1: Display Employee Records Using the Wizard...337
12.3 Displaying Data in a DataGridView 340
12.4 Accessing Data Items from a DataGridView...346
12.5 Displaying Data in Details View 349
12.6 Finding the Record for a Selected Key 351
Tutorial 2: Find Employee Records Using the Wizard..352
12.7 Selecting a Group of Records Using the Binding Source 356
12.8 Selecting Records Using the Table Adapter 362
Tutorial 3: Parameter Query to Select Records..363
12.9 Untyped Datasets: Displaying Records..365

Tutorial 4: Display Records Using Untyped Datasets 366
12.10 Untyped Datasets: Selecting Records 375
Tutorial 5: Select Records Using Untyped Datasets 377
12.11 Untyped Datasets: Add/Edit/Delete Records 384
Tutorial 6: Add/Edit/Delete Using Untyped Datasets 385
12.12 Command Object and DataReader 394
Tutorial 7: SqlCommand and SqlDataReader to Work with Tables 396
Exercises 405
Rentals Assignment 3 407
Help Desk Assignment 3 407

Chapter 13: Object-Oriented Programming 414

13.1 Introduction to Objects and Classes 415
13.2 Classes versus Objects 415
13.3 Information Hiding (Encapsulation) 417
Tutorial 1: Create an Employee Class 419
13.4 Properties 423
Tutorial 2: User Interface and Properties 426
13.5 Calling Methods (Sending Messages to Objects) 434
Tutorial 3: Calling Methods and Passing Data between Forms 435
Exercises 448

Chapter 14: Inheritance 450

14.1 Introduction to Inheritance 451
14.2 Implementing Inheritance 452
Tutorial 1: Creating Subclasses 454
14.3 Using Subclasses and Super Classes 458
14.4 Overriding Methods 459
Tutorial 2: Implementing the GrossPay Method 462
14.5 Polymorphism 467
Tutorial 3: Dynamic Dispatch and the ToString Method 468
14.6 Abstract Classes and Interfaces 472
Tutorial 4: Interfaces 473
14.7 Using Interfaces to Sort Collections 480
Tutorial 5: Sorting Employees 483
14.8 Access Modifiers in C# 486
Exercises 488

Chapter 15: Introduction to Web Applications Development 491
15.1 Introduction to Web Applications 491
15.2 Creating a Single-Page ASP.NET Website 496
Tutorial 1: Developing a Simple Web Page: Ice Cream Cost 496
15.3 Introduction to HTML 502
15.4 Working with Web Server Controls 506
15.5 Validating Data Using Validation Controls 517
Exercises 522

Chapter 16: Web Development: Multipage and Database Applications 523
16.1 Accessing Databases from Web Forms 523
Tutorial 1: Accessing a Database: Theater Tickets Application 524
16.2 Preserving Data during Postbacks Using View State 538
16.3: Multipage Websites and Session State 543
16.4 Setting a Page to Its Previous State 551
Exercises 557

Appendix A: Binary Files 559
A.1 Writing to Binary Files 559
Tutorial: Working with Binary Files 559
A.2 Reading Binary Files 560

Appendix B: Creating a Database 562
Tutorial: Creating HR Database 562

Appendix C: Answers to Review Questions 564

Appendix D: HTML Elements 584

Index 586

Preface

Mission of the Text

Welcome to *Fundamentals of C# Programming for Information Systems*. This book teaches the fundamentals of programming in C# to provide a solid foundation to build business and other real-world applications. Programming concepts are discussed in the context of familiar practical applications that use graphical interfaces.

New to This Edition

A key goal of the new edition was to add significant content so that the book could be used in a two-course sequence in programming. Four new chapters have been added: Two chapters that provide a comprehensive introduction to web applications development, and two other chapters on the concepts and practice of object-oriented programming, including inheritance.

Several other updates were made based on feedback from instructors: a second set of three comprehensive assignments (in chapters 3, 8, and 12), additional end-of-chapter exercises, and learning objectives for each chapter.

Target Audience

This book is designed for introductory programming courses in IS/MIS, CIS and IT. This book also would fit into a computer science curriculum with an introductory course that uses a GUI-based application-oriented approach to teach programming concepts. The breadth and depth of coverage makes this book suitable for a two-course sequence, particularly when students come to the first course with no programming background and a slower pace is desired. An approach in a two-course sequence would be to do in-depth coverage of topics like collections, databases, object-oriented programming, web development, and others presented in later chapters only in the second course.

Key Features

A key feature of the book is that programming concepts are introduced in small chunks through examples and illustrations accompanied by hands-on tutorials. The tutorials, which are interspersed with the concepts, help students apply and explore what they learn immediately. Additionally, review questions and exercises within the chapters enhance student interest and learning.

Although the book is written for beginners, it is thorough and concise. Graphical illustrations and screenshots are used throughout the book to enhance learning for both beginners and experienced students.

Windows forms are used from the beginning to provide GUI-based as opposed to console-based interface.

This book builds graphical user interfaces and code in the .Net environment using Visual Studio. You are encouraged to use the current free version, Visual Studio Community 2017. Earlier versions, including Visual Studio 2015, Visual Studio Express and Visual Studio 2013, generally work well, except that you won't be able to run a few programs that use certain features available only in C# version 7.1 or greater.

Supplements

For Students: Tutorial_Starts.zip file that contains

- o Partially completed projects for tutorials
- o Data files/databases used in projects

You may download the Tutorial_Starts.zip file from
http://www.prospectpressvt.com/titles/c-sharp-programming/student-resources/

For Instructors: Instructor resources include

- o Completed tutorials
- o PowerPoint slides for all chapters
- o Test bank
- o Partially completed projects for tutorials
- o Data files/databases used in projects

To access instructor resources, please complete the request form at
http://www.prospectpressvt.com/faculty-resources/instructors-material/

Installing Visual Studio

You may download Visual Studio Community 2017 from the following website:
https://www.visualstudio.com/downloads/

To install Visual Studio, open the downloaded .exe file and run it.

Overview of the Content and Organization

Every possible sequence of topics seems to put constraints on the quality of illustrative examples and applications that can be used in a chapter. The organization of chapters in this book attempts to minimize such problems and to enhance the ability to build on prior chapters. However, except for the foundational Chapters 1–5, there is significant flexibility in choosing specific topics and the depth of coverage. As suggested by the dependencies summarized below, there is some flexibility in the sequencing too.

Chapters 1–5 cover the Visual Studio environment and introductory programming concepts, including methods. These chapters, which provide the foundational knowledge, should be covered in sequence before other chapters, though certain topics like working with dates and times (section 2.4) may be postponed or skipped.

Chapter 6 presents the application of the GUI controls ScrollBars, RadioButtons, CheckBoxes, ComboBoxes and ListBoxes.

Chapter 7 provides a detailed presentation of one-dimensional arrays, and Chapter 8 presents accessing sequential files and using arrays in combination with files. Chapter 6 is not a prerequisite for Chapters 7 or 8. GUI is presented early on in Chapter 6 to motivate students with more interesting graphical interfaces. It should be noted that the comprehensive assignment (Assignment 2) specified at the end of Chapter 8 requires the use of several GUI controls.

Chapter 9 introduces collections, and discusses the List and Dictionary collections in more detail. Chapter 8 ("Sequential Files and Arrays") is a prerequisite for this chapter. Because of the close relationship between collections and arrays, collections are presented in this book immediately following Chapter 8 on arrays.

Chapter 10 discusses the application of ListView and TabControl. The dependency of this chapter on Chapter 9 ("Collections") is very low. The prerequisite for this chapter includes Chapters 6 and 8.

Chapter 11 presents multiform applications, Menus and ToolStrips. This chapter has some dependency on previous chapters, except Chapter 10.

Chapter 12 provides in-depth coverage of accessing databases from C# programs. Chapter 6 is a prerequisite for this chapter. In addition, the ListView control presented in Chapter 10 is used in an example in the last part of this chapter, and it is required in the third comprehensive assignment (Assignment 3) at the end of this chapter. Assignment 3 also requires the use of MainMenu control discussed in Chapter 11. Other than that, the dependence of Chapter 12 on Chapters 7–11 is relatively low.

Chapter 13 provides an introduction to object-oriented programming (OOP) principles and techniques. The initial part of this chapter may be used for an introduction to OOP early in the semester.

Chapter 14 describes the concept of inheritance, implementation of inheritance, subclasses and super classes, overriding methods and polymorphism. Chapter 13 is a prerequisite for this chapter, and it also relies on collections from Chapter 9.

Chapter 15 presents the concepts of web applications development, and develops simple web applications using the ASP.Net platform in the Visual Studio environment. Only the basic programming knowledge presented in Chapters 1–3 is required for this chapter.

Chaper 16 describes how to access SQL Server databases and develop multipage web projects. Basic programming concepts presented in Chapters 1–6, the concept of collections from Chapter 9, and basic database concepts including binding controls to a database and filtering records (Chapter 12) are used in building the application.

Acknowledgments

I am thankful for the valuable assistance provided by many people in the preparation of this book. I wish to thank Dr. Jakob Iversen, The University of Wisconsin–Oshkosh, for authoring Chapters 13 and 14 on Object-Oriented Programming.

I was fortunate to work with Beth Lang Golub, editor and president of Prospect Press, who was flexible and supportive of my goal to offer a good quality programming textbook at a reasonable price. Special thanks go to Susan Hegedus, Kathy Bond Borie and Rachel Paul for their painstaking attention to detail in editing this book, and to Annie Clark for the cover design.

I wish to acknowledge the contributions of the following reviewers for their valuable guidance in improving the presentation and contents of this book:

> Janet Bailey, University of Arkansas at Little Rock
> Wei Kian Chen, Champlain College
> Clinton Daniel, University of South Florida
> Silvana Faja, University of Central Missouri
> Joni L. Jones, University of South Florida
> David Pumphrey, Colorado Mesa University
> Manonita M. Ratwatte, University of Oklahoma (Retired)
> Theadora Ross, University of Arkansas at Little Rock
> David M. Weber, Northern Arizona University

Thanks are also due to the instructors who provided valuable feedback on the first edition of this book through user surveys:

> Janet Bailey, University of Arkansas, Little Rock
> Jeff Dickson, Oregon Institute of Technology
> Ruth Lamprecht, Virginia Union University
> Panos Linos, Butler University
> Ziping Liu, Southeast Missouri State University
> Robert Pilgrim, Murray State University
> Kris Rosenberg, Oregon Institute of Technology
> Doug Titus, University of North Florida

About the Author

Dr. George Philip is Professor Emeritus of Information Systems at the College of Business, The University of Wisconsin–Oshkosh. He has more than twenty-five years of teaching and consulting experience in the information systems field, including computer programming in multiple languages. He also served as chair of the Information Systems Team, and director of the M.S. in Information Systems program. He has published numerous articles in this field.

Chapter 1

Introduction to Visual Studio and Programming

Welcome to programming in C# language. In this chapter, you will learn to develop simple programs in the Visual Studio development environment, and to work with different types of data.

Learning Objectives

After studying this chapter, you should be able to:

- Identify the inputs, processes and outputs of a software system.
- Describe the steps involved in developing a computer program.
- Describe the terms: syntax, logic and runtime errors, machine language, low-level language, high-level language, compiler and interpreter.
- Develop a simple form in Visual Studio to accept user input, do calculations and display formatted output.
- Work with Label, TextBox, Button, ListBox and MessageBox.
- Develop simple programs that use constants, variables and expressions.
- Use try-catch method to catch errors.

Topics

1.1	Introduction to Programming	1.8	Working with Variables
1.2	Introduction to Visual Studio	1.9	Doing Calculations
1.3	Visual Studio Environment	1.10	Formatting and Displaying Output in a Label
1.4	Adding Controls to a Form	1.11	Additional Concepts: Constants, Scope of Variables and ListBox
1.5	The Code Editor Window		
1.6	Introductory Programming Concepts	1.12	Exception Handling: Catching Errors
1.7	Data Types		

1.1 Introduction to Programming

Programming is the process of developing computer programs. If a computer program seems like a mystery to you, it is just a set of instructions telling the computer how to do a task, like looking up the price of an item or finding the Chinese restaurants in a city.

Unfortunately, computers cannot understand normal English. So, programs have to be written using special commands and statements according to strict rules. A key aspect of programming is breaking down what you want the computer to do, into detailed instructions. Like the directions that a GPS gives you to get to a place, the instructions in a program need to be precise.

Typically, a program uses one or more data items to produce some results. For example, a program that processes an order might use the item number and order quantity to compute the subtotal, sales tax and total cost, as represented in Figure 1-1.

Figure 1-1: Inputs, process and outputs of a program

The program, represented by the block "Process Order," may include multiple subtasks like look up the unit price, check inventory and compute results.

The data that are used by a program are called the **input** to the program, and the results produced are called the **output** of the program. In addition to processing input data to produce the output, a program might write data to and read data from **storage** devices like a flash drive or a hard drive, as represented in Figure 1-2.

Figure 1-2: A general representation of a software system

How Do I Go About Developing a Program?

To understand the process of developing a program, let's use a simplified order-processing system as an example. To develop good programs and do so efficiently, follow these steps:

1. Define the purpose, and identify the input, process and output of the program.
2. Design and develop the graphical user interface (GUI).
3. Identify the components and logic of the program.
4. Design and develop files/databases, if any.
5. Write and test the code.

Let's look at these steps in more detail.

1. Define the purpose, and identify input/process/output

Before you can write the program, you need to lay some groundwork. In this step, you identify what the user wants the program to do, including the input, process and output of the program.

Depending on the size and complexity of the program, this could involve extensive analysis of the requirements, including interviews with users; examination of current forms, reports and transactions; and identifying processes like checking inventory and looking up price in an order.

Here is an example of a simplified statement of the **purpose** of the order processing program:

    Purpose: Compute and display the subtotal, sales tax and total cost for an order

The **output** of the program often follows from the purpose. For this example, the output would be

    Output: subtotal, sales tax, total cost

The **process** specifies not only **what** the program should do (e.g., compute total cost) but also **how** it should be done (e.g., how to compute total cost), as follows:

    Process: (What?)   Look up unit price, look up sales tax rate,
                             compute subtotal, sales tax and total cost
        (How?)   subtotal: unit price * order quantity
                             sales tax: subtotal * sales tax rate
                             total cost: subtotal + sales tax

Specifying the process also would include identifying the sources of data, like the product file to get the unit price and sales tax file to get the sales tax rate.

The **input** specifies the data items that are needed to carry out the process to produce the output. The input for this order-processing system would be

    Input: item number, order quantity

Note that unit price and sales tax rate are not included in the input because the program looks them up. A real-world system would be a lot more complex. Typically, the process would include additional subtasks like handling orders when inventory is insufficient, and output may include various reports. In such systems, graphical methods like **Data Flow Diagrams** and **UML diagrams** are used to represent the processes and the data accessed by them.

2. Design and develop the graphical user interface (GUI)

After identifying the input, process and output of a program, you design and create the user interface—that is, how the user would interact with the program, and how the program would communicate with the user. This is the fun part where you bring in your creativity.

Typically, you use forms to interact with the program. As you will learn in the next section, forms have various types of objects, called **controls**, such as Button, TextBox and Label. In this step, you identify the type of controls to be used, specify their names and captions as appropriate, and design the layout. Figure 1-3 shows an example of the GUI for a simplified order-processing system where the user doesn't provide the unit price and sales tax rate.

Figure 1-3: GUI for an order-processing system

3. Identify the components and logic of the program

This step identifies the major subtasks of the program. For example, in order to process an order, the program needs to do the following subtasks:

> Get Item# and quantity
> Look up unit price and tax rate
> Compute subtotal, sales tax and total cost
> Display subtotal, sales tax and total cost

Again, a real-world program may have to do additional subtasks, like checking the inventory to make sure there is sufficient quantity on stock.

For relatively simpler programs, after identifying the subtasks, you may go directly to writing the program for each subtask. However, for tasks involving more complex logic, it might help to develop an outline of the logic of performing the subtasks. The representation of the logic of a program in plain English is called **pseudo code**. You also may represent the logic graphically using a **flowchart**, as discussed in Chapter 3.

4. Design and develop files/databases

If data is read from or written to files and/or databases, these are designed and developed prior to writing the program. Depending on the application, this step may have to be done in parallel with previous steps.

5. Write and test the code

The final step is to write and test the code. You can program in a variety of languages. C#, Java, Visual Basic, Python and PHP are among the popular languages. You will use C#, which is a popular language for developing desktop and web applications.

Programing may involve iteratively developing an application by going through the above steps multiple times.

Syntax, Logic and Runtime Errors

The programming statements you write have to follow strict rules of the language, called the **syntax**. The program wouldn't **compile** if it had any **syntax error**, like a missing semicolon at the end of a statement or a misspelled key word. Compiling is the process of translating the program you write, called the

source code, to another language before running a program, as described in the next section. So, your first task is to make sure that there are no syntax errors. The good news is that Visual Studio provides a lot of help in identifying syntax errors.

After the syntax errors are eliminated, the program may run. But, it's still too early to celebrate because the results could be incorrect due to errors in the program logic, just like you can write a grammatically correct sentence that doesn't convey the intended message. Errors that cause a program to produce incorrect or unintended results are called **logic errors**. A tax-filing software using the wrong tax rate and a billing software overcharging a customer are examples of logic errors.

There are errors other than logic errors that can occur at runtime. These are called **runtime errors**. Runtime errors cause the program to crash (unless the program catches and handles such errors) because the program asks the computer to do something it is unable to do, like accessing a file with an invalid path or dividing a number by zero.

The process of identifying errors (bugs) is called **debugging**. Testing programs to identify and eliminate errors is an extremely important part of developing software.

Review Questions

1.1 Consider Google as a software system. What would be the input, process and output for Google?

1.2 Consider a software system that enrolls students into classes. Identify some key inputs that the system needs every time a student enrolls in a class, and the subtasks (process) that need to be performed. What are some outputs the system should produce for students and instructors?

1.3 List the major steps in developing a program.

1.4 Incorrect punctuation in a program is an example of what type of error?

1.5 A payroll program uses the wrong formula to compute overtime pay. What type of error is it?

1.6 True or false: A program that doesn't have any syntax errors should produce the correct results.

1.2 Introduction to Visual Studio

In this section, you will learn how to use Visual Studio (VS) to create the user interface and write C# programs. Visual Studio is an integrated development environment (IDE) for developing applications in a variety of languages, including C#, Visual Basic, C++, Python and HTML/JavaScript. VS supports development of desktop and web applications for Windows, Android and iOS. A major strength of Visual Studio is that it provides a user-friendly environment for developing applications.

Installing Visual Studio

You may download Visual Studio Community 2017 from the following website:
    https://www.visualstudio.com/downloads/

To install Visual Studio, *open* the downloaded **vs_community.exe** file. The installation software will guide you through the process. The first time you start Visual Studio, you will be asked to choose the *Development Settings* and a *Color Theme*. Choose *Visual C# for* settings. You may change these settings later, as described in Tutorial 1.

Components of Visual Studio

Though you have the choice to write programs in a variety of languages, the CPU can understand only **Machine Language,** which is extremely difficult for humans to understand. Machine Language requires detailed instructions that consist of patterns of bits (0 and 1), like 10001010, and are dependent on the machine (the specific type of computer). Because of the need to write detailed machine-dependent instructions, Machine Language is called a **low-level language**.

Except in special cases, programs are written in **high-level languages,** like C#, which require fewer statements, are less dependent on the hardware and are easier to understand because they use words rather than patterns of bits.

Programs written in high-level languages are translated to Machine Language before they are run. Different languages use different methods to translate and run programs. Many languages use a special software called a **compiler** to translate the source code to Machine Language. The compiler typically produces a separate executable Machine Language program that can be run any number of times without having to compile every time it is run.

Some programming languages use an **interpreter** that translates each statement to Machine Language and runs it without producing an executable program. So, every time the program is run, it needs to be translated to Machine Language.

Visual Studio uses a compiler, but it translates the source code to an **intermediate language**, which is translated to Machine Language and run using another software, as described later in this section. Thus, Visual Studio translates your statements to Machine Language in two steps:

First, a **compiler** converts the source code into a language called Microsoft Intermediate Language (IL). The compiled code, along with references to prebuilt programs (called classes), is stored in an executable file called **Microsoft Intermediate Language (IL) Assembly**. Such files have the extension .exe or .dll.

Next, another software called **Common Language Runtime (CLR)** translates the assemblies from Intermediate Language to Machine Language and executes the programs. The process of translating and running the source code may be represented as follows:

The products that support developing and running programs within the Visual Studio family include the following:

1. An **Integrated Development Environment (IDE)**
   An IDE provides an environment to develop programs, which includes code editors for **Visual C#, Visual Basic, Visual J#, Visual C++, HTML and XML,** and designers for **Windows forms** and **web forms.**
   In Visual Studio, a software application typically is organized into **Projects** that may contain one or more **forms.**
   Forms provide the **user interface** that allows users to input data for the program, to interact with the program and to display results.

2. **A compiler** that translates the source code into **Microsoft Intermediate Language (MSIL)**

3. **.Net Framework**, which includes
    a. the **Common Language Runtime (CLR)** that translates the assemblies from Intermediate Language to Machine Language and executes the programs, and
    b. .Net Framework **Class Library** that includes a large number of prebuilt programs called **classes**.

Running a program, as described earlier, consists of (1) the compiler translating the source code (the project) into Microsoft Intermediate Language Assembly and (2) the Common Language Runtime translating the assemblies from Intermediate Language to Machine Language and executing the program to produce the output. Next, we will look at how to work with Visual Studio to develop C# programs.

Review Questions

1.7   What is the only programming language that the CPU can understand?

1.8   What is a compiler?

1.9   What is an interpreter?

1.10 What is Microsoft Intermediate Language Assembly?

1.11 What is the function of Common Language Runtime in Visual Studio?

Creating an Application in Visual Studio

To become familiar with the Visual Studio environment, let's create an application for an ice cream parlor, consisting of a simple Windows form to compute and display the cost for an order of ice cream. We will follow the five steps identified earlier to develop it:

1. Define the purpose

    Purpose: Compute and display the total cost for an order of ice cream
    Input: Unit price, number of scoops
    Process: Compute total cost
                Total cost: unit price * number of scoops
    Output: Total cost

2. Design and develop the Graphical User Interface (GUI)

Figure 1-4 presents the design of the form that shows the user interface to let the user enter the number of scoops and unit price to display the cost. You will create the form in Tutorial 1.

Figure 1-4: Windows form to compute ice cream cost

```
┌─────────────────────────────────────────┐
│  ▫      Ice Cream Cost      —  ☐   ✕   │
│                                         │
│    # of Scoops:  [2      ]              │
│    Unit Price:   [2.95   ]  Cost: [ $5.90 ] │
│                                         │
│    [Compute Cost]  [ Clear ]  [ Exit ]  │
│                                         │
└─────────────────────────────────────────┘
```

3. Identify the components and logic of the program

The major subtasks include

      Get number of scoops and unit price
      Compute cost
      Display cost

4. Design and develop files/databases

This application doesn't involve the use of files or databases.

5. Write and test the code

To understand the process of developing the entire application, including the code, let's create the form presented in step 2 and write the code for the tasks identified in step 3.

Creating a Windows Form

In Visual Studio, forms are created within a **Project**. Typically, all forms within a Project relate to a common task. Creating a project also creates a **Solution,** which is a container for one or more projects. Each project we discuss in this book is in a separate solution that has the same name as the project.

Tutorial 1 creates the project and the form to compute the cost of an order of ice cream.

Tutorial 1: Creating a Form: Ice Cream Cost

Step 1-1: Open Visual Studio Community 2017. You will see the start page, shown in Figure 1-5.

Figure 1-5: Visual Studio Start Page

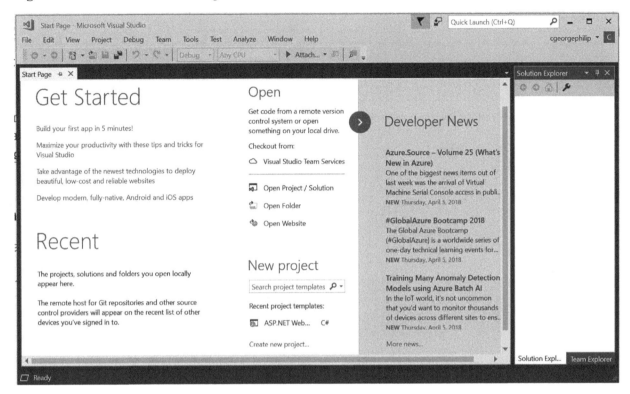

First, you will change some settings.

Changing default settings in Visual Studio

Step 1-2: Change the default folder to save your projects:

> Select *Tools, Options*. Expand *Projects and Solutions* and select *Location*.
>
> For *Projects Location*, select a folder (e.g., your flash drive) where you will save your projects.

Step 1-3: Change the setting to automatically save projects when they are created:

> Select *General* from *Projects and Solutions*.
> Make sure that the CheckBox for *Save new projects when created* is checked. Click *OK*.
>
> You may change the *color theme* by selecting *Environment, General* from the *Options* window. Similarly, you may reset your settings by selecting *Tools, Import and Export Settings, Reset all Settings*.
> Close *Options* window.

Creating a new project

Step 1-4: Create a new project:

Click *Create New Project* on the Start page.
You will see the New Project window displayed in Figure 1-6.

Figure 1-6: *New Project* window

Select *Visual C#* from *Templates, Other Languages*. For the type of application, select *Windows Forms Application*, as shown in Figure 1-6.

For *Name*, enter Ch1_IntroProgramming.

Select the *Location*, if different from the default folder.

Leave the CheckBox for *Create directory for solution* unchecked so that both the Project and the Solution will be created in the same folder. (If there are multiple Projects within a Solution, it would be better to have separate directories for the Solution and Projects.) Note that the default name for the Solution is the same as that of the Project.) Click *OK*.

The Visual Studio development environment appears, as shown in Figure 1-7.

1.3 Visual Studio Environment

This section introduces the various windows and the environment within Visual Studio.

Figure 1-7 shows the default form named **Form1** in the **Designer** window, the **Toolbox** on the left, and the **Solution Explorer** window and the **Properties** window on the right.

The appearance of the form will vary with the version of Visual Studio and Windows you use. For example, the title of the form is centered in Windows 8 and 10, whereas it is left justified in Windows 7.

The **Designer** window allows you to create the user interface by adding controls to a form.

The **Toolbox** contains the components to build the user interface. The items within the Toolbox (e.g., ComboBox, TextBox and Label) are called **controls**.

The **Solution Explorer** window shows the name of the Solution, which is a container of projects; the name of the project (or projects, if there are multiple projects); and the names of the files within each project. If the name of the Solution is not displayed, select

> *Tools, Options, Projects and Solutions, General* and check *Always Show Solution*

The **Properties** window displays and lets you set the properties of the currently selected object in the *Designer* window. For example, the *Name* field in the Properties window shows that the current name of the form is **Form1**.

Figure 1-7: Visual Studio environment

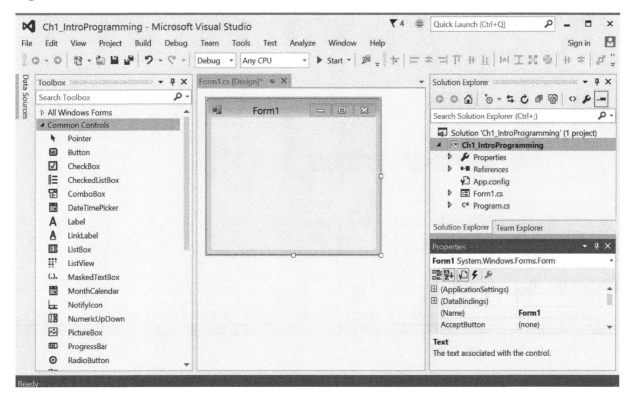

A brief description of each property is displayed at the bottom of the properties window when you click on the property, as shown in Figure 1-8. Make sure you expand the description area to make the description visible. If the description is not displayed, right click the property and check *Description*.

Figure 1-8: Description of properties

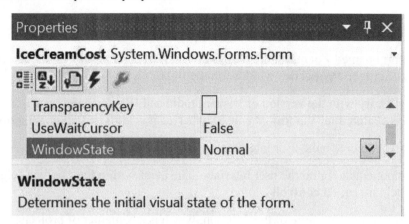

Displaying a closed window

If any window is not displayed, select *View* from the menu, and select the corresponding window.

Changing the docking position of a window

You may change the docking position of a window, such as the Properties window, by grabbing it with the pointer and then dropping it in one of the four positions (top, bottom, left, right), shown in Figure 1-9.

Figure 1-9: Docking positions

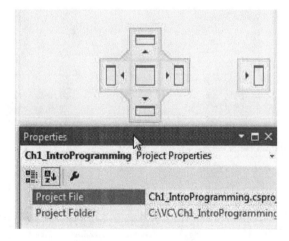

Undocking (floating) a window

By default, all windows are attached to a side of the Visual Studio window. To undock a window, right click its title bar, and select float. To dock a floating window, grab it with the pointer and drop it in one of the positions shown in Figure 1-9.

Autohide

The pushpin icon on a window, as shown in Figure 1-10, allows you to hide the window and show it as a tab on the side.

Figure 1-10: Pushpin for autohide

Click the pushpin to turn autohide on and reduce the window to tab, as shown in Figure 1-11. Note that the Solution Explorer is now a vertical bar to the right of the Properties window.

Figure 1-11: Solution Explorer with autohide on

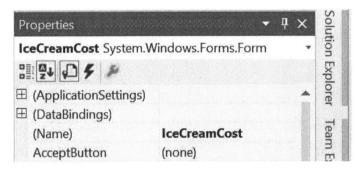

To turn autohide off, click on the tab and click the pushpin.

Menu and toolbar items

The menu and toolbar, shown in Figure 1-12, provide access to an extensive set of functions.

Figure 1-12: Menu and toolbar

The function of each toolbar item is displayed in a **ToolTip** box when you hover the mouse pointer over the item. For a general understanding of the functions, you may view the descriptions of the Toolbar items by displaying the Tooltips. You will become more familiar with them through the projects in this book.

Adding and deleting forms

You may add a form to the project as follows:

> Select: *Project, Add Windows Form*. You will see the *Add New Item* window.
>
> Make sure Visual C# Items and Windows Forms are selected.
>
> Type in a name for the form. Click *Add*.

To delete a form,

> Right click the name of the file (.cs) in the Solutions window, and select *Delete*.

Deleting a file deletes the form and all files associated with the form permanently.

Excluding a form from the project

You may exclude a form from the Solution Explorer but still keep the files associated with the form, in the project folder, as follows:

> Right click the file name (.cs) in the *Solution Explorer*, and select *Exclude from Project*.

To bring back a form that was excluded from the project,

> Select *Project, Add Existing Item*.
>
> Select the file (.cs) from the project folder.

The form will be added to the Solution Explorer.

Changing the name and title of a form

The *Name* field in the Properties window shows that the current name of the form is **Form1**.

The Solution Explorer also displays a file named **Form1.cs**, which is the file that contains the code. Expanding this file, by clicking the triangle to its left, displays two other files,

> **Form1.Designer.cs** that contains the generated code representing the design of the form and
> **Form1.res** that contains additional resources associated with the form,

as shown below:

Guidelines to name a form

Names of forms (and classes, in general) must start with a letter, and can contain only letters, numbers or the underscore character (_). However, the use of the underscore character is not recommended.

The convention in C# is to use the **Pascal Case** to name forms: The first letter of the form name and the first letter of subsequent words, if any, are capitalized, as in IceCreamCost and FinancialPlanner.

Use a meaningful name and avoid using abbreviations or acronyms that are not widely recognized. A noun or noun phrase is recommended for a form name, though you will find a few form names in this book that violate this guideline.

Now that you are familiar with the environment, let's develop the application using the default form, Form1, in the project that you created in Step 1-4.

Chapter 1: Introduction to Visual Studio and Programming

Step 1-5: Change the form name and title:

> Right click *Form1.cs*, and rename it **IceCreamCost.cs**. Notice that the name of the form in the Properties window and the names of all three files under Form1.cs are changed.
>
> The title of the form still remains Form1. To change the title, select the Text property of the form in the Properties window and change it to **Ice Cream Cost**.

Specifying the startup form

A project typically contains multiple related forms. The Program.cs file contains the startup program called Main that includes the following statement, where *FormName* represents the name of the startup form.

> Application.Run(new *FormName*());

Step 1-6: Double click on Program.cs in the Solution Explorer window, and make sure that the name of the startup form is IceCreamCost, as shown below:

> `Application.Run(new IceCreamCost());`

Running a form

Step 1-7: Run the form by clicking the *Start* button (shown below) from the Toolbar.

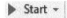

> Visual Studio displays the empty form. Because you haven't added any controls to the form, there is not much you can do with it. Close the form by clicking the Close button ("x") at the top right of the form.

Additional properties of forms

Let's look at a couple of properties that apply to the form only when it is run.

Click on the *start position* property of IceCreamCosts.cs, and view its description displayed at the bottom of the Properties window. If the description is not visible, expand the description area to make it visible.

Click the dropdown arrow to the right of StartPosition to view the options. Select *Center Screen* from the list. Run the form and see how the property impacts the position of the form. Change the property back to *WindowsDefaultPosition*.

Repeat for *WindowState* property, setting it to *Maximized*.

Review Questions

1.12   What property of a form would you use to change the title displayed on the title bar of the form?

1.13   When you run a project containing multiple forms, which form does C# run?

1.4 Adding Controls to a Form

The term **control** means objects like TextBoxes, Labels and Buttons. Typically, **TextBoxes** are used to enter data when you run the form, **Labels** to display a label or computed results and **Buttons** to run a program segment.

Labels and TextBoxes

As stated above, a TextBox allows you to enter data into it when you run the form, whereas a Label control can only display data. Generally, TextBoxes are used to enter input data, and Labels to display output or labels. The Label and TextBox controls are represented in the Toolbox, as follows:

A Label   **abl** TextBox

Text property of Label

To display text in a Label in the Design window, type the text in the box to the right of the **Text property** of the Label.

AutoSize property of Label

When the AutoSize property of a Label is set to its default value, True, you cannot adjust the size of the Label; the size of the Label is determined by what is displayed in it. If you want to adjust the size, you need to set AutoSize to False.

Naming controls

To make programs easy to understand, it is important to give meaningful names to all controls that are used in the code. In addition, use standard prefixes for naming controls on a form. Table 1-1 shows standard prefixes for some commonly used controls.

Table 1-1: Prefixes for controls

Control	Prefix	Example
TextBox	txt	txtCourseNumber
Label	lbl	lblTotalSales
Button	btn	btnComputeTotal
ComboBox	cbo	cboProduct
ListBox	lst	lstIngredient
ListView	lvw	lvwPrice

We will introduce prefixes for other controls later when we use them.

To create the form, IceCreamCost, we need to add several controls from the Toolbox window. First, you need to make sure the Toolbox is displayed.

Step 1-8: Display controls in the Toolbox as follows:

> If the Toolbox tab is not displayed, select *View, Toolbox*.

If the Toolbox appears as a tab, click on the Toolbox tab, and then click the Auto Hide button at the top-right of the window.
Expand *Common Controls*. See Figure 1-13.

Figure 1-13: Toolbox controls

Next, you will add the controls shown in Figure 1-14 to the form. Note that if you use Windows 7, the title of the form won't be centered.

Figure 1-14: The Designer window with the form

You may want to **resize** a form before or after adding controls to it. To resize a form, move the cursor to the border of the form, and click and drag the cursor when it turns to a double-headed arrow.

You may add a control by dragging and dropping it on the form, or by double clicking it.

Step 1-9: Add a TextBox to enter the number of scoops, as follows:

Drag and drop a TextBox from the Toolbox menu to the form and position it at the location shown in Figure 1-14, and resize it as necessary.
Click the ellipsis (…) to the right of Font property to set the FontSize to 10 and the BorderStyle to *Fixed3D*.
Change Name property of the TextBox to **txtScoops**.

Step 1-10: Add a Label to display the label, "# of Scoops":

1.4 Adding Controls to a Form

Drag and drop a Label and position it to the left of the TextBox.
Change its AutoSize and Text properties, as follows:

AutoSize: False
Text: # of Scoops

Step 1-11: Add the remaining seven controls and set their properties, as specified in Table 1-2:
Unless specified otherwise, set the FontSize to 10 for all controls, and BorderStyle to None for Labels, and Fixed3D for TextBoxes. Note that the only Label that is given a name (lblCost) is the one that is used in the code to dislay the cost. Other Labels that are not used in the code keep their default names.

Table 1-2: Controls on the form IceCreamCost and their properties

Purpose	Type of control	Property	Setting
Enter number of scoops	TextBox	Name	txtScoops
Display label for scoops	Label	Text	# of Scoops:
Enter unit price	TextBox	Name	txtUnitPrice
Display label for unit price	Label	Text	Unit Price:
Display cost	Label	Name	lblCost
		BorderStyle	FixedSingle
		Text	(blank)
Display label for cost	Label	Text	Cost:
Compute cost	Button	Name	btnComputeCost
		Text	Compute Cost
Clear controls	Button	Name	btnClear
		Text	Clear
Close the form	Button	Name	btnExit
		Text	Exit

Closing and opening the Designer window and the project

To save the changes to the form, click the **Save** button on the Toolbar. The **Save All** button would save changes to all forms within a project. The Save and Save All buttons appear on the Toolbar as follows:

To close the Designer window, click the **Close** button (x) on the tab above the form (see Figure 1-14), or select **File, Close**.

To open the Designer window, double click the file name IceCreamCost.cs in the Solution Explorer, or right click the file name and select **View Designer**.

To close the project, Select: **File, Close Solution.**

To open the project and the solution from within Visual Studio,

Select File, Open, Project/Solution,
Open the project folder, Ch1_IntroProgramming,
Select Ch1_IntroProgramming.csproj (or, Ch1_IntroProgramming.sln), as shown below, and
Click Open.

If file extensions are not shown along with file names, please see the next subsection, **Display File Extensions**.

To open the project from outside Visual Studio, open the project folder Ch1_IntroProgramming from File Explorer (in Windows 10), or Windows Explorer (in earlier versions of Windows). The project folder displays the following folders and files:

Double click the file Ch1_IntroProgramming.csproj or Ch1_IntroProgramming.sln to open the project and the solution.

Display file extensions

In File Explorer (Windows 10), select View, and check the CheckBox, *File name extensions*.

In Windows Explorer (Windows 8.1), select *Organize, Folders and search options, View*.
Uncheck the CheckBox, Hide extensions for known file types.

Phases completed

In Steps 1-1–1-11 of Tutorial 1, you developed the Graphical User Interface, which is phase 2 of the program development process. Earlier, you also identified the subtasks of the program in phase 3:

     Get number of scoops and unit price
     Compute cost
     Display cost

This application doesn't involve the development of files or databases (phase 4). The next phase of program development is phase 5: Write and test the code.

Review Questions

1.14 What control would you use to let the user enter a data item when you run a form?

1.15 What property of a Label do you set to display a data item in a Label?

1.5 The Code Editor Window

Before you write and test the code, let's take a look at the Code Editor window (or Code window) that includes some initial code already generated by C#.

Step 1-12: Open Code window, as follows:

Select IceCreamCost.cs in Solution Explorer, and
click *View Code* button (<>) on the top right.
(or right click IceCreamCost.cs and select *View Code*).
See Figure 1-15.

Figure 1-15: Opening Code window

Figure 1-16 shows the Code window.

Chapter 1: Introduction to Visual Studio and Programming

Figure 1-16: The Code window

```
 1  using System;
 2  using System.Collections.Generic;
 3  using System.ComponentModel;
 4  using System.Data;
 5  using System.Drawing;
 6  using System.Linq;
 7  using System.Text;
 8  using System.Threading.Tasks;
 9  using System.Windows.Forms;
10
11  namespace Ch1_IntroProgramming
12  {
13      public partial class IceCreamCost : Form
14      {
15          public IceCreamCost()
16          {
17              InitializeComponent();
18          }
19      }
20  }
```

You may **display line numbers** in Code Editor window as follows:

Select *Tools*, *Options*.
Expand *Text Editor* by clicking the triangle on its left.
Expand *C#*.
Select *General* and check the CheckBox *Line numbers* as shown below:

Let's take a look at what the statements mean.

Lines 1–9: The statements in lines 1–9 that start with the key word *using* (e.g., using System.Windows.Forms) are called directives. Each *using* directive specifies a namespace, like **System.Windows.Forms**. A **namespace** is just a container for a group of prebuilt programs (called **classes**) from the .Net Framework Class Library.

For example, the **System.Windows.Forms** is a namespace that contains the classes that define the TextBox, Label, Button and other controls that are commonly used in a form. The directive, using System.Windows.Forms, makes those classes available within the form. The C# compiler looks in the specified namespaces for classes used in the program.

Classes and Objects

Without going into a lot of detail, a **class** may be viewed as a group of program units called **methods,** and data items that are represented by **properties**. A method is a program unit that performs a particular task. For example, a CustomerOrder class representing customer orders may have one method (program unit) to compute the sales tax and another method to compute the total cost. The properties of the class may include product number and order quantity.

To represent an individual order, an application creates an instance of the CustomerOrder class in memory, called an **object**. So, the object is an *implementation* or an *instantiation* of the class. Thus, the same CustomerOrder class can be used to create multiple objects, each object representing a different order. The class is often compared to the blueprint of a house, and the objects compared to multiple houses that are built from the same blueprint.

Thus, namespaces contain classes, and classes contain methods. A method consists of a group of statements that perform a task.

Line 11: `namespace Ch1_IntroProgramming`

This statement specifies the project's namespace, which is essentially the group of classes within the project. The beginning brace on line 12 and ending brace on line 20 indicate that everything in between is part of the project's namespace.

Line 13: `public partial class IceCreamCost : Form`

This statement shows the name of the class IceCreamCost that specifies the form. The key word **Public** means the code units (methods) and properties of this form are accessible to other classes, if the code units are public. **Partial** means that the class may be split to different files.

Everything between the starting brace (line 14) and the closing brace (line 19) is part of this class.

Line 15: `public IceCreamCost()`

This statement is the header of a method named IceCreamCost within the class IceCreamCost.

Developing the Code

As you observed in the above code, C# uses a pair of left (opening) and right (closing) **braces**, {}, to enclose a block of statements.

Before we write the code, let's look at the input, process and output structure of the program, shown in Figure 1-17. The output that the user wants is the cost of ice cream. The cost is computed using unit price and the number of scoops, which are the inputs. The processing consists of computing the cost (= unit price x number of scoops).

Figure 1-17: Input, process and output of the program

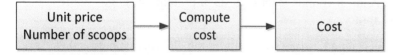

Thus, developing the code involves the following subtasks that were identified in Phase 3, earlier:

    Get number of scoops and unit price
    Compute cost
    Display cost

You need to develop code to do each of the above subtasks.

Event-Driven Programs and Event Handlers

In the .Net environment, code often is connected ("attached" or "wired") to specific events, like clicking a button, loading a form or pressing a key. Therefore, programs developed in .Net are called **event-driven programs**. A method that runs in response to an event such as clicking a button is called an **event handler**.

For example, to compute cost when the user clicks a button, we create a program unit (method) that is connected to the click event of the button. Because the method runs in response to the click event, it is called the click event handler of the button.

Creating event handlers

Let's look at how an event handler is created.

Step 1-13: **Create click event handler method for btnComputeCost:**

Switch back to the Design view by clicking the designer tab, IceCreamCost.cs[Design].

Double click on the Compute Cost button to reopen the Code window and to create the **click event handler**, shown below:

```
private void btnComputeCost_Click(object sender, EventArgs e)
{


}
```

The first line is the header of this method, which shows the method's name, **btnComputeCost_Click**. The key word **private** indicates that this event handler method can be accessed only from this form, and **void** means that this method doesn't return any value to the calling program. These terms and the concept of parameters within the parentheses are explained in later chapters.

Deleting event handlers

Suppose you created an event handler that you don't need. For example, you created the Click event handler of a button and decided to delete the button. Before you delete the button, you need to delete the event handler to avoid an error.

If you manually delete the event handler method from the code window, you will get an error when you try to open the form in the Design window because of the reference to the event handler from the generated code. For example, if you create the click event handler for btnExit, and delete the code for the event handler, you will get the following error message when you try to display the form.

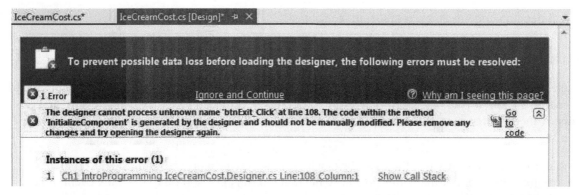

To remove the error, click on the "Go to Code" button on the error window. The **Designer.cs** file, which contains the generated code, opens with the pointer at the line that refers to the click event handler, as shown below:

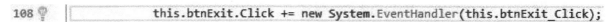

You may manually delete this statement from the Designer.cs file to get rid of the error and display the form. **Be careful not to change or delete any other line**. Close the Designer.cs file to display the form.

How do you properly delete such an event handler? Use the following steps:

Select the button on the form.

Click the **Events** button (⚡) in the Properties window to display the events for the button, as shown below:

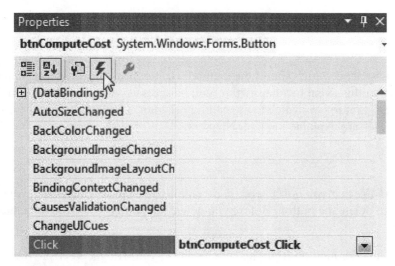

Select the Click event.

Click the dropdown arrow and select the event handler you want to delete (you need to do it even if the event handler already is displayed).

Right click the event handler in the Properties window and select Delete.

This process will delete the event handler if you haven't entered any code within the event handler. If you have entered any code, you still need to select Delete in the Properties window and then manually delete the event handler from the code window.

Step 1-14: Double click btnExit to create the Click event handler. Then delete the event handler by first deleting it from the Properties window, and then deleting the code for the event handler, including the header, the braces (and code, if any, within the braces) from the Code window.

Now you are ready to move to phase 5: Write and test the code.

Review Questions

1.16 What is a namespace?

1.17 What does the *using* directive specify?

1.18 What is a method?

1.19 What is an event handler?

1.6 Introductory Programming Concepts

To help you understand some basic programming concepts, let's first look at the completed code, shown in Figure 1-18, to compute the cost for an order of ice cream.

Figure 1-18: Code to compute cost

```
20          private void btnComputeCost_Click(object sender, EventArgs e)
21          {
22              // Compute the total cost for an order of icecream:
23
24              // Declare variables to hold input data and computed results
25              int scoops;
26              double unitPrice, cost;
27
28              //Get input data from Text Boxes:
29              scoops = int.Parse(txtScoops.Text);
30              unitPrice = double.Parse(txtUnitPrice.Text);
31
32              // Compute result
33              cost = unitPrice * scoops;
34
35              // Display Result
36              lblCost.Text = cost.ToString("C");
37          }
```

As you can see from the code, the method includes three segments corresponding to the three subtasks that were identified earlier:

Get number of scoops and unit price (lines 29 & 30)
Compute cost (line 33)
Display cost (line 36)

Note that the line numbers are not part of the code. They are inserted using a setting in Visual Studio for ease in reference to a statement (select *Tools, Options, Text Editor, C#* and check *Line Numbers*).

Let's look closely at the statements in this program.

Comments

Line 22: `//Compute the total cost for an order of ice cream:`
Lines starting with // are comments for the programmers and are ignored by the compiler.
You may comment multiple lines by placing the lines between /* and */, as in Figure 1-19.

Figure 1-19: Commenting multiple lines

```
22              /*Compute the total cost for an order of icecream
23                 using the formula: cost = unitPrice * NumberOfScoops:
24                 Declare variables to hold input data and computed results*/
```

You may comment out a set of lines by highlighting the lines (drag the mouse pointer over them) and then clicking the Toolbar button *Comment out the selected lines*, shown in the following image:

The button to the right would uncomment the selected lines.

1.7 Data Types

Lines 25 and 26 declare variables and specify what types they are:

Line 25: `int scoops;`
Line 26: `double unitPrice, cost;`

In order to understand the above statements and the rest of the program, we will take a closer look at data types.

Data used in a program can be grouped broadly into the following types:

Integer: An integer is a whole number (e.g., 255677, -54).
Nonintegers or reals: A real number has one or more decimal places (e.g., 255677.75; -0.255).
Text: Text data include **character** type that consists of an individual character (e.g., 'A'; 'B') and **string** type that consists of a group of characters, including alphabetical characters, numeric digits and special characters (e.g., "John Smith"; "P410-2023-3180-05").
Notice that a character is enclosed in single quotes, and a string is enclosed in double quotes.
Boolean: A Boolean has the value *true* or *false*.
Date/Time: Date and time are enclosed in pound signs, as in #05/25/2016 02:30:00 PM#.

Let's look at the different C# types for integers and reals.

Integer types

Table 1-3 shows the different integer types in C#. Note that C# is case sensitive. That means the key words for data types, like short and int, are case sensitive. So, "Short" and "Int" are invalid key words.

Table 1-3: Built-in C# data types for Integers

C# key word	Size (memory used)	Range of values
sbyte	8 bits	-128 to 127
short	16 bits	-32,768 to 32,767
int	32 bits	-2,147,483,648 to 2,147,483,647
long	64 bits	-9,223,372,036,854,775,808 to 9,223,372,036,854,775,807
char	16 bits	0 to 65535

There are additional "unsigned" data types, **byte, ushort, uint** and **ulong,** which can store only positive, but larger, integers with maximum size close to double that of signed integers. For example, **uint** type variables can hold values up to 4,294,967,295, whereas the maximum for **int** type is only 2,147,483,647.

The size shows the memory needed to store the data, and the range shows the range of values that can be stored in the variable. Larger data types take more memory, as reflected in the size.

Note that **character** data (denoted by the C# key word **char**) is represented by integers using one of the popular standards called Unicode that allows representation of 65,535 different characters from different languages.

C# versus .Net Types

The names of basic data types in C# are different from those in .Net Framework. However, there is direct correspondence between them. For example, **int** type in C# is the same as **System.Int32** in .Net, and **long** in C# is the same as **System.Int64**.

Noninteger (real) types

Table 1-4 shows the data types for non-integers.

Table 1-4: C# noninteger types

C# key word	Size	Range of values	Precision	How the number 123456789012.345678901234567890 may be represented
float	32 bits	$\pm 1.5 \times 10^{-45}$ to $\pm 3.4 \times 10^{38}$	7 digits	0.1234568×10^{12}
double	64 bits	$\pm 5.0 \times 10^{-324}$ to $\pm 1.7 \times 10^{308}$	15 digits	$0.123456789012346 \times 10^{12}$
decimal	128 bits	$\pm 1.0 \times 10^{-28}$ to $\pm 7.9 \times 10^{28}$	28–29 digits	123456789012.345678901234567800

As indicated by the ranges, float and double can store larger numbers than decimal type. They also take less memory. But float and double cannot represent all noninteger numbers exactly (due to the method used by them to represent numbers), making them less suitable than decimals for financial calculations.

Decimal type, in general, can represent real numbers exactly, and it has better precision than float and double. Hence, as stated earlier, decimal is better suited for financial calculations where exact representation is important. However, decimal type takes more memory and is slower to work with than float and doubles. Because decimal has a lower range, it may not be appropriate for some scientific calculations.

The C# types, float and double, use a method of representation called **floating point** method to approximate a wide range of values to a fixed number of significant digits. Therefore, float and double also are referred to as floating point types.

string type

The C# type **string** represents a combination of zero or more characters. "John Smith" and "P410-2023-3180-05" are examples of string type data. Note that string type data is enclosed in quotes.

bool type

The C# type bool represents Boolean data that include the values **true** and **false**.

DateTime type

C# doesn't have a data type for dates and times. The .Net base type DateTime is used to represent dates and/or times.

Review Questions

1.20   What is the appropriate data type to represent student GPA?

1.21   Which real data type is best suited for financial calculations where exact representation of numbers is important?

1.22   What are the advantages of float and double types compared to decimal?

1.8 Working with Variables

Programs typically use variables to hold data. A simple variable holds one piece of data, and more complex types like an array can hold a set of data. In general, we will use the term "variable" to denote a simple variable.

A variable is a location in computer memory where it holds data while the program is running. The program accesses the location using the name given to the variable.

For variables, C# requires that you specify what type of data will be stored in each variable.

Creating (Declaring) Variables

Variables are created by specifying the type and a name of the variable. The statement shown below from Line 25 in Figure 1-18 creates a variable named **scoops** that is of **int** type:

```
int scoops;
```

When the program is run, the above statement creates a variable named *scoops* in computer memory to hold an int type integer data. Because we haven't assigned any value to scoops, by default, the value zero is stored in the variable, as represented below:

```
┌─────┐
│  0  │
└─────┘
 scoops
```

You may declare two or more variables of the same type in a statement, as in line 26 in Figure 1-18:

```
double unitPrice, cost;
```

This statement creates two variables, unitPrice and cost, to hold double type **floating point** data.

Note that a **semicolon** is required to denote the end of each C# statement.

Guidelines for naming variables

Variable names must start with a letter, and can contain only letters, numbers or the underscore "_" character.

The convention in C# is to use the **Camel Case:** The first letter of a variable is lowercase, and the first letter of subsequent words, if any, are capitalized, as in totalCost and averageRetailPrice.

Use meaningful names for variables to make programs easy to read. You should avoid using acronyms or abbreviations that are not widely recognized. For example, use totalQuantitySold rather than tqs.

Variables to hold other types of data

The example of code in Figure 1-18 that we are looking at does not include variables to hold other types of data, like character and string data. Every type of variable is declared using the same syntax. Here are examples of statements to create character and string variables:

```
char middleInitial;
string lastName;
```

Assigning Values to Variables

After you create a variable, you can assign a value to it using a statement of the form

variable name = value (or an expression that computes a value)

For example, the variable scoops declared in the program as type **int** may be assigned the value 3, using the statement

```
scoops = 3;
```

That is, the value 3 will be stored in a memory location called scoops, as represented below:

```
┌─────┐
│  3  │
└─────┘
 scoops
```

Similarly, the statement

```
scoops = (10/5) + 1;
```

would assign the value 3 to scoops.

Lifetime of a variable

You cannot use a variable that doesn't exist! That means you cannot use a variable before it is declared, because the variable exits only after it is created using the declaration statement.

Similarly, a variable ceases to exist after completion of execution of the method where it is declared. Thus, the **lifetime** of a variable that is declared within a method is the time period between its creation in memory during execution of the method and the time when the execution of the method is completed. In general, the lifetime represents the time period between the creation of a variable and the destruction of the variable from memory.

A related term is **scope** of a variable, which is the section of a program where the variable can be accessed. The scope of a variable is the block of statements between the variable's declaration and the next right curly brace. The scope of variables is discussed further in section 1.10.

Initializing a variable

You may declare a variable and also assign it a value (initialize the variable) in one statement, using the syntax,

    type variable name = value

Here are some examples:

```
int scoops = 3;
double unitPrice = 3.95;
char middleInitial = 'C';
string lastName = "Philip";
bool validData = true;
```

Caution! The statement "3 = scoops" will not do the same thing as "scoops=3"; it is meaningless in a program. The variable name must be on the left side of the equal sign, and the value on the right side.

Default Values of Variables

If a variable is declared without initializing it, the variable will have no value until a value is assigned to it. However, if you use the **new** key word to create a variable, as in

    int scoops = new int();

the variable will have a default value depending on its type.

The use of the default constructor, new, is explained in Chapters 2, 7 and 13. The default values of a few different types of variables are shown in Table 1-5.

Table 1-5: Default values of data types

Type	Default value
Integers and reals, except decimal types	0
decimal	0m ("m" represents decimal type)
bool	false
DateTime	# 01/01/0001 12:00:00AM #
String	null reference

Thus, the statement `int scoops = new int()` has the same effect as `int scoops = 0`.

Representing Numbers

Scientific notation

You may represent very large numbers using scientific notation:

    3.95e8

e8 represents 10 to the power 8. Thus, 3.95e8 is equal to 3.95 times 10^8—that is, 395,000,000.

Similarly, 3.95e-8 represents 3.95 times 10 to the power -8—that is, 0.0000000395.

Numbers can have only numeric digits, decimal point, minus and plus signs, and the letter "e" or "E" when using scientific notation. Commas and dollar signs ($) are invalid within a number. So, the following statements are invalid:

```
double unitPrice = $1.50;  //Invalid; dollar sign is not allowed in a number
double cost = 3,520.00;    //Invalid; comma is not allowed in a number
double amount = "15.45"    //Invalid; "15.45" is a string
```

Caution! In general, the compiler does not like it when you assign the wrong type of value to a variable. The following statement would result in an error because scoops is declared as type int, which is a type of integer, but 1.5 is a real number.

```
int scoops = 1.5;    //Invalid; 1.5 is not an integer
```

We will look at specific rules on mixing different types of integer and real types in Chapter 2.

Assigning the Value of One Variable to Another Variable

You may assign the value of one variable to another variable, as in the following example:

```
int scoops = 3;
int temp;
temp = scoops;
```

The statement "**temp = scoops;**" copies the value of scoops into the variable temp. After execution of this statement, the storage locations for both variables will contain the same value 3:

What happens if you change the value of scoops by adding the following statement after the above code?

```
scoops = 5;
```

Does it change the value of temp? The answer is no, because each variable has its own storage location. After execution of the statement "**scoops = 5,**" the two variables will have the following values:

Caution! The statement

```
temp = scoops;
```

which assigns the value of scoops to temp, is not the same as

```
scoops = temp;
```

which assigns the value of temp to scoops.

Tutorial 2: Developing the Code

In this tutorial, you enter the code shown in Figure 1-18 into the click event handler of the button that computes the cost.

Step 2-1: Declare variables scoops, unitPrice and cost:

Make sure the Code window is open.
Enter the following two statements after the opening brace within the btnComputeCost_Click method that was created in Step 1-13, as shown in Figure 1-20. If you don't have this method, double click the btnComputeCost button to create it.

```
int scoops;
double unitPrice, cost;
```

Figure 1-20: Click event handler for btnComputeCost button

```
15      public IceCreamCost()
16      {
17          InitializeComponent();
18      }
19
20      private void btnCompteCost_Click(object sender, EventArgs e)
21      {
22          int scoops;
23          double unitPrice, cost;
24      }
```

As stated earlier in secton 1.6, C# is case sensitive. That means the key words for data types, like int and double are case sensitive. So, if you type "Int scoops" instead of "int scoops," you will get an error when you run the project.

IntelliSense

As you may have noticed, when you type code, the IntelliSense list box pops up with a list of items you can choose from to help you complete the code. For example, when you type the letter "i" at the beginning of a statement, the following list box is displayed.

```
private void btnComputeCost_Click(object sender, EventArgs e)
{
    i
}
```

- InstallerTypeAttribute
- InstanceCreationEditor
- InsufficientExecutionStackException
- InsufficientMemoryException
- **int**
- Int16

You may select "int" by pressing the Tab key.

Next, do Step 2-2 to see what happens if you assign the wrong type of data to a variable.

Step 2-2: Experiment with assigning different types of data to the variables:

Add each of the following statements, identify the error in each (hover the cursor over the number), write down in your own words what the problem is and how to correct it, and delete the statement.

```
scoops = 1.5;
cost = $3.95;
unitPrice = 1,575.50;
cost = "3.95";
```

What's special about double types?

If you are wondering why we used only double type (not float or decimal) for variables that store real numbers like unit price and cost, it is because all real numbers, by default, are treated as double type. For example, 3.5, 1575.0 and 0.6 are double types. As we will see in Chapter 2, it is invalid to assign such double type numbers to a float or decimal type variable without appropriate conversion of the double type.

Getting Values from a TextBox

The data you type into a TextBox, whether it is a number or a string, is represented as a string. So, if you enter 3.95 into the TextBox, txtUnitPrice, the Text property of the TextBox, txtUnitPrice.Text, holds the string "3.95," not the number 3.95.

Because the variable unitPrice is of double type, the value that is assigned to it cannot be a string; the value (txtUnitPrice.Text) must be converted to double before assigning it to the variable unitPrice.

Parse Method: Converting String Type Data to a Numeric Type

Every data type has a set of built-in programs called **methods** to do different tasks related to the data type. Numeric data types have a method called **Parse** that converts a string data to the numeric data type by parsing each character.

Let's look at the statements in lines 29 and 30 in Figure 1-18. The statement in line 29,

```
29      scoops = int.Parse(txtScoops.Text);
```

uses the **Parse()** method of int type to convert the string, txtScoops.Text, to an int type number. The converted number is assigned to the variable scoops.

Similarly, in line 30,

```
30      unitPrice = double.Parse(txtUnitPrice.Text);
```

double.Parse(txtScoops.Text) converts the string value txtUnitPrice.Text to double data type, which is assigned to unitPrice.

The Parse method would result in error at runtime if the string to be converted is empty. For example, if the TextBox, txtUnitPrice, is empty, the program would break at line 30 with an error message. Further, the Parse method can covert only a string to numeric data types. It cannot convert one numeric data type, like int, to another numeric data type, like double.

Convert Class: An Alternative to Parse Method

Convert is a .Net class that has different methods that convert to/from base .Net data types. As discussed earlier under "C# versus .Net Types," the data types in C# are different from the base .Net data types. However, there is direct correspondence between them. For example, the int type in C# is the same as **Int32** in .Net, and double in C# is the same as **Double** in .Net. Because Convert is a .Net class, it can be used in other .Net languages like Visual Basic as well.
The following statement converts the string, txtScoops.Text, to int type number, using the ToInt32() method.

```
scoops = Convert.ToInt32(txtScoops.Text);
```

You may use the above statement in place of the current statement in line 29 that uses the Parse method.

Similarly, the ToDouble() method of Convert class converts the string, txtScoops.Text, to Double type.

```
unitPrice = Convert.ToDouble(txtScoops.Text);
```

The Parse method can convert only a **string** to numeric data types, but the methods of Convert class also can be used to covert a numeric data type to another numeric data type.

Step 2-3: Add the following code to convert the strings, txtScoops.Text and txtUnitPrice.Text, to numeric values, and to store them in variables:

```
scoops = int.Parse(txtScoops.Text);
unitPrice = double.Parse(txtUnitPrice.Text);
```

Review Questions

1.23   What is the Camel Case convention to name variables in C#?

1.24   What are the requirements for variable names?

Correct errors, if any, in the following statements:

1.25   int orderQty = 4.5;

1.26   char lastName = "Jones";

1.27   string firstName = 'Jack';

1.28   double totalAmount = "145.50";

1.29   double totalCost = 2,145.00;

1.30 double unitPrice = $15.50;

1.31 int quantity;
      15 = quantity;

1.32 int quantity = txtQuantity.Text; (txtQuantity is the name of a TextBox)

1.33 int units = double.Prase(txtUnits.Text); (txtUnits is the name of a TextBox)

What would be the values of the variables currentSales and previousSales after execution of the following set of statements?

1.34 int currentSales = 110;
      int previousSales;
      currentSales = previousSales;

1.35 int currentSales = 110;
      int previousSales;
      previousSales = currentSales;

1.9 Doing Calculations

Common mathematical operators used in calculations are shown in Table 1-6.

Table 1-6: Math operators to do calculations

Operator	Operation	Example
*	multiply	Multiply 3.95 by 3 (3.95*3)
/	divide	Divide 11.85 by 3 (11.85/3)
+	add	Add 11.85 and 0.65 (11.85+0.65)
-	subtract	Subtract 1.25 from 12.5 (12.5-1.25)
-	unary minus, for negative numbers; e.g., -350.50	Negative 2.5 (-2.5)

The statement shown below from Line 33 in Figure 1-18 multiplies the unit price and number of scoops to compute the cost.

33   cost = unitPrice * scoops;

Step 2-4: Type in the code to compute cost:

    cost = unitPrice * scoops;

Creating Expressions: Precedence of Operators

In an expression, multiplications and divisions are done before additions and subtractions, and then calculations take place from left to right. Consider the following expression:

    5+4*3-2/2

The above expression would evaluate to 5+(4*3)-(2/2), yielding 16.

Here is another example:

    12/4*2

This expression would be evaluated from left to right as (12/4)*2, yielding 6.

It is always a good practice to use parentheses to make the order of calculations clear and to avoid unintended errors.

Review Questions

Evaluate the following expressions:

1.36   10/2+3*4

1.37   12-6/2*3+1

1.10 Formatting and Displaying Output in a Label

What you see in a Label is the value of its Text property. Earlier in Step 1-10, you displayed text in a Label in the Designer window using its Text property. Now we look at how to display a text in a Label using code at runtime. To do this, you assign the text to the Text property of the Label.

The statement in line 36,

```
lblCost.Text = cost.ToString("C");
```

displays the value of cost in the Label, lblCost, by assigning it to the Text property of lblCost.

The value of cost, which is a double type, is converted to a string using the ToString() method of the variable, described next.

Formatting Numbers Using the ToString() Method

Each numeric data type has a method (program unit) called ToString() that converts and formats the number to a string. For example,

```
cost.ToString()
```

would convert the value of cost to a string so that it can be displayed in a Label.

The ToString() method also allows a format code (string) to be specified, as in

```
cost.ToString("C")
```

where "C" specifies currency format.

Examples of using additional ToString() format codes are presented in Table 1-7.

Table 1-7: ToString() format codes

Number	Format string	ToString() would format the number as
1145.857	null	1145.857
1145.857	"C"	$1,145.86
1145.857	"N"	1,145.86
1145.857	"F"	1145.86
0.156	"P"	15.60%

Chapter 1: Introduction to Visual Studio and Programming

Optionally, you may specify the number of decimal places. Here are some examples:		
1145.857	"C0"	$1,146
-1145.857	"C1"	($1,145.9)
-1145.857	"N1"	-1,145.9
1145	"N3"	1,145.000
Format code for integers only		
157	"D5"	00157
157	"D"	157
157	"D1"	157

If the format code doesn't specify the number of decimal places, as in **cost.ToString("C")**, the default number of decimal places is two.

Step 2-5: Add the statement to display cost:

```
lblCost.Text = cost.ToString("C");
```

The btnComputeCost_Click event handler should look as shown in Figure 1-21.

Figure 1-21: Click event handler of btnComputeCost button

```
20          private void btnComputeCost_Click(object sender, EventArgs e)
21          {
22              /*Compute the total cost for an order of icecream
23                  using the formula: cost = unitPrice * NumberOfScoops:
24                  Declare variables to hold input data and computed results*/
25              int scoops;
26              double unitPrice, cost;
27
28              //Get input data from Text Boxes:
29              scoops = int.Parse(txtScoops.Text);
30              unitPrice = double.Parse(txtUnitPrice.Text);
31
32              // Compute result
33              cost = unitPrice * scoops;
34
35              // Display Result
36              lblCost.Text = cost.ToString("C");
```

Step 2-6: Run the form:

Double click **program.cs** to open the code editor window.
Specify IceCreamCost as the name of the form to be run, as shown:

Application.Run(new IceCreamCost());

Click the *Start* button on the Toolbar to run the form.
Type in 3 for scoops and 3.95 for unitPrice. Click the Compute Cost button to see the result.
Close the window.

Return to the Code Editor window by clicking on the IceCreamCost.cs tab.
Change the format string "C" to "C0," run the program and observe the difference in format.

Repeat with format strings "C1," "N" and "F," and observe the format in each case.

Run the form with no data for unit price. The program breaks at line 30, with an error message as shown below, because txtUnitPrice.Text is an empty string.

```
unitPrice = double.Parse(txtUnitPrice.Text);

// Compute result                    ⚠ FormatException was unhandled
cost = unitPrice * scoops;           An unhandled exception of type 'System.FormatException' occurred in mscorlib.dll

// Display Result                    Additional information: Input string was not in a correct format.
```

Click the **Stop Debugging** button (■) on the Toolbar to stop the program.
Click **IceCreamCost.cs[Design]** tab to open the Designer window.

Clearing Controls

TextBoxes have a Clear() method that clears their content. The Labels, however, are cleared by setting the Text property to null (which is nothing) or an empty string represented by two quotes with nothing in between ("").You also may use this method for clearing TextBoxes.

Step 2-7: Double click the Clear button to create the click event handler of the button, and type in the code to clear the controls:

```
private void btnClear_Click(object sender, EventArgs e)
{
        txtScoops.Clear();
        txtUnitPrice.Clear();
        lblCost.Text = null;
}
```

Note that Clear is the name of a method (a program unit), and methods require the pair of parentheses at the end of their names.

Closing a Form

To close a form that is running, use the Close() method of the form.

Step 2-8: Double click the Close button, and type in the code within the click event handler:

```
private void btnExit_Click(object sender, EventArgs e)
{
        this.Close();
}
```

In the above code, "this" refers to the current form.

The Complete Code for the IceCreamCost Form

Figure 1-22 shows the complete code.

Figure 1-22: The complete code for IceCreamCost

```csharp
using System;
using System.Collections.Generic;
using System.ComponentModel;
using System.Data;
using System.Drawing;
using System.Linq;
using System.Text;
using System.Threading.Tasks;
using System.Windows.Forms;

namespace VCintro
{
  public partial class IceCreamCost : Form
  {
    public IceCreamCost()
    {
      InitializeComponent();
    }
    private void btnComputeCost_Click(object sender, EventArgs e)
    {
            //Compute the total cost for an order of ice cream:

            //Declare variables to hold input data and computed results
            int scoops;
            double unitPrice, cost;

            //Get input data from Text Boxes:
            scoops = int.Parse(txtScoops.Text);
            unitPrice = double.Parse(txtUnitPrice.Text);

            // Compute result
            cost = unitPrice * scoops;

            // Display Result
            lblCost.Text = cost.ToString("C");
    }
    private void btnExit_Click(object sender, EventArgs e)
    {
            this.Close();
    }
    private void btnClear_Click(object sender, EventArgs e)
    {
            txtScoops.Clear();
            txtUnitPrice.Clear();
            lblCost.Text = null;
    }
  }
}
```

Step 2-9: Run the form:

Test the Clear and Exit buttons.

Experimentation with Expressions

To see for yourself how C# evaluates an expression, do Step 2-10.

Step 2-10: To give $1 off on each scoop, change the expression for computing the cost as follows:

```
cost = unitPrice-1 * scoops;
```

Click the Start button to run the program. Enter 2 for scoops and 3.5 for unit cost.
Before clicking the button, compute the value of cost manually.
Note: If you manually computed 3.5-1 * 2 to be 5, you are wrong.

Next, click the Compute Cost button to display the actual value.
Why is it 1.5? Because multiplication is done before addition, the expression is evaluated as 3.5-(1*2), which yields 1.5.

To specify that the expression, (unitPrice-1), should be multiplied by scoops, use parentheses around the expression, as in (unitPrice-1)*scoops. Run the program again with the same values for scoops and unitPrice (2 and 3.5, respectively). Check the value of cost. Now, it should be evaluated as (3.5-1)*2, yielding 5.

Change the expression back to cost = unitPrice * scoops.

Invalid numbers

To see for yourself errors due to invalid data, do Step 2-11.

Step 2-11: Run the program, and enter 2 for scoops and $3.95 for unitPrice.

Click the Compute Cost button.
The program breaks and highlights the statement that contains the error
    unitPrice = double.Parse(txtUnitPrice.Text);
and displays the message

The error occurs because $3.95 is not a valid number; hence, it cannot be converted to a decimal.

The Parse method, however, accepts numbers with commas. For example 2,450 is valid. To test it, run the program, enter 2 for scoops and 2,450 for unitPrice (it is some expensive ice cream!) and compute the cost. But, remember, it is invalid to assign a number that contains a comma to a variable, as in

    cost = 2,450;   // invalid!

It's time to practice! Do Exercise 1-1 (uses integer data) and Exercise 1-2 (uses real type data), provided at the end of the chapter.

1.11 Additional Concepts: Constants, Scope of Variables and ListBox

Now we expand on the concepts you learned in previous sections and introduce constants, scope of variables, and formatting multiple lines of output using ListBox. We use an expanded version of the application developed in Tutorials 1 and 2 to help understand these topics.

Tutorial 3: Ice Cream Cost with Sales Tax

This tutorial expands the application developed in Tutorials 1 and 2 by adding sales tax to compute the total cost. The subtotal, sales tax and total cost are formatted and displayed in a ListBox on a new form called **IceCreamTotal,** as shown in Figure 1-23.

Figure 1-23: IceCreamTotal form with sales tax

Figure 1-24 shows the input, process and output of the program. The output now consists of the subtotal and sales tax in addition to the cost.

Figure 1-24: Input, processing and output for IceCreamTotal

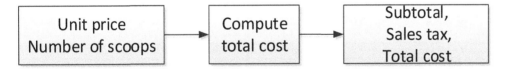

The process, compute total cost, involves computing the following expressions:

    Subtotal = Unit price * Number of scoops
    Sales tax = Subtotal * Sales tax rate
    Total cost = Subtotal + Sales tax

Note that the sales tax rate, which is needed to compute sales tax, is not a part of the input. Then where does the program get it from? The value of sales tax rate is "hard-coded" into the program, as shown in Figure 1-25—that is, it is a part of a program statement, which keeps the example simple but makes it difficult to change the rate.

Step 3-1: To create this application, add a new form to the project, Ch1_IntroProgramming:

    Select: *Project, Add Windows Form*

Change the name of the file form1.cs in Solution Explorer to **IceCreamTotal.cs**.
Add the controls shown in Figure 1-23, as follows:

Rather than adding the Labels, TextBoxes and Buttons to the form, you may copy the following controls from IceCreamCost and paste them to the current form:

Labels and TextBoxes for number of scoops and unit price
Buttons for compute total, clear and exit

To copy multiple controls from a form, you may select the controls by dragging the pointer over an area that contains the controls, or you may select individual controls while the shift key is pressed. Right click any of the selected controls, and select *Copy*.

To paste the control on IceCreamTotal, right click the form, and select *Paste*.

In addition, add a ListBox and rename it **lstOutput**.

Step 3-2: Double click btnComputeTotal to create the Click event handler of the button, as shown below:

```
private void btnComputeTotal_Click(object sender, EventArgs e)
{
}
```

The complete code required to compute and display the total cost is shown in Figure 1-25. You will enter each segment of the code after we discuss the code.

Figure 1-25: Code to compute and display total cost

```
20        private void btnComputeTotal_Click(object sender, EventArgs e)
21        {
22            // Compute the total cost for an order of icecream, including sales tax.
23
24            // Declare variables to hold input data and computed results
25            int scoops;
26            double unitPrice, subTotal, salesTax, totalCost;
27
28            //Declare a constant to hold the tax rate
29            const double TAX_RATE = 5.5;
30
31            //Get input data from Text Boxes:
32            unitPrice = double.Parse(txtUnitPrice.Text);
33            scoops = int.Parse(txtScoops.Text);
34
35            // Compute results
36            subTotal = unitPrice * scoops;
37            salesTax = subTotal * TAX_RATE / 100;
38            totalCost = subTotal + salesTax;
39
40            // Display Results with decimals aligned
41            lstOutput.Items.Add(string.Format("{0,12}{1,10:C}", "Sub Total:", subTotal));
42            lstOutput.Items.Add(string.Format("{0,12}{1,10:C}", "Sales Tax:", salesTax));
43            lstOutput.Items.Add("");   // Add a blank line
44            lstOutput.Items.Add(string.Format("{0,12}{1,10:C}", "Total Cost:", totalCost));
45        }
```

Lines 25 and 26 declare variables as in Tutorial 2.

Constants

Line 29 declares a **constant**. The key word **const** in line 29,

```
const double TAX_RATE = 5.5;
```

specifies that TAX_RATE is a constant (not a variable).

Constants are used to represent literal values that do not have to be changed at runtime. The use of a constant for tax rate assumes that the tax rate is a fixed value. However, if the tax rate is to be read from a file or a TextBox at runtime, a variable would be more appropriate.

The constant has the following characteristics:

1. The value of a constant cannot be changed at runtime.
   The following code that assigns a value to the constant, numberOfScoops, at runtime, is invalid:
   ```
   const int numberOfScoops = 0;
   numberOfScoops = 2;   // Invalid
   ```

2. A constant must be initialized as part of the declaration by assigning a literal value or a constant. For example, the statement,
   ```
   const int numberOfScoops = int.Parse(txtScoops.Text);   // Invalid
   ```

   is invalid because the value assigned to the constant is not a constant or a literal value. Similarly, the following statement also is invalid because the constant is not initialized as part of the declaration:
   ```
   const int numberOfScoops;   // Invalid
   ```

3. A constant is replaced by its value at compile time.

Step 3-3: Double click btnComputeTotal to create its click event handler, and add the comments and statements to declare the variables and a constant, as shown in Figure 1-26.

Figure 1-26: Code to declare variables and constant

```
22              // Compute the total cost for an order of icecream, including sales tax.
23
24              // Declare variables to hold input data and computed results
25              int scoops;
26              double unitPrice, subTotal, salesTax, totalCost;
27
28              //Declare a constant to hold the tax rate
29              const double TAX_RATE = 5.5;
```

Figure 1-27 shows the code to get input data from TextBoxes (the same as in Tutorial 2), and to compute subtotal, sales tax and total cost.

Figure 1-27: Code to get input data and compute results

```
31                      //Get input data from Text Boxes:
32                      unitPrice = double.Parse(txtUnitPrice.Text);
33                      scoops = int.Parse(txtScoops.Text);
34
35                      // Compute results
36                      subTotal = unitPrice * scoops;
37                      salesTax = subTotal * TAX_RATE / 100;
38                      totalCost = subTotal + salesTax;
```

The statement in line 32 converts the data typed into txtUnitPrice to double type because unitPrice is a double. Similarly, in line 33, data from txtScoops is converted to int type because scoops is an int type variable.

Note that when an expression has multiplication and division, as in line 37, the two operations are treated as equal in deciding the order of computation. The calculations take place from left to right, as in (subTotal * TAX_RATE) / 100, which is okay here because it doesn't make any difference whether multiplication is done first or division is done first.

Step 3-4: Enter code from Figure 1-27.

Displaying Data in a ListBox

In this application, the output is displayed in a ListBox in two columns, the first column for the labels and the second column for the numbers, as shown in Figure 1-28. The decimals are aligned automatically by right justifying the numbers in the second column and formatting them with two decimal places.

Figure 1-28: Formatted output

A ListBox can be used to display items and also let users select one or more items from the list. Here we use the ListBox just to display output. We will discuss ListBox in more detail in Chapter 5.

string.Format() method

First, we look at formatting data using the Format() method of string type, which has the following syntax:

    string.Format(format string, data items separated by commas)

The Format method formats multiple data items and combines them into a single string. For example, the following statement formats the first line of the output shown in Figure 1-28:

```
string.Format("{0,12}{1,10:C}", "Subtotal:", subTotal)
```

This statement lists two data items, a literal and a variable, to be formatted: "Subtotal:" and subTotal. These two items are identified by indices 0 and 1, respectively.

The format string, "{0,12}{1,10:C}," specifies that

> parameter 0 ("Subtotal") should be displayed in a column that is 12 characters wide, and parameter 1 (subtotal) should be displayed in a column that is 10 characters wide.

The "C" in {1,10:C} specifies currency format (default is two decimal places).

By default, each data item is right justified within its column.

Displaying output in a ListBox

Each line (item) in a ListBox is a member of the **Items collection** of the ListBox. The **Add() method** adds a new line to the Items collection of the ListBox. For example,

```
lstOutput.Items.Add("")
```

would add a blank line to the ListBox.

The following statement would display a title in the ListBox:

```
lstOutput.Items.Add("Ice cream cost with sales tax");
```

We will discuss collections and its methods in detail in Chapter 9.

To display the first line of output, you use the string.Format method, as follows:

```
lstOutput.Items.Add(string.Format("{0,12}{1,10:C}", "Subtotal:", subTotal));
```

The following statements from lines 43 and 44 use the same format string as the one used to display the subtotal. Because all three values (subtotal, sales tax and total cost) are right justified (by default) and displayed using the same column width, the decimal points of the three numbers are automatically aligned.

```
43  lstOutput.Items.Add(string.Format("{0,12}{1,10:C}", "Sales Tax:", salesTax));
44  lstOutput.Items.Add(string.Format("{0,12}{1,10:C}", "Total Cost:", totalCost));
```

Additional formatting codes presented earlier in Table 1-7 for the ToString() method also apply to string.Format(). These include N, F, and P and others.

Font for ListBox

For the decimals to align, it also is important that the font for the ListBox is set to a fixed-width font like Courier New.

Step 3-5: **Change the font type for the ListBox to *Courier New*.** Add the following code from Figure 1-29 to display results in the ListBox.

Figure 1-29: Code to display formatted output in ListBox

```
40              // Display Results with decimals aligned
41              lstOutput.Items.Add(string.Format("{0,12}{1,10:C}", "Sub Total:", subTotal));
42              lstOutput.Items.Add(string.Format("{0,12}{1,10:C}", "Sales Tax:", salesTax));
43              lstOutput.Items.Add("");   // Add a blank line
44              lstOutput.Items.Add(string.Format("{0,12}{1,10:C}", "Total Cost:", totalCost));
```

The complete code should look as shown in Figure 1-25.

Step 3-6: Run the form:

Double click **program.cs** to open the code window.
Specify IceCreamTotal as the name of the form to be run:

```
Application.Run(new IceCreamTotal());
```

Click the *Start* button to run the form and test the code you entered. Run the form with no data for number of scoops. Run it again with $3.90 for unit price.
Note that in both cases, the program crashes with an error message, and you will have to rerun the program. Later in this chapter you will learn how to have the program deal with errors more gracefully.

Clearing a ListBox

The Clear method of the Items collection clears all lines (items) from the ListBox. Figure 1-30 shows the code to clear the TextBoxes and the ListBox.

Figure 1-30: Code to clear controls

```
private void cmdClear_Click(object sender, EventArgs e)
{
        txtScoops.Clear();
        txtUnitPrice.Clear();
        lstOutput.Items.Clear();
}
```

Step 3-7: Add the code from Figure 1-30 to the click event handler of Clear button.

Scope of Variables

Variables like scoops and unitPrice that are declared within a method are called **local variables.** A local variable is available for use only within the method where it is declared, and only in the statements that follow the declaration.

As discussed earlier, the part of a program where a variable is accessible is called the **scope** of the variable. The scope of a variable is the block of statements between the variable's declaration and the next right curly brace. For example, the scope of the variable unitPrice is lines 27–44, which is part of the btnComputeTotal_Click event handler method following the declaration of unitPrice.

A variable is not known outside its scope. That means, it would be an error to use scoops in any method other than btnComputeTotal_Click where it is declared

Step 3-8: Use the variable scoops inside btnClear_Click method as follows:

```csharp
private void cmdClear_Click(object sender, EventArgs e)
{
    txtScoops.Clear();
    txtUnitPrice.Clear();
    lstOutput.Items.Clear();
    scoops = 0;   // Error! - Invalid use of scoops
}
```

Note that the use of scoops is invalid, as indicated by the error.

Fields: Variables That Are Declared Outside Methods

Variables that are declared within a class but outside all methods are called fields. For example, the code shown in Figure 1-31 declares scoops as a field (line 19) within the class IceCreamTotal that represents the form, rather than as a local variable inside btnComputeTotal_Click method (line 29). The rest of the code remains the same as what was previously discussed. Fields are declared at the form level rather than inside a method.

Figure 1-31: Declaring fields

```csharp
13   public partial class IceCreamTotal : Form
14   {
15       public IceCreamTotal()
16       {
17           InitializeComponent();
18       }
19       int scoops;    // scoops declared outside the methods
20       private void btnComputeTotal_Click(object sender, EventArgs e)
21       {
22           // Compute the total cost for an order of icecream, including sales tax
23
24           // Declare variables to hold input data and computed results
25           // int scoops;
26           double unitPrice, subTotal, salesTax, totalCost;
27
28           //Declare a constant to hold the tax rate
29           const double TAX_RATE = 5.5;
30
31           //Get input data from Text Boxes:
32           unitPrice = double.Parse(txtUnitPrice.Text);
33           scoops = int.Parse(txtScoops.Text);
34
35           // Compute results
36           subTotal = unitPrice * scoops;
37           salesTax = subTotal * TAX_RATE / 100;
38           totalCost = subTotal + salesTax;
39
40           // Display Results with decimals aligned
41           lstOutput.Items.Add(string.Format("{0,12}{1,10:C}", "Sub Total:", subTotal));
42           lstOutput.Items.Add(string.Format("{0,12}{1,10:C}", "Sales Tax:", salesTax));
43           lstOutput.Items.Add("");   // Add a blank line
44           lstOutput.Items.Add(string.Format("{0,12}{1,10:C}", "Total Cost:", totalCost));
45   }
```

Because scoops is declared at the class level, it is available within each method and outside the methods. That means the variable can be assigned a value within btnComputeTotal_Click event handler method and set to zero within the btnClear_Click method. Variables declared at the form level exist until the form is closed.

Step 3-9: Declare scoops at the class level, and comment out its declaration within the method, as shown in Figure 1-31.

Note that the statement scoops = 0, within btnClear_Click method, does not give any error because scoops is declared at the class level. (It is not necessary to set scoops to zero; it is done to demonstrate the use of fields.)

Now, comment out the class level declaration in line 19, and uncomment the declaration of unitPrice in line 27. Comment out the statement unitPrice = 0 that you inserted into the btnClear_Click event handler.

When to use fields in a form

A variable may be declared as a field (i.e., declared at the class level) if it is to be used in more than one event handler method. We will see such use of fields in later chapters.

Why not declare every variable at the class level (as fields)?

First, a variable declared as a field stays in the memory until the form is closed. If the variable is needed only in one method, declaring it as a local variable inside the method will remove it from memory when execution of that method is completed—that is, fields consume memory for longer periods of time.

Second, when a variable is declared at the class level when it is needed only in one method, any other method may inadvertently change its value.

So, if a variable is to be used only in one method, then it should be declared inside that method—that is, declare it as a local variable, except in some special cases.

Review Questions

1.38   What would the Label lblCost display after execution of the following set of statements?

doubleextendedCost = 4525.457;
lblCost.Text = extendedCost.ToString("N");

1.39   A program needs to read the sales tax rate at runtime from a database and use it to compute the sales tax for customer orders. Would you use a constant or a variable to hold the sales tax rate while the program is running?

1.40   What are the special requirements of a constant?

1.41   What is a local variable?

1.42   How is a field different from a variable?

1.12 Exception Handling: Catching Errors

It would be annoying if you have to rerun a program just because you made a simple mistake like including a "$" sign in the dollar amount for unit price, as you found out in Step 3-6. How do you prevent the program from crashing if it encounters an error when it is run?

There is special code, called **exception handler**, which you can insert to catch such errors (exceptions). When a program encounters an error, it creates an object that contains information about the error and hands it to the system that runs the program. This process is called throwing an exception. Let's look at how to use the **try-catch** statement to catch exceptions.

Figure 1-32 shows how you would use a simple form of the try-catch method in the btnComputeTotal_Click event handler.

An error in any statement within the **try** clause will cause the method to throw an exception, skip the rest of the statements and jump to the catch clause. If there are no errors, the statements in the catch clause, called the catch block, are skipped.

Figure 1-32: Try-catch statement

```
private void btnComputeTotal_Click(object sender, EventArgs e)
{
    try
    {
        // Compute the total cost for an order of ice cream, including sales tax.
        // Declare variables to hold input data and computed results
        int scoops;
        …
        lstOutput.Items.Add(string.Format("{0,12}{1,10:C}", "Total Cost:",
        totalCost));
    }
    catch (Exception err)
    {
        MessageBox.Show(err.Message);
    }
}
```

You assign the name **err** (or any meaningful name) to the exception object using the catch clause as follows:

    catch (Exception err)

A commonly used property of the exception object is the **Message** property, which contains the error message that describes the error that caused the exception—that is, **err.Message** contains the specific error message.

MessageBox

MessageBox is an object available in .Net to display messages. The **Show** method of the MessageBox allows any string to be displayed as the message. For example, a message could be displayed using the statement
    `MessageBox.Show("Invalid data; please re-enter");`

We use MessageBox to display the message contained in err.Message. In the catch clause, we use the statement

1.12 Exception Handling: Catching Errors

```
MessageBox.Show(err.Message)
```

to display err.Message whose value is a string that describes the error.

Step 3-10: Incorporate a try-catch statement in the btnComputeTotal_Click event handler, as shown in Figure 1-32.

Step 3-11: Run the form. Click the Compute Total button after entering unit price but no data for number of scoops. Note the error message. Repeat with both data entered, but include a "$" sign in unit price.

Figure 1-33 shows the complete code that includes computing the total, clearing the controls and closing the form.

Figure 1-33: Complete code for IceCreamTotal form

```csharp
20      private void btnComputeTotal_Click(object sender, EventArgs e)
21      {
22          try
23          {
24              // Compute the total cost for an order of icecream, including sales tax.
25
26              // Declare variables to hold input data and computed results
27              int scoops;
28              double unitPrice, subTotal, salesTax, totalCost;
29
30              //Declare a constant to hold the tax rate
31              const double TAX_RATE = 5.5;
32
33              //Get input data from Text Boxes:
34              unitPrice = double.Parse(txtUnitPrice.Text);
35              scoops = int.Parse(txtScoops.Text);
36
37              // Compute results
38              subTotal = unitPrice * scoops;
39              salesTax = subTotal * TAX_RATE / 100;
40              totalCost = subTotal + salesTax;
41
42              // Display Results with decimals aligned
43              lstOutput.Items.Add(string.Format("{0,12}{1,10:C}", "Sub Total:", subTotal));
44              lstOutput.Items.Add(string.Format("{0,12}{1,10:C}", "Sales Tax:", salesTax));
45              lstOutput.Items.Add("");   // Add a blank line
46              lstOutput.Items.Add(string.Format("{0,12}{1,10:C}", "Total Cost:", totalCost));
47          }
48          catch (Exception err)
49          {
50              MessageBox.Show(err.Message);
51          }
52      }
```

```
53
54      private void cmdClear_Click(object sender, EventArgs e)
55      {
56          txtScoops.Clear();
57          txtUnitPrice.Clear();
58          lstOutput.Items.Clear();
59          //unitPrice = 0;   // Error! - Invalid use of unitPrice.
60      }
61
62      private void cmdExit_Click(object sender, EventArgs e)
63      {
64          this.Close();
65      }
```

It's time to practice! Do Exercise 1-3 (uses constant and try-catch) and Exercise 1-4 (uses ListBox and try-catch).

Exercises

Exercise 1-1: Maximum heart rate

The maximum heart rate during exercise is given by the formula: (220 - age). Create a form named HeartRate, within a project named Exercises_Ch1, to let the user enter an integer age into a TextBox, and compute and display the maximum rate in a Label by clicking a button.

Exercise 1-2: MPG

The miles per gallon is given by dividing the miles driven by the gallons of gas used. Assume both are nonintegers. Create a form named MilesPerGallon, within a project named Exercises_Ch1. The form should let the user enter the miles driven and gallons used into two TextBoxes, and compute and display the miles per gallon in a Label by clicking a button.

Exercise 1-3: Total cost

Create a form, TotalCost, within a project named Exercises_Ch1, to let the user enter the labor and material costs of a job into two TextBoxes, and display the subtotal, sales tax, and total as follows:

Use a try-catch statement to catch unexpected errors, like a missing data item. Use a constant for sales tax rate (6%). Subtotal = labor cost + material cost. Total cost is the sum of subtotal and sales tax.

Exercise 1-4: Total cost with ListBox

Create a form, TotalCostWithListBox, within a project named PrExercises_1, to let the user enter the labor cost and material cost of a job, and display the subtotal, sales tax and total cost in a ListBox as follows:

Use a try-catch statement to catch unexpected errors, like a missing data item. Assume 6% sales tax. Subtotal = labor cost + material cost. Total cost is the sum of subtotal and sales tax.

Chapter 2

Data Types

In Chapter 1, we looked at different types of integer, noninteger (real) and text data that are used in programs. There are certain rules you need to follow when you mix different types of numbers, like a decimal type and double type, in an expression or statement.

In this chapter, first we look at the differences between different types of numeric data and the rules on mixing them. Next, we discuss string type, followed by DateTime type.

Learning Objectives

After studying this chapter, you should be able to:

- Use literals, constants and variables of different data types.
- State the rules for casting one data type to another.
- Do common operations on strings and characters.
- Do computations with dates/times and format them.
- Use the Locals window and break points to debug programs.

Topics

2.1	Working with Different Numeric Data Types	2.3	Working with Strings and Characters
2.2	Casting	2.4	Working with Dates and Times

2.1 Working with Different Numeric Data Types

To refresh your memory, Table 2-1 shows commonly used integer types of numbers, their sizes and range of values. You are already familiar with the int type that you used in Chapter 1.

Table 2-1: Integer types

C# key word	Size (memory used)	Range of values
sbyte	8 bits	-128 to 127
short	16 bits	-32,768 to 32,767
int	32 bits	-2,147,483,648 to 2,147,483,647
long	64 bits	-9,223,372,036,854,775,808 to 9,223,372,036,854,775,807

As stated in Chapter 1, there are additional "unsigned" data types, **byte, ushort, uint** and **ulong,** which can store only positive, but larger, integers with maximum size close to double that of signed integers. For example, **uint** type variables can hold values up to 4,294,967,295, whereas the maximum for **int** type is only 2,147,483,647. There is also a special integer type, denoted by C# key word char, which can represent 65,535 characters from different languages.

The size shows the memory needed to store the data, and the range shows the range of values that can be stored in each type. For integers, larger size also means larger range.

One of the basic rules discussed in the next section is that **you cannot store a data type with a larger range in a variable of data type with a smaller range without explicitly converting the larger type to the smaller type**.

For example, the second statement in the following code is invalid:

```
int count = 150;      //valid; 150 is well within the range for int type
short recordCount = count; //invalid; count is int type, which is larger than short type
```

Note that the actual value of count is well within the range of short type, but the assignment is still invalid because count is of a larger type than recordCount. You will learn how to convert data item of one type to another type in the next section.

Next, we will look at the noninteger types, shown in Table 2-2.

Table 2-2: Real types

C# key word	Size	Range of values	Precision	How the following number may be represented: 123456789012.345678901234567890
float	32 bits	$\pm 3.4 \times 10^{38}$	7 digits	0.1234568×10^{12}
double	64 bits	$\pm 1.7 \times 10^{308}$	15 digits	$0.123456789012346 \times 10^{12}$
decimal	128 bits	$\pm 7.9 \times 10^{28}$	28–29 digits	123456789012345678901234567800

In addition to size and range, noninteger types have a precision property, with float type having only 7 digits of precision.

It is important to note that although the decimal type uses more memory (i.e., it has a larger size) than float and double, it has a smaller range (10^{28}). Decimals also are slower to work with. Hence, decimals are less suited than float and double for scientific applications involving extremely large numbers.

Why does decimal type take more memory although its range is smaller? Decimal type uses a method of internal representation that allows exact representation of real numbers with higher precision than float and double.

Although float and double have larger ranges and still take less memory, they cannot represent all noninteger numbers exactly (due to the method used by them to represent numbers), making them less suitable than decimals for financial calculations. For example, you would expect the following calculation to yield a value of 84.83 for cost:

```
float cost = (float) 4.99 * 17;
```

However, the actual value would be 84.82999. The meaning of "(float)" on the right side will be explained later in this section.

So, decimals are better suited for financial calculations than float and double when exact representation of numbers is important.

The C# types, float and double, use a method of representation called the **floating point** method to approximate a wide range of values to fixed number of significant digits. Therefore, float and double also are referred to as floating point types.

In the following sections, we examine the use of different types through a modified version of the IceCreamTotal form that you developed in Chapter 1.

Tutorial 1: Ice Cream Total with Data Types

This tutorial creates a new project named Ch2_DataTypes, and a modified version of IceCreamTotal form from Chapter 1. The modified form is named IceCreamTotalDataTypes. Figure 2-1 shows the form, which looks identical to IceCreamTotal.

The only change is that in place of double type variables, we use **decimal** type that is more suitable for financial calculations.

Figure 2-1: The IceCreamTotalDataTypes form

The primary goal of this tutorial is to learn how to work with different numeric data types.

Step 1-1: Copy the zipped file, Tutorial_Starts.zip. Extract the files (right click and select *Extract All*). Open the project, Ch2_DataType. Open the form **IceCreamTotalDataTypes.**

Because there are multiple forms in this project, you need to specify which form you want to run. Double click the Program.cs file to open it, and specify IceCreamTotalDataTypes as the form to be run:

```
Application.Run(new IceCreamTotalDataTypes());
```

Step 1-2: Open the code window to view the code that computes the total cost, as shown in Figure 2-2.

The code shown in Figure 2-2 differs from the code in the **IceCreamTotal** form only in lines 26 and 29. In line 26, we use **decimal** type variables to represent the unit price, subtotal, sales tax and total cost. Similarly, in line 29, we use decimal type constant for tax rate.

First, let's look at the statement in line 29 that assigns a value to the constant TAX_RATE,

```
const decimal TAX_RATE =  (decimal) 5.5;
```

What does "**(decimal) 5.5**" mean? The word **decimal** within the parentheses converts the number 5.5 to decimal type, as explained in the next section.

Figure 2-2: Modified code to compute total cost

```csharp
20  private void btnComputeTotal_Click(object sender, EventArgs e)
21  {
22      //Compute the total cost for an order of icecream:
23
24      //Declare variables to hold input data and computed results
25      int scoops;
26      decimal unitPrice, subTotal, salesTax, totalCost;
27
28      //Declare a constant to hold the tax rate
29      const decimal TAX_RATE =  (decimal) 5.5;
30
31      //Get input data from Text Boxes:
32      unitPrice = decimal.Parse(txtUnitPrice.Text);
33      scoops = int.Parse(txtScoops.Text);
34
35      // Compute results
36      subTotal = unitPrice * scoops;
37      salesTax = subTotal * TAX_RATE / 100;
38      totalCost = subTotal + salesTax;
39
40      // Display Results with decimals aligned
41      lstOutput.Items.Add(string.Format("{0,12}{1,10:C}", "Sub Total:", subTotal));
42      lstOutput.Items.Add(string.Format("{0,12}{1,10:C}", "Sales Tax:", salesTax));
43      lstOutput.Items.Add("");   // Add a blank line
44      lstOutput.Items.Add(string.Format("{0,12}{1,10:C}", "Total Cost:", totalCost));
45  }
```

2.2 Casting

The process of explicitly converting one data type to another type is called **casting.** Let's look at when casting is required and what the rules are.

Literals

Numbers like 5.5 that are used in programs are called **literal values** or **literals**. **All noninteger literal values, by default, are treated as double type.** Thus, 5.5 is double type. But, TAX_RATE is decimal type. Recall that double type has a larger range (10^{308}) than decimals (10^{28}).

Rule for Assigning Literal Values to Variables and Constants

You cannot store a literal of a data type that has a larger **range** in a variable of a data type that has a smaller range without explicitly converting (casting) the literal to the smaller type. Here are the data types sorted by increasing range of values:

> **sbyte, short, int, long, decimal, float, double**

Thus, a data type cannot be assigned to any of the data types on its left. For example, double type cannot be assigned to any of the other types without explicit conversion. This is true even if the actual value of the data item is small; it is the **range of the data type** that matters.

So, the following code will result in an error because 5.5 is double type, which has a larger range than decimal, although double uses less memory than decimal:

```
const decimal TAX_RATE = 5.5;    // invalid
```

Similarly, float cannot be assigned to decimal or integer types, but it can be assigned to double, which has a larger range.

To correct the problem, the literal 5.5 needs to be explicitly converted to decimal type by specifying the data type enclosed in parentheses as follows:

```
const decimal TAX_RATE = (decimal) 5.5;    // valid
```

Another way to cast a literal to decimal type would be to add the letter "m" or "M" to the literal:

```
const decimal TAX_RATE = 5.5m;    // valid
```

Except for decimals, data types of smaller range can be stored in data types of larger range. So, the following assignment is valid because the literal 5 is an integer. All integer types have ranges smaller than that of decimal:

```
const decimal TAX_RATE = 5;    // valid
```

decimal type: The exception

The exception is that **decimal type cannot be stored in float and double (though they have large ranges) without explicitly converting it.** Remember, decimal needs more memory than float and double, though it has a smaller range. So, the following is invalid, because decimal type cannot be assigned to double type:

```
const double TAX_RATE = (decimal) 5.5;    // invalid
```

Step 1-3: Change the statement in line 29 to: `const decimal TAX_RATE = 5.5;`
Hover the pointer over the number to view the error message, "Literal of double type cannot be implicitly converted to type 'decimal'..."

Step 1-4: Change the statement in line 29 to

```
const decimal TAX_RATE = (decimal) 5.5;
```

Note that the error disappears.
Next, change the statement to

```
const decimal TAX_RATE = 5;
```

Still there is no error.
Now, you may change it back to

```
const decimal TAX_RATE = 5.5m;
```

The above rules imply that **casting is needed when storing a double type in any other type,** because all other types have smaller ranges than double.

Casting is needed also when storing a decimal type in any other type, because of the special nature of decimals—decimal needs more memory than float and double, though it has a smaller range.

Rule for Casting

You can cast a literal or variable of one type to another type only if the value of the literal/variable is within the range of the type it is cast to.

For example, casting 40000 to short in the following statement would fail because it is outside the range (32767) for short:
```
short sales = (short)40000;   // Cast would fail
```

Casting a real to integer

You may cast a real to an integer type if the value of the real is within the range of the integer type, but you will lose the decimal part of the value. For example, the following statement is valid, but it will result in losing the decimal part (0.5) of the number:
```
int quantity = (int)10.5;
```

Casting to float

Caution! When you cast a real number with more than 7 significant digits to float, you lose some precision. Consider the following:
```
float price = 123456789.55F;
```

The number 123456789.55, which is a double by default, is cast to a float and assigned to the float type variable price. The value stored in unitPrice would have only 7 digits of precision, because float type has only 7 digits of precision.

Integer literals

With integer numbers, the value of the number determines its type. For example, the literal 120 is considered by the compiler as sbyte, because it is less than the sbyte's limit of 127. Similarly, 32000 is short type, and 2,000,000,000 is int type.

Debugging Programs: Putting Break Points in a Program

If you encounter unexpected results or errors in a program, a good way to determine what is going on is to examine the values of variables at different statements of the program. To do this, you can stop execution of a program at a statement by putting a break point at that statement.

To put a break at a statement, click on the margin to the left of the statement. You will see a break represented by a red dot in the margin.

Locals window

When the program breaks, you can view the values of all **local variables** (variables declared within a method) by displaying the Locals window (select Debug, windows, Locals). You also can move the pointer over each variable to see its value.

Suppose you inadvertently declared tax rate as int type, which results in truncation of the number 5.5 to 5, and the wrong output. Let's look at how to identify the problem.

Step 1-5: Check the effect of casting a double to int by putting a break in the program:
  Change the statement in line 29 to

```
const int TAX_RATE = (int) 5.5;
```

and run the form with 1 scoop and a unit price of 10. The sales tax should be 0.55 (10 x 0.055), but the output shows 0.50.

To identify the problem, put a break point at the closing brace "}" on line 45 by clicking on the margin to the left of the brace. C# inserts a red circle in the margin, as shown below:

```
44            lstOutput.Items.Add(string.Format("{0,12}{1,10:C}", "Total Cost:", totalCost));
45       }
```

Run the program. Program execution stops at the brace, indicated by a yellow arrow over the red circle.

Display the Locals window by selecting Debug, Windows, Locals. The Locals window appears, as shown in Figure 2-3.

Figure 2-3: Locals window

Name	Value	Type
this	{Ch2_DataTypes.IceCreamTota	Ch2_DataTypes.
sender	{Text = "Compute Total"}	object {System.\
e	{X = 20 Y = 15 Button = Left}	System.EventAr
scoops	1	int
unitPrice	10	decimal
subTotal	10	decimal
salesTax	0.5	decimal
totalCost	10.5	decimal
TAX_RATE	5	int

Examining the values of variables helps identify that TAX_RATE has the wrong value of 5. You also may move the pointer over TAX_RATE or any variable to display its value, one at a time.

Click the Continue button on the Toolbar to continue running.

Close the form, and correct the problem by changing the statement in line 29 back to the original form: `const decimal TAX_RATE = (decimal) 5.5;`

Next we will use the break to examine the effect of casting a double to float.

Step 1-6: Insert the following statement in line 39, which is currently blank:

```
float testNumber = 123456789.45F;
```

Run the form. Enter values for scoops and unit price, and click the Compute Total button. When the program breaks, hover the pointer over testNumber to view its value, 123456792.0, which is different from 123456789.45 due to the lower precision (7 digits) of floats compared to doubles. Close the form, and delete the newly inserted statement.

Rule for Assigning One Variable to Another Variable of a Different Type

The rule for assigning a variable of one type to a variable of another type is essentially the same as the rule for assigning a literal to a variable: You cannot store a variable (or constant) of a data type that has a larger **range** in a variable of a data type that has a smaller range. Here are the data types sorted by increasing range of values:

sbyte, short, int, long, decimal, float, double

The second statement shown below is invalid because price is of double type, which has a large range (10^{308}) compared to decimal (10^{28}).

```
double price = 25;    // valid; all integer types have ranges smaller than double
decimal currentPrice = price; // invalid; double has a larger range than decimal
```

To correct the problem, you need to cast price to decimal as follows:

```
decimal testNumber = (decimal) price;    // valid
```

Except for decimals, data types of smaller range can be stored in data types of larger range. Decimal type cannot be stored in float or double (though they have large ranges) without explicitly converting it.

The second statement shown below is invalid because price is of decimal type, which cannot be assigned to any other type.

```
decimal price = 45500;    // valid
double currenPrice = price; // invalid
```

Again, to correct the problem, you need to cast price to double, as follows:

```
double currentPrice = (double) price;    // valid
```

Thus, both double and decimal cannot be assigned to any other type without explicit conversion.

Step 1-7: To observe the role of casting when a variable of larger type is assigned to another variable of smaller type, enter the following two statements into the click event handler of ComputeTotal button, and note the error in the second statement:

```
decimal price = 2.95m;    // "m" represents decimal
double testNumber = price; // invalid
```

To correct the error, cast price to double.

```
double testNumber = (double) price;    // valid
```

Delete both statements. Repeat the process by adding the following statements and observing the error:

```
double price = 2.95;
decimal testNumber = price; // invalid!
```

To correct the error, cast price to decimal.

```
decimal testNumber = (decimal) price;   // valid
```

Casting: Narrowing and Widening Conversion

Explicit casting and narrowing conversions

As discussed above, **explicit casting** is needed when you store a data type with a larger range in a data type with a smaller range, and also when you store a decimal in other types. Such a conversion is called a **narrowing conversion.**

Implicit casting and widening conversion

Explicit casting is not needed when storing any data type, other than decimal, in another data type that has a larger range.

Thus, literal or variable of any type other than decimal can be assigned to a double type variable without explicit casting. In such cases, C# performs an **implicit cast** to do a **widening conversion**.

Consider the following code that assigns the literal 1234567890 of int type (determined by the value) to a float type variable, costPerUnit, and then assigns costPerUnit to another variable, price.

```
float costPerUnit = 1234567890;    // valid
double price = costPerUnit;        // valid
```

In the first statement, C# performs an implicit cast from integer to float that has a larger range to do widening conversion. The second statement assigns the value of costPerUnit, which is float, to unitPrice, which is a double. Again, C# performs an implicit cast from float to the larger double.

Caution! Float has only 7 digits of precision. Hence, the value of costPerUnit will have only 7 digits of precision.

Step 1-8: Make sure that there is break at the closing brace. Add the following two statements to the click event handler of Compute Total button, run the program and note the value of costPerUnit and price.

```
float costPerUnit = 1234567890;    // valid, but loses precision.
double price = costPerUnit;        // valid
```

Delete the statements.

Note that the values of costPerUnit and price are different from 1234567890 due to lack of precision for float.

Mixing data types in an expression

When an operation involves two different data types, then the data type with the smaller range, except for decimal, is implicitly converted to the larger type, and the result would be of the same type as the larger type.

So, mixing different integer types in an expression, as in the following code, is valid:

```
short weight = 60;
int score = 1000;
int weightedScore = weight * score;  // valid to mix different integer types
```

The short type is implicitly converted to the larger int type.

Similarly, mixing integer types with a real type like float, double or decimal is valid, as in the following code:

```
float weight = 60;
int score = 1000;
float weightedScore = weight * score;// valid to mix integer types with any real
```

Again, the int type is implicitly converted to the larger float type.

Mixing float and double types also is valid. The float type will be implicitly converted to the larger double type.

```
float weight = 60;
double score = 950.50;
double weightedScore = weight * score;//valid to mix float with double and integer
```

Decimal type: Again, the troublemaker!

While mixing float and double types is valid, **mixing decimal with float or double is invalid.** Decimal is not implicitly converted to float or double. Similarly, double and float are not implicitly converted to decimal. The expression in the following statement is invalid because subTotal is declared as decimal, but the literal 5.5 is double:

```
salesTax = subTotal * 5.5 / 100;   // invalid!
```

So, the literal 5.5 must be explicitly converted to decimal, as follows, so that the result will be decimal type:

```
salesTax = subTotal * (decimal) 5.5 / 100;
```

or

```
salesTax = subTotal * 5.5m / 100;
```

Caution! Division of an integer by another integer would be an integer, resulting in loss of the fractional part, if any. For example, the value of average in the following statement would be 2, though average is float type, because division of 25 by 10 is represented as an integer:

```
float average = 25/10;     // yields the result 2
```

Casting at least one of the two integers to float type would yield the correct result:

```
float average = 25f/10;
```

The result of 25f/10 would be float type, which is the larger of the two types, yielding the correct result, 2.5.

So, when you divide an integer by another integer, at least one of the two must be cast to a real type to get the result with the fractional part, if any.

Step 1-9: Add the following statement to the click event handler of Compute Total button, run the program and observe the value of average:

```
float average = 25 / 10;    // yields 2
```

Change the expression to 25f/10f, and see the difference in value (2.5).
Change the expression to 25f/10, which yields 2.5.
Delete the statement.

Multiplication with large integers

Multiplication with large integers can result in unexpected consequences:

```
int price = 3000;
int quantity = 1000000;
long cost = price * quantity;
```

The product 3000 * 1000000 yields 3 billion. However, because the product of the two int types (price * quantity) is represented as int, whose limit is approximately 2.15 billion, the result is outside the limit and yields incorrect results.

Step 1-10: Add the following statements, run the program and note the value of cost.

```
int price = 3000;
int quantity = 1000000;
long cost = price * quantity;
```

Delete the statements.

The value of cost is incorrect because the result is int type, whose limit is about 2.15 billion, which is too small for the result of 3000,000,000 (3 billion).

An oddity in C#

When an operation involves short or sbyte types, these types are automatically converted to int type before performing the operation. So, the result will be int type.

```
short exam1 = 80;
sbyte exam2 = 90;
short total = exam1 + exam2;  // Error! The result is int type (odd!), whereas
total is short type
```
You may cast the result to short, as in
```
short total = (short)(exam1 + exam2);
```

Because of this odd behavior when using sbyte and short, in general, **you may use int type for all integers that are within the range for int**, except in applications where memory usage is of concern.

It's time to practice! Do Exercise 2-1.

Review Questions

2.1   Which real data type is best suited for financial calculations where exact representation of numbers is important?

2.2   What are the advantages of float and double types compared to decimal?

2.3   Which data type has the largest range?

2.4   Which data type has the largest size (memory)?

Correct errors, if any, in the following statements:

2.5   double sales = 2560.50;

2.6   float cost = 2.50;

2.7   float units = 45000;

2.8   sbyte quantity = 200;

2.9   decimal price = 300;
      float currentPrice = price;

2.10 short units = 2500;
      int quantity = units;

2.11 int qty1 = 1000000000;
      int qty2 = 2000000000;
      long total = qty1 + qty2;

2.12 float total = 255.50 + 500;

2.13 The following code is intended to compute the average:
int totalScore = 25;
int count = 4;
float average = total/count;

2.14 decimal price = 100;
float quantity = 5.5;
decimal netCost = price*quantity - 10;

2.3 Working with Strings and Characters

In this section, you will learn how to work with string type and character type data and variables. The C# type **string** represents a combination of zero or more characters. "John Adams" and "P410-2023-3180-05" are examples of string type data. Note that strings are enclosed in quotes.

Value and Reference Types

The integer and real types we discussed in Chapter 1 are called **value types** because each variable of these types holds the actual value t in the memory location for that variable.

By contrast, strings (and other objects like arrays) are called **reference types** because each variable of these types holds only a reference to the data—that is, the memory location or address of the data, not the actual data. Reference type is discussed later in Chapter 7.

To understand working with strings, let's look at an application that creates a user name by combining the following three pieces of string type data:

1. First 5 characters of the last name
2. First character of the first name
3. Last two digits of the id

Figure 2-4 shows the user interface form named UserNameString.

Figure 2-4: UserNameString form

Tutorial 2: String Operations

This tutorial creates the UserNameString form shown in Figure 2-4. The primary objective is to learn string operations.

Step 2-1: Open the project, Ch2_DataTypes, from Tutorial_Starts folder. Open the form, UserNameString. To make it easy to understand the code, the names of controls that are used in the code are shown in Table 2-3.

Table 2-3: Controls on the form, UserNameString

Purpose	Type of control	Property	Setting
Enter last name	TextBox	Name	txtLname
Enter first name	TextBox	Name	txtFname
Enter id	TextBox	Name	txtId
Create user name	Button	Name	btnUserName
Display user name	Label	Name	lblUserName

Figure 2-5 shows the code to create the user name.

Figure 2-5: Code to create the user name

```
20      private void btnUserName_Click(object sender, EventArgs e)
21      {
22          string lName = txtLname.Text;    // Get input data
23          string fName = txtFname.Text.Trim();
24          string id = txtId.Text.Trim();
25          string userName;
26          // Combine three strings to get the user name
27          userName = lName.Substring(0, 5).ToLower() + fName.Substring(0, 1).ToLower() +
28                                  id.Substring((id.Length - 2), 2);
29          // Alternate way using the Concat method;
30          userName = string.Concat(lName.Substring(0, 5).ToLower(), fName.Substring(0, 1).ToLower(),
31                                  id.Substring((id.Length - 2), 2));
32          lblUserName.Text = userName;
33      }
```

Let's take a look at what the statements do.

Lines 22–24 get the input data from TextBoxes and save the data in variables.

Line 27 extracts the required substrings from the last name, first name and id; converts them to lowercase; and combines them to form the user name.

The string type has a large number of methods and properties to help work with strings. We first look at the methods/properties used in the above program.

Trim() method

The Trim() method removes all leading and trailing blank (white) spaces from the string. For example,
```
"   George   ".Trim();
```
would return the string, "George".

In the statement in line 22,
```
string lName = txtLname.Text.Trim();
```

the Trim() method of the string txtLname.Text removes trailing and leading blank spaces, if any, inserted by the user into the TextBox.

Additional options to trim include

TrimStart()—removes leading blanks, if any
TrimEnd()—removes trailing blanks, if any

Step 2-2: Double click the Create User Name button, and add lines 22 through 25 to the click event handler of the button. Put a break on the closing brace of the event handler. Run the form after entering the data with leading and trailing blank spaces. Make sure the last name is at least 5 characters long, because the code doesn't allow for shorter names.

Substring(i, n) method

The Substring(i, n) method returns n characters from the string, starting with index position i. Index position 0 refers to the first character, 1 refers to the second character, 2 refers to the third character and so on.

For example, in line 27, **lName.Substring(0, 5)** returns a string that consists of 5 characters, starting with the character in position zero—that is, it gives the first 5 characters of the last name.

Similarly, **fName.Substring(0, 1)** returns a string that consists of the first character of the first name.

id.Substring((id.Length - 2), 2) gives the two digits of the id starting with the second position from the end—that is, the last 2 digits of id. For example, for a 7-digit id, (id.Lenth - 2) would be 5, thus making the expression equivalent to **id.Substring(5,2)**, which gets the 2 characters starting with index position 5—that is, starting with the sixth character.

Length property

The Length property gives the number of characters in the string.

In line 27, **id.Length** uses the Length property of the string, id, to get the number of characters in the id to help extract the last two digits of id:

```
27      userName = lName.Substring(0, 5).ToLower() + fName.Substring(0, 1).ToLower() +
                  id.Substring((id.Length - 2), 2);
```

ToLower() method

The ToLower() method returns a copy of the string in lowercase.

lName.Substring(0, 5).ToLower() converts the first five characters of the last name to lowercase. Similarly, **ToUpper()** converts to uppercase.

Concatenating strings

Two or more strings may be concatenated together using the "+" operator, as in

```
String fullName = "G " + "C " + "Philip";
```

The value of fullName would be "G C Philip." This is the method used in computing username in Line 27:

```
userName = lName.Substring(0, 5).ToLower() +
           fName.Substring(0, 1).ToLower() + id.Substring((id.Length - 2), 2);
```

Step 2-3: Add the statement that computes user name in line 27 & 28. Run the form to test it.

Change the parameters of Substring method for fName to 3 and 1, as in

```
fName.Substring(3, 1)
```

Observe the change in user name.

Type in the last name with uppercase and with leading and trailing blanks, and observe the output. Type in id of a different lengths.

Another option to concatenate strings is to use the **Concat()** method of string, as in

```
String fullName = String.Concat("G ", "C ", "Philip");
```

which yields "G C Philip."

Line 30 uses the Concat method to combine strings:

```
30    userName = string.Concat(lName.Substring(0, 5). ToLower(),
              fName.Substring(0, 1).ToLower(),id.Substring((id.Length - 2), 2));
```

Step 2-4: Add lines 30 through 32 from Figure 2-5 to complete the program, and test the code.

Make each of the following changes, run the program, and describe what impact it has on the output and why.

Change the parameters (0, 5) in lName.Substring(0, 5) to (0,4).

Change the same set of parameters as above to (2, 3).

Change the parameters ((id.Length - 2), 2)) to ((id.Length - 1), 1)).

Additional Methods of Strings

Replace (string, string) method: Replaces the first string, if any, with the second string.

Example: `lName = lName.Replace(" ", null);`

The above statement would replace blank spaces, if any, by null, in essence removing any spaces within the name.

Step 2-5: Insert the statement
```
lName = lName.Replace(" ", null);
```
immediately after line 22 that has the statement
```
string lName = txtLname.Text.Trim();
```

Test the program by including spaces in the last name that you type in.

Some of the commonly used methods and properties of string are given in Table 2-4.

Table 2-4: Methods and properties of string type variables

String methods & properties	Description
Methods	
IndexOf(*ch*)	Returns the index of the first occurrence of the character, *c*, if it is found; if not, -1 is returned
IndexOf(*str*)	Returns the index of the first occurrence of the string, *str*, if it is found; if not, -1 is returned.
Insert(*i, str*)	Inserts the string, *str*, at index position i, and returns the resulting string
Remove(*i, n*)	Removes *n* characters starting at index position i, and returns the resulting string
Replace(*str1, str2*)	Replaces every occurrence of *str1* by *str2* and returns the resulting string
SubString(*i,n*)	Returns *n* characters starting with index position *i*
ToLower()	Returns a copy of the string in lowercase
ToUpper()	Returns a copy of the string in uppercase
Trim()	Removes all leading and trailing blank (white) spaces, and returns the resulting string
Property	
Length	Gives the number of characters in the string

Additional methods of string variables include

    Contains(*ch*), Contains(*str*)
    IndexOf(*string, i*), IndexOf(*string, I, n*), IndexOf(*ch, i*), IndexOf(*ch, i, n*), LastIndexOf(*ch*)
    Insert(i, str)
    StartsWith(*str*), EndsWith(*str*)
    SubString(i)
    TrimStart(), TrimEnd()

String versus Character

A string that consists of only one character is not the same as a character type value. As stated earlier, a string variable is a reference type, but a character is a value type.

The string methods are applicable to strings of all sizes, including a size of one character, **but they are not applicable to character type.**

Combining strings with characters

If you concatenate a char type to a string, C# will implicitly convert the char type to string, as in the following statement:

```
char initial = 'g';
lblUserName.Text = "Phili" + initial;
```

Step 2-6: Insert the above two statements to the end of the click event handler. Comment out the rest of the code, and run the form. The string "phili" and character 'g' are combined to display the string "Philig." Delete the two statements.

The methods of string type variables, like ToUpper and ToLower, are not valid for character type variables. Hence, the following statement would give an error:

```
char initial = 'g';
lblUserName.Text = "Phili" + initial.ToUpper(); //error
```

You may convert the character type variable, initial, to string type and then use the ToUpper method of string variables.

Step 2-7: Add the above two statements to the end of the code in the click event handler, and observe the error message.

Correct the error by converting the value of initial to string type using the ToString() method of char type, as shown below:

```
lblUserName.Text = "Phili" + initial.ToString().ToUpper();
```

You also may use the ToUpper method of the char type, as in

```
lblUserName.Text = "Phili" + char.ToUpper(initial);
```

An alternative is to declare initial as a string instead of a character, and then use the ToUpper method of the string.

Step 2-8: Add the two statements shown below to declare initial as string, and then use the ToUpper() method. Run the code. Observe that there is no error.
```
string initial = "g";
lblUserName.Text = "Phili" + initial.ToUpper();
```

Delete the above two statements.

It's time to practice! Do Exercise 2-2.

Review Questions

2.15 A string variable named firstName contains the first name, and another string variable named lastName contains the last name of a person. Write an expression to create a string that consists of the first 2 characters of the first name and the entire last name, with a space between them, and convert the string to all capital letters.

2.16 A student id consisting only of numeric digits is stored in a string variable named studentId. Write the code to display the last 2 digits of the id in a ListBox named lstDisplay.

2.4 Working with Dates and Times

Working with dates and times is important in a variety of applications, like computation of due dates, late fees, duration of events and rental fees. This section takes a detailed look at how to work with dates and times. However, the materials presented in this section are not prerequisites for the chapters that follow.

As stated in Chapter 1, C# does not have a data type for dates and times. The .Net base type DateTime is used for dates and/or times. DateTime data are enclosed in pound signs, as in #12/25/2017 10:30:00 AM#.

Tutorial 3: DateTime Operations

To explore various operations with dates and times, you will create an application that uses your graduation date and time. Different pieces of information related to the graduation date/time are displayed on a form named GraduationDate, as shown in Figure 2-6.

Figure 2-6: GraduationDate form

Step 3-1: Open the form named GrduationDate from Ch2_DataTypes projects in Tutorial_Starts folder.

Step 3-2: The controls on the form are shown in Table 2-5.

Table 2-5: Controls on the form GraduationDate

Purpose	Type of control	Property	Setting
Select date	DateTimePicker	Name	dtpGradDate
		Format	short
		Value	Delete the time, leaving just the date
Select time	DateTimePicker	Name	dtpGtadTime
		Format	Custom
		CustomFormat	hh:mm tt (use HH:MM tt for 24-hour military time)
		ShowUpDown	True
Display prompt for GradDate	Label	Text	Graduation Date:
Display prompt for GradTime	Label	Text	Graduation Time:
Compute results	Button	Text	Graduation Info
Clear controls	Button	Text	Clear
Display results	ListBox	Name	lstDisplay
		Font	Courier New, 8.25pt.

Using DateTimePicker to select date and times

Rather than hardcoding dates as strings and converting to DateTime, a DateTimePicker may be used to let users select a date or time. By default, the DateTimePicker, shown below with Format set to **short**, lets the user select a date.

The selected date, which also includes the current time, is returned by the **Value** property. To remove the current time, delete the time from the Value property. C# will add the default time 12:00:00 to the selected date. In the GraduationDate form, the current time has been removed from the Value property of dtpGradDate. So, the date includes the default time 12:00:00.

To let users select a Time from DateTimePicker, you need to set the Format, CustomFormat and ShowUpDown properties, as shown in Table 2-6.

Table 2-6: Settings for DateTimePicker to display time

Format	Custom
CustomFormat	hh:mm tt
ShowUpDown	True

The time selected by the user is returned by the Value property that also includes today's date. The above properties of dtpGradTime are already set to the values specified in Table 2-6.

Step 3-3: Double click the Graduation Info button to open the click event handler of the button.

Working with Dates

Here is an example of creating a date and storing it in a DateTime variable by hard-coding the date:
```
DateTime gradDate = DateTime.Parse("05/12/17");
```

You also may specify a time:
```
DateTime gradDateTIme = DateTime.Parse("05/12/17 02:30 PM");
```

Or, you may specify a time with no date:
```
DateTime gradTime = DateTime.Parse("02:30 PM");
```

When no date is specified, today's date is added to the time, by default.
Similarly, if no time is specified, the default time 12:00 AM is added to the date.

The date must be specified in one of the standard date formats, 05-12-2017, May 12, 2017 or 2017-05-12.

The general syntax to create a DateTime is
DateTime variableName = DateTime.Parse(string);

Step 3-4: Enter the following statements in the click event handler of Graduate Info button to convert strings to DateTime.

```
DateTime gradDate = DateTime.Parse("05/12/17");
DateTime gradDateTime = DateTime.Parse("05/12/17 02:30 PM");
DateTime gradTime = DateTime.Parse("02:30 PM");
```

Put a break at the last line of the click event handler that has the closing brace.
Run the program, click on the Graduation Info button and observe the value each of the three variables by moving the cursor over it. Pay particular attention to the default value of date and time.
Delete the statements. Remove the break from the closing brace.

Creating a date/time using the "new" key word

Alternatively, you may use the **new** key word, using the syntax

DateTime variableName = new DateTime(year, month, day, hour, minute, second, millisecond)

where hour, minute, second and millisecond are optional.

The **new** key word creates a structure of DateTime type in memory with the default value of DateTime (#01/01/0001 12:00:00 AM#). The default value is assigned to the variable specified on the left side of the = operator. The new key word, which is called the default constructor, is discussed in more detail in Chapters 7 and 13.

Here is an example:

```
DateTime gradDate = new DateTime(2017, 5, 12);
// The value of gradDate would be 5/12/2017
```

Alternatively, you can include the time 2:30 PM represented by 14, 30, 0:
```
DateTime gradDateTime = new DateTime(2017, 5, 12, 14, 30, 0);
// The value of gradDateTime would be 5/12/2017 2:30 PM.
```

Step 3-5: Specify date/time using the new key word, by adding the following statements:

```
DateTime gradDate = new DateTime(2017, 5, 12);
DateTime gradDateTime = new DateTime(2017, 5, 12, 14, 30, 0);
```

Run the program, and observe the values of gradDate and gradDateTime.
Delete the statements.

Step 3-6: Add the following code to declare variables and get graduation date and time from DateTimePickers.
Run the form, and check the values of the variables gradDate and gradTime.

```
22
23              DateTime gradDate, gradDateTime, gradTime;
24              float numberOfDays;
25
26              //Get date of graduation:
27              gradDate = dtpGradDate.Value;
28              // Get time of graduation:
29              gradTime = dtpGradTime.Value;
```

Properties of DateTime

DateTime structure has a variety of properties. Some of the commonly used properties are shown in Table 2-7. Column 3 shows the value of the property for gradDate created as follows:
```
DateTime graduationDate = DateTime.Parse("5/12/17 02:30 PM");
```

Table 2-7: Properties of DateTime structure

Property	Description	Value of property for graduationDate (5/12/2017 02:30 PM)
Day	Returns the day as an integer	graduationDate.Day: 12
Month	Returns the month as an integer	graduationDate.Month: 5
Year	Returns the year as an integer	graduationDate.Year: 2017
Now	Returns the current date and time	DateTime.Now: (today's date and time)
Today	Returns the current date	DateTime.Today: (today's date with the default time 12:00:00 AM)
Hour	Returns the hour as an integer	graduationDate.Hour: 2
TimeOfDay	Returns the duration that has elapsed since midnight, as a TimeSpan (Hours, Minutes, Seconds, …)	TimeOfDay.Hours: 2 TimeOfDay.Minutes: 30 TimeOfDay.Seconds: 0
Additional properties include Minutes, Seconds, DayOfWeek and DayOfYear		

Today versus Now

The property **Today** gives today's date with the default time 12:00:00 AM, whereas **Now** gives today's date with the current time.

Formatting DateTime

Table 2-8 shows commonly used formatting options for DateTime.

Table 2-8: Formatting options for DateTime

Format	5/12/2017 will be displayed as
ToShortDateString	5/12/2017
ToLongDateString	Friday, May 12, 2017
(No Format)	5/12/2017 12:00:00 AM
ToString	5/12/2017 12:00:00 AM

Step 3-7: Type in the code shown below to display current date & time and the graduation date:

```
31    // Format and display current date & time and graduation date:
32    lstDisplay.Items.Clear();
33    lstDisplay.Items.Add("Current Date & Time:    " + DateTime.Now);
34    lstDisplay.Items.Add("Date of graduation:     " + gradDate.ToShortDateString());
```

Run the program and observe the output. Change the format for gradDate to ToLongDateString, run the program and observe the output.
Now remove the format, and see the output with the default format.

Formatting date/time using ToString() method

Just like formatting numbers using ToString() method, you can format dates and times using ToString method of DateTime. For example, you can display the name of the day or name of the month for a date. Table 2-9 shows format codes available for ToString() method.

Table 2-9: Format codes for ToString method

Code	5/12/2017 2:30 PM will be displayed as
"dd"	12
"ddd"	Fri
"dddd"	Friday
"MM"	5
"MMM"	May
"MMMM"	May
"yy"	17
"yyyy"	2017
"hh"	02
"mm"	30
"ss"	00

You may combine the above codes to create custom formats as follows:
    "dd MMM yyyy hh:mm:ss tt"   12 May 2017 2:30:00 PM
    "MM/dd/yy hh:mm tt"   5/12/17 2:30 PM

Step 3-8: Type in the code to display day, month, day of the month from gradDate, and time of graduation from gradDate, as follows:

```
What day is the graduation?        Friday
What month is it?                  May
What day of the  Month?            12
What time?                         2:30:00 PM
```

Here is the code to be added. Note how the format codes "dddd" and "MMM" are used to display the day and month from a date:

```
36    // Display the month & day of graduation:
37    lstDisplay.Items.Add("");
38    lstDisplay.Items.Add("What day is the graduation?     " + gradDate.ToString("dddd"));
39    lstDisplay.Items.Add("What month is it?               " + gradDate.ToString("MMM"));
40    lstDisplay.Items.Add("What day of the month?          " + gradDate.Day);
41
42    //Format and display the time of graduation:
43    lstDisplay.Items.Add("What time?                      " + gradTime.ToLongTimeString());
```

Modify line 38 so that the three-letter abbreviated name of day (e.g., Fri) is displayed.

Step 3-9: You may experiment with the different format codes from Table 2-8 and observe the outputs. Next, change the statement in line 43 as follows so that no format is used:
`lstDisplay.Items.Add("What time?          " + gradTime)`

Run the form. Note that the time is displayed along with today's date, which is the default date.

Doing Computations with Date/Times

The DateTime structure has a number of methods to find the difference between two date/times and to add intervals to a date/time. Table 2-10 shows such methods.

Table 2-10: Methods of DateTime

Method	Description	Value returned by the method
AddDays(days)	Returns the DateTime obtained by adding the specified number of days to a DateTime	graduationDate.AddDays(3): Returns 5/15/17 02:30 PM
AddMonths(months)	Returns the DateTime obtained by adding the specified number of months to a DateTime	graduationDate.AddMonths(3): Returns 8/12/17 02:30 PM
AddHours(hours)	Returns the DateTime obtained by adding the specified number of hours to a DateTime	graduationDate.AddHours(3): Returns 5/12/17 05:30 PM
Note: Additional methods with the same syntax are available to add years, minutes and seconds.		
Subtract(DateTime)	Returns a TimeSpan obtained by subtracting a specified DateTime	graduationDate.Subtract(DateTime.Now): Returns a TimeSpan with the number of days, hours, and so on between now and graduation date.
Add(TimeSpan)	Returns the DateTime obtained by adding the specified TimeSpan to a DateTime	DateTime.Now.Add (fillTime)*: Returns the DateTime value obtained by adding 3 days, 2 hours and 30 minutes to current date and time. *fillTime: (3, 2, 30, 0, 0)

Computing the difference between two dates

.Net Framework uses a **TimeSpan** structure to represent the time interval between two DateTime values.

TimeSpan structure

To get the difference between two DateTime values, you may subtract one from the other. For example, the time to graduate may be computed and stored in a TimeSpan variable as follows:

```
TimeSpan timeToGraduate = gradDate-DateTime.Now;
```

Table 2-11 shows commonly used properties of TimeSpan Stucture.

Table 2-11: Properties of TimeSpan stucture

Properties of TimeSpan	Description
Days, Hours, Minutes, Seconds, Milliseconds, Ticks	The number of whole days and remaining whole hours, minutes, seconds, milliseconds and ticks in a time interval. For example, the TimeSpan representing (#12/19/2017 4:40:45 PM# - #12/18/2017 1:10:30 PM#) would have following values for its properties: Days - 1 Hours - 3 Minutes - 30 Seconds - 15
TotalDays, TotalHours, TotalMinutes, TotalSeconds, TotalMilliSeconds	Each property returns the entire interval expressed in a particular unit like days, hours or minutes. For example, the TimeSpan representing (#12/19/2015 4:30:45 PM# - #12/18/2015 1:20:40 PM#) would have following values for its properties: TotalDays - 1.146 (0.146 represents the fractional days) TotalHours - 27.504 (includes 27 days and .504 fractional hours) TotalMinutes - 1650.25

A TimeSpan also may be created using the new key word:

```
TimeSpan variableName = new TimeSpan(days, hours, minutes, seconds, milliseconds);
```

Here is a specific example of a TimeSpan that represents the time it takes to fill an order: 2 days, 3 hours and 30 minutes:

```
TimeSpan fillTime = new TimeSpan(2, 3, 30, 0, 0);
```

Step 3-10: Insert the following statements that compute the number of days to graduate:

```
46      // Compute the interval between today's date and graduation date
47      TimeSpan tsGraduation = (gradDate - DateTime.Today);
48      /* today includes current date, but not current time.
49         You may also use the Subtract() method of DateTime
50         TimeSpan tsGraduation = gradDate.Subtract(DateTime.Today);     */
51
52      numberOfDays = (float)tsGraduation.Days;
53      // Since graduation date and DateTime.Today both have the same default time (12:00:00),
54      // the remaining hours, minutes and seconds of the difference will be zero.
55      lstDisplay.Items.Add("");
56      lstDisplay.Items.Add("How long?");
57      lstDisplay.Items.Add("You have only  " + numberOfDays + " more days to graduate");
```

Note that in the expression (gradDate - DateTime.Today), both dates have the same default time (12:00:00). Hence, the remaining hours, minutes and seconds of the difference will be zero. So, the TotalDays property of the TimeSpan would give the same value as the Days property.

Later, Step 3-14 introduces an example where there is a time difference between two DateTime values.

Step 3-11: This is an **optional** addition to the code. Skipping this code won't affect the rest of the program. Add the code to compute and display the number of weeks, months and years to graduation, as follows:

```
59      // Computing weeks, months and years:
60      float numberOfWeeks = numberOfDays / 7;
61      float numberOfMonths = ((gradDate.Year - DateTime.Today.Year) * 12) +
62                                      (gradDate.Month - DateTime.Today.Month);
63      float numberOfYears = (float) (numberOfDays / 365.25);
64      // The above calculation of month and year may be imprecise under certain conditions.
65
66      // Display number of weeks, months & years
67      lstDisplay.Items.Add("That is, about " + numberOfWeeks.ToString("N0") + " more weeks.");
68      lstDisplay.Items.Add("That is, about " + numberOfMonths + " more months.");
69      lstDisplay.Items.Add("That is, over   " + numberOfYears + " year.");
```

Adding years, months, days, hours, and so on to a DateTime

Table 2-10 presented the methods of DateTime to add years, months, days, hours, and so on to a DateTime. Let's use one of the methods to compute the date when the student loan is due, assuming it is 60 days after graduation.

Step 3-12: Add the following statements to compute and display the date when the student loan is due:

```
71      // Compute the day when the loan is due - Add 60 days to graduation date:
72      lstDisplay.Items.Add("");
73      lstDisplay.Items.Add("When is my loan due? " + gradDate.AddDays(60).ToLongDateString());
74      // To subtract days, use a negative number
```

Combining a date and time

A date from a DateTime variable and the time from another DateTime variable can be combined into a single variable. For example, the date from gradDate and the time from gradTime can be combined and stored into the single variable, gradDateTime. Because you cannot add two DateTimes, the time from gradTime is converted to a TimeSpan using the TimeOfDay property of DateTime and then added to gradDate, as shown in Step 3-13.

Step 3-13: Add the code to combine the date from gradDate and time from gradTime into a single variable gradDateTime. Here is the code to do it:

```
76    // Combine the date and time and store in a single variable:
77    // Convert graduation time to a TimeSpan and add to date:
78    TimeSpan tsGradTime = gradTime.TimeOfDay;
79    // or,                = new TimeSpan(gradTime.Hour, gradTime.Minute, gradTime.Second);
80    gradDateTime = gradDate.Add(tsGradTime);
81    lstDisplay.Items.Add("");
82    lstDisplay.Items.Add("Again, graduation date & time is: " + gradDateTime);
```

Step 3-14: Add the code to compute the number of days to graduation, taking into account the difference between graduation time and current time.

```
84    // Compute the number of days to graduation, including the fractional part
85    tsGraduation = (gradDateTime - DateTime.Now);   //Now includes current date and time
86    lstDisplay.Items.Add("Just " + tsGraduation.TotalDays.ToString("N") + " more days");
```

Run the program, and observe that the number of days includes a fractional part due to the difference between graduation time and current time.

Alternatively, you may use the Days (instead of TotalDays) and Hours properties to display the whole days and hours using the following statement:

```
lstDisplay.Items.Add("Just " + tsGraduation.Days.ToString("N1") + " more days" + " and "
    + tsGraduation.Hours + " hrs" );
```

Stop the program.

Now you have the complete code that should look as follows (a copy of the code can be found at Tutorial_Starts/Ch2_Data Types/Ch2_GraduationDate_Code.txt):

```
20    private void btnDisplayGradInfo_Click(object sender, EventArgs e)
21    {
22
23        DateTime gradDate, gradDateTime, gradTime;
24
25        float numberOfDays;
26
27        //Get date of graduation:
28        gradDate = dtpGradDate.Value;
29        // Get time of graduation:
30        gradTime = dtpGradTime.Value;
31
32        // Format and display current date & time and graduation date:
33        lstDisplay.Items.Clear();
34        lstDisplay.Items.Add("Current Date & Time:    " + DateTime.Now);
35        lstDisplay.Items.Add("Date of graduation:     " + gradDate.ToShortDateString());
36
37        // Display the month & day of graduation:
38        lstDisplay.Items.Add("");
39        lstDisplay.Items.Add("What day is the graduation?      " + gradDate.ToString("dddd"));
40        lstDisplay.Items.Add("What month is it?                " + gradDate.ToString("MMM"));
41        lstDisplay.Items.Add("What day of the  Month?          " + gradDate.Day);
42
43        //Format and display the time of graduation:
44        lstDisplay.Items.Add("What time?                       " + gradTime.ToLongTimeString());
```

```
45
46          // Compute the interval between today's date and graduation date
47          TimeSpan tsGraduation = (gradDate - DateTime.Today);
48          /* today includes current date, but not current time.
49             You may also use the Subtract() method of DateTime
50             TimeSpan tsGraduation = gradDate.Subtract(DateTime.Today);    */
51
52          numberOfDays = (float)tsGraduation.Days;
53          // Since graduation date and DateTime.Today both have the same default time (12:00:00),
54          // the remaining hours, minutes and seconds of the difference will be zero.
55          lstDisplay.Items.Add(" ");
56          lstDisplay.Items.Add("How long?");
57          lstDisplay.Items.Add("You have only  " + numberOfDays + " more days to graduate");
58
59          // Computing weeks, months and years:
60          float numberOfWeeks = numberOfDays / 7;
61          float numberOfMonths = ((gradDate.Year - DateTime.Today.Year) * 12) +
62                                  (gradDate.Month - DateTime.Today.Month);
63          float numberOfYears = (float) (numberOfDays / 365.25);
64          // The above calculation of month and year may be imprecise under certain conditions.
65
66          // Display number of weeks, months & years
67          lstDisplay.Items.Add("That is, about   " + numberOfWeeks.ToString("N0") + " more weeks.");
68          lstDisplay.Items.Add("That is, about " + numberOfMonths + " more months.");
69          lstDisplay.Items.Add("That is, over  " + numberOfYears + " year.");
70
71          // Compute the day when the loan is due - Add 60 days to graduation date:
72          lstDisplay.Items.Add("");
73          lstDisplay.Items.Add("When is my loan due? " + gradDate.AddDays(60).ToLongDateString());
74          // To subtract days, use a negative number
75
76          // Combine the date and time and store in a single variable:
77          // Convert graduation time to a TimeSpan and add to date:
78          TimeSpan tsGradTime = gradTime.TimeOfDay;
79          // or,             = new TimeSpan(gradTime.Hour, gradTime.Minute, gradTime.Second);
80          gradDateTime = gradDate.Add(tsGradTime);
81          lstDisplay.Items.Add("");
82          lstDisplay.Items.Add("Again, graduation date & Time is: " + gradDateTime);
83
84          // Compute the number of days to graduation, including the fractional part
85          tsGraduation = (gradDateTime - DateTime.Now);  //Now includes current date and time
86          lstDisplay.Items.Add("Just " + tsGraduation.TotalDays.ToString("N") + " more days");
87      }
```

Step 3-15: Double click the Clear button, and enter the following code into the Click event handler of the Clear button. The code shows three different statements to reset the date. Run the form and observe the effect of each statement by commenting out the others.

```
89          private void btnClear_Click(object sender, EventArgs e)
90          {
91              //Clear ListBox
92              lstDisplay.Items.Clear();
93
94              //Reset the DateTimePicker for date:
95              dtpGradDate.Value = DateTime.Parse("01/01/1900");
96              // Or,
97              dtpGradDate.Value = new DateTime(1900, 01, 01);
98              //Or,
99              dtpGradDate.Value = DateTime.Today;
100
101             // Reset the DateTimePicker for Time:
102             dtpGradTime.Value = DateTime.Parse("12:00 AM");
103         }
```

It's time to practice! Do Exercise 2-3 and 2-4.

Review Questions

2.17   The DateTime structure has two related properties—**Today** and **Now**. What is the difference between them?

Refer to the following figure and code for the next two questions:

```
//Get rental date from DateTimePicker named dtpRentalDate
        DateTime rentalDate = dtpRentalDate.Value;
// Get time of rental DateTimePicker named dtpRentalTime
        DateTime rentalTime = dtpRentalTime.Value;
```

2.18   When a rented item is returned, it is desired to compute and display the whole number of days (without the fractional part due to time difference, if any) the item was rented. Write the code to display the number of days in the ListBox, lstDisplay.

2.19   Write the code to combine the rental date and time, and store in a single variable named rentalDateTime declared as **DateTime rentalDateTime**

2.20   Write the code to compute the due date for payment by adding 30 days to the invoice date declared as **DateTime invoiceDate = DateTime.Today;**

Exercises

Exercise 2-1: Cost of oil change

The service department of an auto dealership computes the cost of an oil change by adding a fixed labor charge to the cost of oil used (quarts used x cost per quart). Develop an application that allows the user to enter the labor charge, quarts used, cost per quart and sales tax rate in percents. The application should compute the total charge and display the subtotal (labor + cost of oil), sales tax and total cost (subtotal +

sales tax). Develop the necessary form within a project named Exercises_Ch2. Use appropriate data types and names for variables. Use decimal type for variables that store dollar amounts.

Exercise 2-2: User name

Make necessary changes in the code in the UserNameString form (line 27 in Figure 2-5) so that the user name consists of the last 5 characters of the last name, the second character of the first name, and the first 2 digits of the id.

Exercise 2-3: Auto rental charge

"Ace Auto Rentals" is a small business that rents low-cost automobiles. Ace would like you to develop an application that lets them compute the rental charges. The application should let the user enter the date rented and date returned (use the DateTimePicker control), enter the charge/day, and compute and display the days rented and the total rental cost (days rented x charge per day). Note that this application doesn't take into account the times of rental and return.

Extra credit (30%): Modify the application so that the user also enters the time of rental and time of return. These times are used in computing the days rented to include fractional days. For fractional days, a customer is charged only for the fraction of the day the vehicle is used.

Exercise 2-4: Late fee for video game rental

Develop an application for a video game rental store to compute the late fee for videos that are returned late. Develop a form named VidioRentals that lets the store to do the following:

Enter a fixed handling charge, late charges per day per item, number of items rented and the due date. The user doesn't have to enter the date returned; use today's date as the date returned.

Compute the number of days late (today's date - due date), late charge (= No of items × No of days late × late charges per day per item), and total charge (= handling charge + late charge). Today's date is given by **DateTime.Today**."

Display the number of days, late charge and total charge in a ListBox.

Chapter 3

Decision Structures and Validation

The programs presented in Chapters 1 and 2 have simple structures, called **sequence structures,** where the statements are processed sequentially, starting with the first statement and ending with the last, processing each statement, exactly one time. In this chapter, we look at **decision (selection) structure**, which lets you execute a block of statements only under certain conditions.

Learning Objectives

After studying this chapter, you should be able to:

- Use flow charts to represent decisions.
- Create Boolean expressions using relational operators and combine Boolean expressions using logical operators.
- Make decisions using *if else, if else if* and *nested if* statements with Boolean expressions.
- Make decisions using the *switch* statement.
- Validate input data before a control loses focus and also after a control loses focus.

Topics

3.1	Flowchart for Decision Structure	3.4	Nested if Statements
3.2	if else Statement	3.5	The switch Statement
3.3	if else if Statement	3.6	Data Validation

3.1 Flowchart for Decision Structure

As an example of decision structure, consider an updated version of Tutorial 3 from Chapter 1 to include a 20% discount for orders that are greater than or equal to 10 dollars, as shown in Figure 3-1.

Figure 3-1: The IceCreamTotalDiscounted form

Because the discount applies only to orders greater than 10 dollars, the program has to check the value of subtotal and select alternate routes depending on the value. Such a structure that allows a program to execute one or more statements only if a certain condition is true is called a **decision (or selection) structure**.

A popular graphical tool to represent program logic is a flowchart. The flowchart that represents the logic for computing the discount percentage is shown in Figure 3-2.

Figure 3-2: Flowchart for a simple selection structure

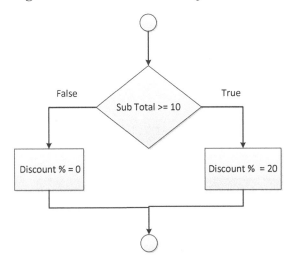

In a flowchart, the diamond is used to check whether a certain condition is true or false. The connector symbol, circle, is used to connect the selection structure to the rest of the program logic that is not shown in the diagram.

A rectangle is used to represent the action to be taken in each path. The action may consist of processing one or more statements with a sequence structure, another selection structure or other structures.

First, we will look at the use of the **if** statement to implement a decision structure.

3.2 if else Statement

This is how you compute the discount via an if statement with an else clause using the decision structure shown in Figure 3-2:

Example 1:
```
if (subTotal >= 10)
        discountPercent = 20;
else
        discountPercent = 0;
```

The above if statement uses the expression (subTotal >= 10), which evaluates to true or false. Such an expression that evaluates to true or false is called a **Boolean expression**.

The Boolean expression (subTotal >= 10) uses the **relational operator** ">=" to compare the value of the decimal type variable subTotal and the literal 10. Table 3-1 presents additional relational operators. Note that when the "< " (less than) operator is used with dates, it means "earlier than." For example, in Table 3-1, the Boolean expression (dateDue < DateTime.Today) evaluates to true if dateDue is earlier than today's date. Similarly, when the ">" (greater than) operator is used with dates, it means "later than."

How does the **if else** statement work? C# first evaluates the Boolean expression (subTotal >= 10).

If it is true, the statement within the **if** clause (discountPercent = 20) is executed, and the else clause (discount = 0) is skipped.

If expression (subTotal >= 10) evaluates to false, the statement within the if clause is skipped, and the statement within the else clause (discount = 0) is executed.

Table 3-1: Relational operators

Relational operator	What it means	Examples
>	Greater than	if (roomCapacity > enrollment)
<	Less than	if (dateDue < DateTime.Today)
>=	Greater than or equal to	if (examScore >= 90)
<=	Less than or equal to	if (dateOfBirth <= DateTime.Parse("12/31/1980")

== (combination of two "=" signs)	Equal to	if (city == "Oshkosh") // city is string type
		if (inState ==true) // inState is bool type
		// The expression if (inState ==true) may be replaced by
		// if (inState), because inState itself is true or false.

!=	Not equal to	if (middleInitial != 'C') // middleInitial is of char type.

Tutorial 1: Ice Cream Total with Discount

This tutorial develops the updated version of Tutorial 3 from Chapter 1 to include a discount, as represented in the flowchart in Figure 3-2.

Step 1-1: Open the project, Ch3_DecisionAndValidation from the Tutorial_Starts folder. Open the form IceCreamTotalDiscounted.

Edit Program.cs file to specify IceCreamTotalDiscounted as the form to be run.

Step 1-2: Modify the code for computing total cost, as shown in Figure 3-3: The changes include

Lines 40–43: Add if else statement to compute discountPercent.

Lines 46–52: Uncomment statements to compute discount, net cost and total cost.

Lines 57–60: Uncomment statements to display subtotal, discount and net cost.

Lines 62–63: Add statements to display sales tax and net cost.

Test the program with combinations of unit price and number of scoops that yield subtotals that are less than 10, equal to 10 and more than 10.

Figure 3-3: Compute total cost with discount

```csharp
private void btnComputeTotal_Click(object sender, EventArgs e)
{
    //Compute the total cost for an order of icecream:

    //Decalre variables to hold input data and computed results

    decimal unitPrice, subTotal, salesTax, totalCost;
    int scoops;
    //Declare constants to hold the discount and tax rates
    const decimal TAX_RATE = 5.5m;

    //Get input data from Text Boxes:
    unitPrice = decimal.Parse(txtUnitPrice.Text);
    scoops = int.Parse(txtScoops.Text);

    // Compute sub-total
    subTotal = unitPrice * scoops;

    // Compute discount percent
    decimal discountPercent, discount;
    if (subTotal >= 10)
        discountPercent = 20;
    else
        discountPercent = 0;

    // Compute discount
    discount = subTotal * discountPercent / 100;

    // Compute Total Cost
    decimal netCost;
    netCost = subTotal - discount;
    salesTax = netCost * TAX_RATE / 100;
    totalCost = netCost + salesTax;

    /* Display Results with decimals aligned
       Important Note: To align characters in a column, the Font type of
       the ListBox must be of fixed size, like Courier New.           */
    string formatString = "{0,-12}{1,10:C}";
    lstOutput.Items.Add(string.Format(formatString, "Sub Total:", subTotal));
    lstOutput.Items.Add(string.Format(formatString, "Discount:", discount));
    lstOutput.Items.Add(string.Format(formatString, "Net Cost:", netCost));
    lstOutput.Items.Add("");   // Add a blank line
    lstOutput.Items.Add(string.Format(formatString, "Sales Tax:", salesTax));
    lstOutput.Items.Add(string.Format(formatString, "Total Cost:", totalCost));
}
```

Mixing Different Data Types in a Boolean Expression

The rules you learned in Chapter 2 about mixing different data types in a formula (mathematical expression) involving numbers also apply to Boolean expressions—that is, you may mix different real and integer data types in a Boolean expression, but you cannot mix a decimal with a float or double without explicit conversion. For example, the expression in the following if statement is invalid:

```
decimal price = 12.5m;
if (price > 10.5)      // invalid; cannot compare a decimal to double
```

The expression should be changed to

```
if (price > 10.5m)     // valid
```

The following comparisons are valid:

```
decimal price = 12.5m;
if (price > 10)    // valid; decimal and integer types may be mixed

float price = 12.5f;
if (price > 10.5)      // valid; float and double types may be mixed
```

An if statement may use one or more of the following **relational operators** to compare data items in a Boolean expression:

if Statement with No else Clause

Example 2: An alternate way to represent the decision structure shown in Figure 3-2 is to remove the statements in the **false** path and place them before the decision structure, as shown below:

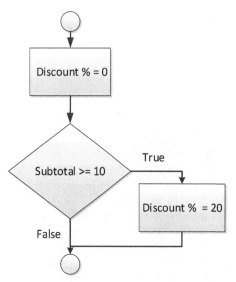

This is how you would translate the flowchart to programming statements:

```
float discountPercent = 0;
if (subTotal >= 10)
     discountPercent = 20;
```

If the value of the Boolean expression (subTotal >= 10) is true, discountPercent is assigned the value 20. What happens if subTotal is less than 10? discountPercent keeps its initial value, zero.

Caution! If there is no else clause, discountPercent must be assigned a value before the variable can be used in the program.

In computing the discountPercent, the above if statement with no else clause will have the same effect as the following code with an else clause:

```
float discountPercent;
if (subTotal >= 10)
        discountPercent = 20;
else
        discountPercent = 0;
```

The two if statements are equivalent because the value assigned in the else clause is the same as the initial value of the variable.

A Block of Statements in the if or else Clause

When the if or else clause includes multiple statements, they are enclosed in a pair of braces. Optionally, you may use braces also when the if or else clause includes only a single statement.

Example 3: if else statement with a block of statements in the if and else clauses.

Assume that the customer gets a free topping if the subtotal is $10 or above. In this case, the if clause assigns values to discountPercent and to the Boolean variable, freeTopping.

```
bool freeTopping;
if (subTotal >= 10)
{
        discountPercent = 20;
        freeTopping = true;
}
else
{
        discountPercent = 0;
        freeTopping = false;
}
```

Note that the block of statements in the if clause is enclosed in a pair of braces.

Step 1-3: Modify the if statement in the click event handler of btnComputeTotal as follows:

```
bool freeTopping;
if (subTotal >= 10)
{
        discountPercent = 20;
        freeTopping = true;
}
else
{
        discountPercent = 0;
        freeTopping = false;
}
```

Step 1-4: Add the following statements at the end of the click event handler to display the message "*** Free Topping ***" if freeTopping has the value, **true**.

```
if (freeTopping)   // same as, if (freeTopping == true)
      lstOutput.Items.Add(" *** Free Topping ***");
```

Note that it is not necessary to use the Boolean expression "(freeTopping = true)" because the value of freeTopping is true or false.

Step 1-5: Run the form to test the code.

The general syntax of the if else statement is

```
if (Boolean expression)
{
      A block of statements  (or, a single statement -- doesn't require braces).
}
else     // else clause is optional
{
      A block of statements (or, a single statement -- doesn't require braces).
}
```

Note that there is no semicolon at the end of the line starting with if (Boolean expression) and the else line.

Logical Operators

If statements may use **logical operators** to combine multiple Boolean expressions that specify the selection criterion. As stated earlier, a Boolean expression is an expression that evaluates to true or false. Table 3-2 shows a list of such logical operators and examples of their use.

Table 3-2: Logical operators

Logical operator	Name	Example	What it means
&&	Conditional And	if((ACT >=20) && (GPA>=2.5)) The second Boolean expression is evaluated only if the first one is **true**.	If **both** expressions are true, "&&" returns **true**; if not, "&&" returns **false**. That is, **true&&true** yields **true**; **true&&false** yields **false**; **false&&true** yields **false**; **false&&false** yields **false**;
&	And	if((ACT >=20) & (GPA >=2.5)) Both expressions are always evaluated.	Same as "&" (i.e., if both expressions are true, "&" returns **true**; if not, "&" returns **false**).
||	Conditional Or	if ((ACT <20) || (GPA <2.5)) The second Boolean expression is evaluated only if the first one is **false**.	If **either** expression is true, "||" returns **true**; if not, "||" returns **false**. That is, **true||false** yields **true**; **false||true** yields **true**; **true||true** yields **true**; **false||true** yields **false**;
|	Or	if ((ACT <20) | (GPA <2.5)) Both expressions are always evaluated.	Same as "|" (i.e., if either expression is true, "|" returns **true**; if not, "|" returns **false**).
!	Not	if (!(ACT >= 20))	if (ACT >= 20), "!" returns **false**; if not, "!" returns **true**.

if statement with logical operators

Table 3-2 presented different logical operators like "&&" (conditional And) and "||" (conditional Or) that can be used in an expression when the decision involves multiple criteria.

Next, we will look at a modified version of the decision structure that uses logical operators. In this version, discount and free topping are given if **the subtotal is at least $10 or the number of scoops is at least 3**.

The following program shows the modified if statement to implement the new decision structure:

```
int scoops=0, subTotal=0, discountPercent;

bool freeTopping;
if ((subTotal >= 10) || (scoops >= 3))
{
    discountPercent = 20;   // processed if either (subTotal >= 10) or (scoops >=3),
    freeTopping = true;     // that is, if one of the two Boolean expressions is true.
}
else
{
    discountPercent = 0;    // processed if both Boolean expressions are not true
    freeTopping = false;    // that is, if both are false.
}
```

Step 1-6: Modify the if statement to include the revised criteria. Run and test the form with different combinations of subtotal and scoops so that (1) both Boolean expressions are false, (2) only the first expression is true, (3) only the second expression is true and (4) both expressions are true. Observe the result in each case.

Change the logical operator from "||" (Or) to "&&" (And), run the form and observe the difference under the four different combinations of subtotal and scoops.
Change the logical operator back to "||"

The expression **((subTotal >= 10) || (scoops >=3))** checks whether subtotal is at least 10 or scoops is at least 3. An alternative to do the same thing is to check whether the expression

```
((subTotal < 10) and (scoops < 3))
```

is **false**—that is, both subtotal and scoops are not below the required limit. Note that the expressions have the "<" operator in place of ">=." Further, "Or" is replaced by "And."

You can use the "!" (Not) operator to check whether the expression is false, as follows:

```
if (!((subTotal < 10) && (scoops < 3)))
```

Step 1-7: Change the if statement to: if (!((subTotal < 10) && (scoops<3))).
Run the form and observe the difference under the four different combinations of subtotal and scoops specified in Step 1-6. Observe the result in each case.
Now, change "&&" to "||" as follows:

```
bool freeTopping;
if (!((subTotal < 10) || (scoops < 3)))
{
    discountPercent = 20;
    freeTopping = true;
}
else
{
    discountPercent = 0;
    freeTopping = false;
}
```

This won't give the desired result. If any one of the two Boolean expressions is false, then the combined expression, ((subTotal < 10) || (scoops < 3)) is true, which makes (!((subTotal < 10) || (scoops < 3))) false, resulting in no discount.

Run the form and observe the difference under the four different combinations of subtotal and scoops. Observe the result in each case.

Review Questions

What would be the output displayed in the ListBox lstOutput by the following programs?

3.1
```
int totalAmt = 80, percentDiscount = 5;
if (totalAmt >= 100)
        percentDiscount = 20;
lstOutput.Items.Add(percentDiscount);
```

3.2
```
int totalAmt = 80;
int percentDiscount = 5;
if (totalAmt >= 100)
        percentDiscount = 20;
else
        percentDiscount = 0;
lstOutput.Items.Add(percentDiscount);
```

3.3
```
int ACT = 23;
float GPA = 3.2F;
if ((ACT >= 24) || (GPA >= 3))
        lstOutput.Items.Add("Admit");
else
        lstOutput.Items.Add("Deny");
```

3.4
```
int ACT = 23;
float GPA = 3.2F;
if ((ACT >= 24) && (GPA >= 3))
        lstOutput.Items.Add("Admit");
else
        lstOutput.Items.Add("Deny");
```

3.5
```
int ACT = 23;
float GPA = 3.2F;
if (!((ACT >= 24) || (GPA >= 3)))
        lstOutput.Items.Add("Admit");
else
        lstOutput.Items.Add("Deny");
```

3.6
```
int ACT = 23;
float GPA = 3.2F;
if (!((ACT >= 24) && (GPA >= 3)))
        lstOutput.Items.Add("Admit");
else
        lstOutput.Items.Add("Deny");
```

3.3 if else if Statement

Consider a grading scale with a minimum of 90, 80 and 70 for A, B and C, respectively (anything below 70 is an F). The flowchart that represents the decision structure would look as follows:

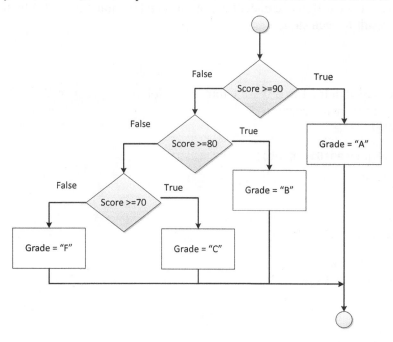

Example 4: The following program uses the **if else if** structure to compute the letter grade based on the decision structure represented by the flowchart. In addition to the grade, the program also computes the grade point.

```
string letterGrade;
int gradePoints;
if (totalScore >= 90)
{
    letterGrade = "A";
    gradePoints = 4;
}
else if (totalScore >= 80)
{
    letterGrade = "B";
    gradePoints = 3;
}
else if (totalScore >= 70)
{
    letterGrade = "C";
    gradePoints = 2;
}
else
{
    letterGrade = "F";
    gradePoints = 0;
}
```

In an if else if statement, in addition to the Boolean expression(s) in the if clause, you may have additional Boolean expressions in else if clauses.

In Example 4, if the expression (totalScore >= 90) is true, the statements in the if clause,

```
letterGrade = "A";
gradePoints = 4;
```

are processed, and the rest of the statements are skipped.

If the expression (totalScore >= 90) is false, the statements in the if clause are ignored, and the next else if clause is processed.

If none of the Boolean expressions are true, the statements in the else clause are executed.

General syntax:
if (Boolean expression)

    one or more statements

else if (Boolean expression)

    one or more statements

...9.5

else

    one or more statements

Review Questions

3.7   What would be the value of letterGrade after execution of the following program?

```
string letterGrade = "F";
int totalScore = 85;

if (totalScore >= 70)
     letterGrade = "C";
else if (totalScore >= 80)
     letterGrade = "B";
     else if (totalScore >= 90)
          letterGrade = "A";
```

3.8   Convert the following **if else if** statement to an **if else** statement:

```
string grade;
int totalScore = 85;

if (totalScore >= 70)
     grade = "Pass";
else if (totalScore < 70)
     grade = "Fail";
```

3.9 What would be the value of discount after execution of the following program?

```
int orderQty = 120, totalAmt = 900, discount =0;

if (totalAmt > 1000)
    discount = 20;
else if (orderQty > 100)
    discount = 10;
else if (totalAmt > 500)
discount = 5;
```

3.4 Nested if Statements

An if statement (with or without the else clause) may be nested within another if statement's if clause or else clause.

Consider a modified version of the flowchart discussed in Example 4 to compute grade. As in the previous example, letter grades A, B and C require a minimum total score of 90, 80 and 70, respectively. In addition, an A requires a minimum of 90 for attendance, as shown in the following flowchart.

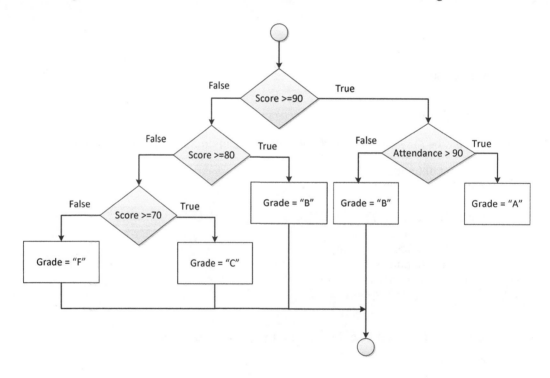

Example 5: The following program uses a nested if statement where the if statement that checks attendance is nested inside the if statement that checks total points.

```
string letterGrade;

if (totalScore >= 90)
    if (attendance >= 90)    // this statement processed only if totalScore is at least 90
        letterGrade = "A";
    else                     // else is processed only if attendance is less than 90
        letterGrade = "B";
else if (totalScore >= 80)   //else if means if totalScore is less than 90
    letterGrade = "B";
else if (totalScore >= 70)
    letterGrade = "C";
else
    letterGrade = "F";
```

If totalScore is greater than or equal to 90, the grade is assigned based on attendance, and the remaining else if clauses and the else clause are skipped.

If totalScore is less than 90, the first else if clause is processed. The process continues until a condition is satisfied.

If no Boolean expression evaluates to true, the final else clause is processed.

Tutorial 2: Compute Grade

In this tutorial, you work with different forms of if statements through an application that computes letter grade based on total score and attendance, as shown in Figure 3-4.

Figure 3-4: The ComputeGrade form

Step 2-1: Open the project, Ch3_DecisionAndValidation from the Tutorial_Starts folder. Open the form ComputeGrade that computes and displays grade based on total points and attendance.

Step 2-2: Review the user interface controls specified in Table 3-3.

Table 3-3: Controls on ComputeGrade form

Purpose	Type of control	Property	Setting
Enter total score	TextBox	Name	txtScore
Enter attendance %	TextBox	Name	txtAttendance
Display grade	Label	Name	txtGrade
Display label for total score	Label	Text	Score:
Display label for attendance	Label	Text	Attendance:
Display label for grade	Label	Text	Grade:
Compute grade	Button	Name	btnGrade
Compute grade	Button	Text	Compute Grade

First, you develop the code to compute grade based on just the total points, ignoring attendance.

Step 2-3: Add the following code to the click event handler of btnGrade to compute grade based only on score.

Figure 3-5: Code to compute grade based only on score.

```
20      private void btnGrade_Click(object sender, EventArgs e)
21      {
22          string letterGrade;
23          float totalScore = float.Parse(txtScore.Text);
24
25          // Compute grade using if else if, based only on score:
26
27          if (totalScore >= 90)
28              letterGrade = "A";
29          else if (totalScore >= 80)
30              letterGrade = "B";
31          else if (totalScore >= 70)
32              letterGrade = "C";
33          else
34              letterGrade = "F";
35
36          // Display grade:
37          lblGrade.Text = letterGrade;
```

Step 2-4: Change the startup form to ComputeGrade, and run the project with values of score in the range 0 to 100, a value above 100 and one negative value. No data for attendance is needed for this version of the program.

Step 2-5: Make changes in the code, as shown in Figure 3-6, to include the requirement that an "A" grade requires an attendance of at least 90%.

The changes include:
Declare attendance in line 26.
Insert the if statement shown in lines 28-31 in place of line 28 in Figure 3-5.

Figure 3-6: Code to compute grade based on score and attendance

```
20      private void btnGrade_Click(object sender, EventArgs e)
21      {
22          string letterGrade;
23          float totalScore = float.Parse(txtScore.Text);
24
25          // compute grade using if else if, based on score and attendance:
26          float attendance = float.Parse(txtAttendance.Text);
27          if (totalScore >= 90)
28              if (attendance >= 90F) // nested within the outer if
29                  letterGrade = "A";
30              else
31                  letterGrade = "B";
32          else if (totalScore >= 80)
33              letterGrade = "B";
34          else if (totalScore >= 70)
35              letterGrade = "C";
36          else
37              letterGrade = "F";
38
39          // Display grade:
40          lblGrade.Text = letterGrade;
41      }
```

Run the project with attendance below 90, and also with attendance greater than or equal to 90, while keeping the score above 90. Verify the results. Run the project when grade is less than 90, and verify that the attendance does not have any effect, in this case.

Review Questions

3.10   A restaurant offers a $10 discount if the total amount is at least $40. If the amount is less than $40, a $5 discount is given if the customer has a coupon and the amount is at least $20. If the amount is less than $20, the coupon gives a 10% discount. Draw a flowchart that represents the decision structure.

3.11   Write an if statement that implements the decision structure described in 3.10. Assume that the variables are declared as follows:

```
float amount = 15.0F;
float discount;
bool coupon = true;
```

3.5 The switch Statement

The switch statement provides a convenient way to branch to one of many paths based on the value of a variable or an expression. How does it differ from an if statement with else if clause? Let's look at the code to determine the discount percent for a hotel reservation based on the discount type ("AAA," "Best Rate," "Government" or "Senior") for a hotel guest.

Here is the if statement with else if clause to determine the discount:

```
float discountPercent;
if (discountType == "AAA")
    discountPercent = 10;
else if (discountType == "Best Rate")
    discountPercent = 5;
else if (discountType == "Government")
    discountPercent = 15;
else if (discountType == "Senior")
    discountPercent = 12.5F;
else
    discountPercent = 0;
```

The corresponding switch statement is

```
float discountPercent;
switch (discountType)
{
    case "AAA":
        discountPercent = 10;
        break;
    case "Best Rate":
        discountPercent = 5;
        break;
    case "Senior":
        discountPercent = 12.5F;
        break;
    case "Government":
        discountPercent = 15;
        break;
    default:
        discountPercent = 0;
        break;
}
```

The switch statement checks the value of the **test variable**, discountType, and branches to one of the first four paths based on its value ("AAA," "Best Rate," "Government" or "Senior"). If the value does not match any one of the four values, the statement under default is processed.

Note that there is a break statement in each path, which causes the program to skip the rest of the switch statement. For example, if the discountType is "Best Rate," then the program assigns the value 5 to discountRate, and the break statement causes it to skip checking for the values "Government" and "Senior" that are specified in the rest of the case statements.

A break statement (or other statements like return) that causes the program to jump out of the switch statement is required in each case and the default statement.

The general syntax of the switch statement is

```
switch (test variable or expression)
{
   case value_1:
      one or more statements
      break;
   case value_2:
      one or more statements
      break;
   case value_3:
      one or more statements
      break;
         ...
   default:
      one or more statements
      break;
}
```

Unlike in an if statement, the value of the test variable or expression in a switch statement must be an integer, string or Boolean. Further, the expression can be compared to the different values only for equality, not for other relational operators like "<" and ">".

For example, the following code, which computes discount when discountCode is equal to 1, is valid:

```
int discountCode = int.Parse(txtDiscountCode.Text);
switch (discountCode)
{
    case 1:     // valid
        discount = 10;
        break;
}
```

The following case statement that uses the ">=" operator is invalid:

```
int score = int.Parse(txtScore.Text);
switch (score)
{
    case >= 90: // invalid use of ">=" operator. Case = 90 also is invalid.
        letterGrade = "A";
        break;
}
```

Tutorial 3: switch Statement

To help understand the switch statement, you will create a form that computes the discount for a hotel room reservation based on the discount type ("AAA," "Best Rate," "Government" or "Senior"), as shown in Figure 3-7:

Figure 3-7: Reservation discount calculator

Step 3-1: Open the project Ch_DecisionAndValidation.

Step 3-2: Open the form Reservation.

Step 3-3: Add the following code to the click event of btnComputeDiscount button.

```csharp
        private void btnComputeDiscount_Click(object sender, EventArgs e)
        {
            string discountType = txtdiscountType.Text;
            float discountPercent;

            // Compute discount percent
            switch (discountType)
            {
                case "AAA":
                    discountPercent = 10;
                    break;
                case "Best Rate":
                    discountPercent = 5;
                    break;
                case "Government":
                    discountPercent = 15;
                    break;
                case "Senior":
                    discountPercent = 12.5F;
                    break;
                default:
                    discountPercent = 0;
                    break;
            }
            // Display discount percent
            lblDiscountPercent.Text = discountPercent.ToString("N");
```

Step 3-4: Make Reservation the startup form, run the project and verify the result for the different discount types.

The following is the equivalent code to compute discount using if else if statement:

```
// Alternative to using if else if statement:
if (discountType == "AAA")
    discountPercent = 10;
else if (discountType == "Best Rate")
    discountPercent = 5;
else if (discountType == "Government")
    discountPercent = 15;
else if (discountType == "Senior")
    discountPercent = 12.5F;
else
    discountPercent = 0;
```

It's time to practice! Do Exercise 3-1.

Review Questions

3.12 In a **switch** statement, what are the valid data types for the test variable (or expression)?

3.13 A university rates its faculty members' performance on a scale of 1 to 3. The annual percent raise in pay is determined by the rating as follows:

Rating	Percent raise
3	0.25%
2	0.1%
1	-1.0%

Write a switch statement that computes the percent raise based on the above rule. Use the following declaration of variables:

int rating = 2;
double raise;

3.6 Data Validation

So far, we assumed that the user would enter valid data into TextBoxes. It is not uncommon that users inadvertently enter invalid data. Such data, if not caught properly, often result in abnormal termination of the program, frustrating the user. A good program should catch such runtime errors and let the user correct them.

Data entered into a TextBox or other controls may be validated before the user moves focus away from the control, or after the focus is moved away from the control. We refer to validation before leaving the control as **immediate validation**, and validation after leaving the control as **late validation.** You will use the **TryParse** method to of numeric data types to do both types of validation.

Consider the ComputeGrade form shown in Figure 3-8, which was presented earlier in Tutorial 2 to compute letter grade.

Figure 3-8: The ComputeGrade form from Tutorial 2

In Tutorial 2, we used the Parse method to convert the total score entered into the TextBox, txtScore, to float, as follows:

```
float totalScore = float.Parse(txtScore.Text);
```

If you use the Parse method to convert a string that contains nonnumeric characters, it will result in an error when you run the program. You can use the TryParse method to validate the data and avoid such errors.

The TryParse method

The TryParse method tries to convert a string to a specific numeric data type. The syntax of TryPrase() method is

numericType.TryParse(string, out resultVariable)

In general, TryParse() tries to convert the string specified as the first parameter to the specific numeric data type (represented by *numericType*). If successful, the resulting number is assigned to the variable specified in the second parameter (represented by *resultVariable*). In addition, TryParse() returns the value true. If not successful, the value zero is assigned to *resultVariable*, and TryParse() returns the value false. The key word *out* specifies that totalScore is an output parameter, which will be discussed in more detail later.

Here is an example:

```
float.TryParse(txtScore.Text, out totalScore);
```

In this example, TryParse() tries to convert the string represented by txtScore.Text to float type. If successful, the resulting number is assigned to the variable totalScore. In addition, TryParse() returns the value true. If not successful, the value zero is assigned to totalScore, and TryParse() returns the value false.

The assignment of a numeric type to the resultVariable must follow the rules of casting discussed in Chapter 2. Thus, you cannot store a data type that has a larger range in a *resultVariable* of a smaller range.

This is how you would use TryParse to validate the total score from the TextBox, txtScore:

```
if (float.TryParse(txtScore.Text, out totalScore) == false)
{
    // Display error message and let user re-enter the data
}
```

How does it work? If conversion of the string to float type is successful, the resulting float type data will be stored in the variable totalScore, and the method will return the value **true,** causing the Boolean expression to have the value, **false**. So, the block within the if statement (to display the error message) would be skipped.

If conversion is unsuccessful, the value zero will be stored in totalScore, and the Parse method will return the value false, causing the Boolean expression to have the value, true. So, the block within the if statement (to display the error message) would be executed.

Late Validation of Data

Late validation validates the data after the user moves focus away from a control. We will use the TryParse method to validate the score and attendance when the user clicks the button, btnGrade.

Figure 3-9 shows the code to validate total score when the user clicks btnGrade.

Figure 3-9: Code to validate the score

```
20          private void btnGrade_Click(object sender, EventArgs e)
21          {
22              float totalScore;
23              // Verify that the score is a number
24              if (float.TryParse(txtScore.Text, out totalScore) == false)
25              {
26                  // Display invalid number message
27                  MessageBox.Show("Please enter a valid number for score");
28                  txtScore.Focus();
29                  txtScore.SelectAll();
30                  return;
31              }
```

In line 27, the statement,

```
MessageBox.Show("Please enter a valid number for score");
```

uses the **Show()** method of MessageBox to display the string ("Please enter a valid number for score") specified as the parameter.
The general syntax is

MessageBox.Show(*string***)**

The second statement in line 28, **txtScore.Focus()**, uses the Focus() method of the TextBox, txtScore, to put the cursor in the TextBox.

The **SelectAll()** method in line 29 highlights the entire string in the TextBox so that the user can type over the data.

Tutorial 4: Late Validation

In this tutorial, you will create a form, **GradeWithLateValidation,** that validates the total score entered into a TextBox, as presented in Figure 3-11. The validation takes place when the user clicks the button to compute the grade. This is called late validation because the data in the TextBox is not validated before the focus leaves the TextBox.

Step 4-1: Open the form **GradeWithLateValidation** from the project Ch3_DecisionAndValidation (within Tutorial_Starts folder), shown in Figure 3-10. This form has the same user interface as the ComputeGrade form.

Figure 3-10: GradeWithLateValidation form

Step 4-2: Open the click event handler of btnGrade, which contains the following code to compute the letter grade. Note that the code to get the total score and attendance from TextBoxes is missing.

```
string letterGrade;

if (totalScore >= 90)
    if (attendance >= 90F) // nested within the outer if
        letterGrade = "A";
    else
        letterGrade = "B";
else if (totalScore >= 80)
    letterGrade = "B";
else if (totalScore >= 70)
    letterGrade = "C";
else
    letterGrade = "F";
// Display grade:
lblGrade.Text = letterGrade;
```

Step 4-3: Add the following code to the beginning of btnGrade click event handler to get the score and validate it:

```
float totalScore;
// Verify that the score is a number
if (float.TryParse(txtScore.Text, out totalScore) == false)
{
    // Display invalid number message
    MessageBox.Show("Please enter a valid number for score");
    txtScore.Focus();
    txtScore.SelectAll();
    return;
}

// Get attendance
float attendance = float.Parse(txtAttendance.Text);
```

Step 4-4: Run the form and test it using valid and invalid score (non-numeric score and no score).

Figure 3-11 shows the complete code for GradeWithLateValidation form.

Figure 3-11: Code for GradeWithLateValidation form

```csharp
20      private void btnGrade_Click(object sender, EventArgs e)
21      {
22          float totalScore;
23          // Verify that the score is a number
24          if (float.TryParse(txtScore.Text, out totalScore) == false)
25          {
26              // Display invalid number message
27              MessageBox.Show("Please enter a valid number for score");
28              txtScore.Focus();
29              txtScore.SelectAll();
30              return;
31          }
32
33          // Get attendance
34          float attendance = float.Parse(txtAttendance.Text;
35
36          // Compute grade:
37          string letterGrade;
38          if (totalScore >= 90)
39              if (attendance >= 90F)
40                  letterGrade = "A";
41              else
42                  letterGrade = "B";
43          else if (totalScore >= 80)
44              letterGrade = "B";
45          else if (totalScore >= 70)
46              letterGrade = "C";
47          else
48              letterGrade = "F";
49
50          // Display grade:
51          lblGrade.Text = letterGrade;
52      }
```

It's time to practice!

Modify the code to verify that the attendance percent is a valid number. If invalid, display an appropriate message, highlight the current data, and allow the user to type over the existing data. If valid, verify that it is within the range 0 to 100. If not, display an appropriate message, highlight the current score and allow the user to type over the existing score.

Immediate Validation of Fields

On the GradeWithLateValidation form, the score and attendance data entered into TextBoxes are validated when the user clicks the button to compute the grade. Next, we look at how to validate data before the user moves away from a TextBox to give immediate feedback. We use the **Validating event** of the TextBox to do this.

Validating event

The Validating event of a TextBox is raised when the user tries to move the focus away from a TextBox—for example, by pressing the Tab key or clicking in another TextBox. This event is raised before another control gets the focus.

To do immediate validation, you validate the data within the Validating event handler of the TextBox and, if invalid, cancel the process of moving focus away from the TextBox.

To create the Validating event handler for txtScore TextBox, select txtScore and select the *events* tab in the Properties window, as shown in Figure 3-12. Double click Validating.

Figure 3-12: Validating event in Properties window

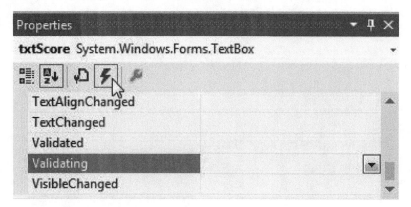

Here is the code to validate the TextBox, txtScore:

```csharp
private void txtScore_Validating(object sender, CancelEventArgs e)
{
    float totalScore;
    // Verify that the score is a number
    if (float.TryParse(txtScore.Text, out totalScore) == false)
    {
        MessageBox.Show("Please enter a valid number for score");
        e.Cancel = true; // Cancel the normal sequence of events
        txtScore.SelectAll();
        return;
    }
}
```

The code is identical to the validating code in the click event handler of the button on the previous form, except for the statement that sets focus back in the TextBox. Instead of txtScore.Focus(), here we use the statement
```
e.Cancel = true;
```

Setting the Cancel property of the parameter e to true has the same effect as txtScore.Focus(). We use this method because setting focus from within the Validating event is not recommended because it can cause the application to stop responding.

Tutorial 5: Immediate Validation Using Validating Event

This tutorial creates the form, GradeWithImmediateValidation, that validates the input data before the focus is moved away from the TextBoxes. The user interface is the same as in GradeWithLateValidation that we discussed previously.

Step 5-1: Open the form named GradeWithImmediateValidation from Ch3_DecisionAndValidation project.

The click event handler of the btnGrade contains the code shown in Figure 3-13, which computes the letter grade with no validation of data.

Figure 3-13: Code from GradeWithImmediateValidation form

```csharp
20      private void btnGrade_Click(object sender, EventArgs e)
21      {
22          // Get totalScore and attendance
23          float totalScore = float.Parse(txtScore.Text);
24          float attendance = float.Parse(txtAttendance.Text);
25
26          // Compute grade:
27          string letterGrade;
28          if (totalScore >= 90)
29              if (attendance >= 90F)
30                  letterGrade = "A";
31              else
32                  letterGrade = "B";
33          else if (totalScore >= 80)
34              letterGrade = "B";
35          else if (totalScore >= 70)
36              letterGrade = "C";
37          else
38              letterGrade = "F";
39
40          // Display grade:
41          lblGrade.Text = letterGrade;
42      }
```

Step 5-2: Create the Validating event handler for txtScore, and add the code from lines 46–54 in Figure 3-14 to the event handler on the line following the first opening brace, "{."
Run the form, enter an invalid score and try moving away by tabbing or clicking in the second TextBox. Repeat with valid score.

Figure 3-14: Validating event handler with MessageBox

```csharp
43
44      private void txtScore_Validating(object sender, CancelEventArgs e)
45      {
46          float totalScore;
47          // Verify that the score is a number
48          if (float.TryParse(txtScore.Text, out totalScore) == false)
49          {
50              // Display invalid number message
51              MessageBox.Show("Please enter a valid number for score");
52              txtScore.Focus();
53              txtScore.SelectAll();
54              return;
55          }
56      }
```

Note that the code that validates the score, as discussed earlier, is the same as in the GradeWithLateValidation form, except that the Cancel property of the parameter **e** is set to true to cancel the process of moving away from the TextBox, thus setting focus back to the TextBox.

An alternative to using a MessageBox to alert users of data errors is to use an **ErrorProvider** control.

ErrorProvider Control

The ErrorProvider indicates an error by placing an error indicator image next to the control. Moving the mouse over the image displays the error message as a tooltip, as shown in Figure 3-15.

Figure 3-15: ErrorProvider control with error message

Unlike the MessageBox, ErrorProvider can be used to indicate errors on multiple TextBoxes simultaneously.

To use an ErrorProvider, add the ErrorProvider control from the Container group in the Toolbox. To indicate error next to a control, use the SetError method of the ErrorProvider object, which specifies the control and the error message, as in

.SetError(controlName, errorMessage);

See the code in Figure 3-16 for an example. To remove the image, set the error message to a null string.

Step 5-3: Drag and drop the ErrorProvider control from the Toolbox to the form. The ErrorProvider appears in the **component tray** at the bottom of the Designer window.
Change the name of the control from ErrorProvider1 to erpAttendance.

Create the Validating event handler for txtAttendance.
Add the code from lines 60–71 in Figure 3-16 to the Validating event handler of txtAttendance on the line following the opening brace "{."
Run the form, enter an invalid data for attendance and try moving away by tabbing or clicking in the first TextBox.
Repeat with valid data.

Figure 3-16: Validating event handler with ErrorProvider

```
57
58      private void txtAttendance_Validating(object sender, CancelEventArgs e)
59      {
60          float attendance;
61          // Verify that attendance is a number
62          if (float.TryParse(txtAttendance.Text, out attendance) == false)
63          {
64              // Display invalid number message
65              erpAttendance.SetError(txtAttendance, "Please enter a valid number for attendance");
66              txtAttendance.Focus();
67              txtAttendance.SelectAll();
68              return;
69          }
70          else
71              erpAttendance.SetError(txtAttendance, "");
72      }
```

Checking for Null Values

The code in the validation event of each TextBox would validate the data entered into it, **if and only if** the user tries to move away from the TextBox. What if the user skips a TextBox and clicks the button to compute grade? That may result in an undesirable abnormal termination of the program.

So, even with the validation codes we have, it is necessary to verify that the TextBoxes are not empty. This can be done in the click event handler of the btnGrade button, as shown in Figure 3-17.

Step 5-4: Add the if statement shown in Figure 3-17 to the beginning of btnGrade_click event handler.

Step 5-5: Run the form, and test the code when attendance data is missing and when invalid attendance data is entered.

Figure 3-17: Checking for nulls in TextBoxes

```
private void btnGrade_Click(object sender, EventArgs e)
{
    // Verify that the TextBoxes are not empty
    if ((txtScore.Text.Length == 0) || (txtAttendance.Text.Length == 0))
    {
        MessageBox.Show("Please enter score and attendance");
        return;
    }

    // Get totalScore and attendance
    ...
```

CauseValidation Property

When you move focus from one control to another, the Validating event for the first control is raised only if the CauseValidation property of both controls are true, which is the default.

Step 5-6: Set CauseValidation for txtAttendance to False. Run the form, enter an invalid score into txtScore and press the Tab key. No message is displayed, indicating that the Validating event for txtScore was not raised.

You can use the CauseValidation property to allow users to move away from a control without entering valid data. For example, you can let the user close the form without entering valid data by clicking the Exit button. To suppress the Validating event when the user moves to the Exit button, set the CauseValidation property of the Exit button to false.

Step 5-7: Add a new button, btnExit, and set its CauseValidation property to false. Enter the code shown in Figure 3-18 to the click event handler of btnExit. Run the form. Enter an invalid score and click the Exit button. No error message is displayed because setting CauseValidation of Exit button to false suppresses the validating event of the TextBox.

Figure 3-18: Click event handler of Exit button

```
73
74          private void btnExit_Click(object sender, EventArgs e)
75          {
76              Environment.Exit(0); // 0 is an exit code passed to the O.S.
77              // Environment.Exit closes all "threads"; but this.Close doesn't
78          }
```

Other Relevant Events

A variety of events are raised when you move from one control to another. These include

1. Enter
2. GotFocus
3. Leave
4. Validating
5. Validated
6. LostFocus

The complete code for computing grade in GradeWithImmediateValidation form is shown in Figure 3-19.

Figure 3-19: Complete code for GradeWithImmediateValidation form

```csharp
20      private void btnGrade_Click(object sender, EventArgs e)
21      {
22          // Verify that the TextBoxes are not empty
23          if ((txtScore.Text.Length == 0 || (txtAttendance.Text.Length == 0)))
24          {
25              MessageBox.Show("Please enter score and attendance");
26              return;
27          }
28          // Get score and attendance
29          float totalScore = float.Parse(txtScore.Text);
30          float attendance = float.Parse(txtAttendance.Text);
31
32          // Compute grade:
33          string letterGrade;
34          if (totalScore >= 90)
35              if (attendance >= 90F)
36                  letterGrade = "A";
37              else
38                  letterGrade = "B";
39          else if (totalScore >= 80)
40              letterGrade = "B";
41          else if (totalScore >= 70)
42              letterGrade = "C";
43          else
44              letterGrade = "F";
45
46          // Display grade:
47          lblGrade.Text = letterGrade;
48      }
49
50      private void txtScore_Validating(object sender, CancelEventArgs e)
51      {
52          float totalScore;
53          // Verify that the score is a number
54          if (float.TryParse(txtScore.Text, out totalScore) == false)
55          {
56              // Display invalid number message
57              MessageBox.Show("Please enter a valid number for score");
58              txtScore.Focus();
59              txtScore.SelectAll();
60              return;
61          }
62      }
```

```csharp
63
64      private void txtAttendance_Validating(object sender, CancelEventArgs e)
65      {
66          float attendance;
67          // Verify that attendance is a number
68          if (float.TryParse(txtAttendance.Text, out attendance) == false)
69          {
70              // Display invalid number message
71              erpAttendance.SetError(txtAttendance, "Please enter a valid number for attendance");
72              txtAttendance.Focus();
73              txtAttendance.SelectAll();
74              return;
75          }
76          else
77              erpAttendance.SetError(txtAttendance, "");
78      }
79
80      private void btnExit_Click(object sender, EventArgs e)
81      {
82          Environment.Exit(0);     // 0 is an exit code passed to the O.S.
83          // Environment.Exit() closes all "threads"; but, this.close() doesn't
84      }
```

It's time to practice! Do Exercise 3-2 and Programming Assignment 1.

Review Questions

Refer to the following code to answer the first three questions:
```
double salary;
double.TryParse(txtSalary.Text, out salary);
```

3.14   What would be the value returned by TryParse if txtSalary contains an invalid number?

3.15   What would be the value stored in salary if txtSalary contains an invalid number?

3.16   What would be the value returned by TryParse if txtSalary contains 245?

3.17   When is the Validating event of a TextBox raised?

3.18   When you move focus from one control to a second control, what is the requirement for the validating event of the first control to be raised?

Exercises

Exercise 3-1

Create a new form named Exercise_3.1 within the project Ch3_DecisionAndValidation. Copy the controls and the code from the form IceCreamTotalDiscounted to the new form Exercise_3.1. Modify the form as shown below, so that it computes discount percent based on the discount code, not based on the subtotal: 20% discount if discount code is "1" and 10% discount if discount code is "2;" otherwise, there is no discount (if there is no discount, the user wouldn't enter any discount code). The discount code is to be entered into a TextBox as shown below. Compute the discount using an if statement and also using a switch statement.

When the user clicks the compute button, the program should validate the data entered into the TextBoxes. The unit price must be a valid number; the number of scoops must be an integer; and if the user enters a discount code, it must be a "1" or "2."

Hint: To check whether a number is an integer, you may compare the number to the truncated number, as in
```
if (scoops == (int)scoops)
```
or,
```
if (scoops == Math.Truncate(scoops))
```

Exercise 3-2

Create a new form named Exercise_3.2 within the project Ch3_DecisionAndValidation. Copy the controls and code from the form IceCreamTotalDiscounted to the new form Exercise_3.2. Modify the form so that the user won't be able to move away from the TextBox for scoops, if the data entered into the TextBox is not an integer. Display a message using the Error Provider. Highlight the data so that the user can type over it.

Hint: To check whether a number is an integer, you may compare the number to the truncated number, as in
```
if (scoops == (int)scoops)
```
or,
```
if (scoops == Math.Truncate(scoops))
```

Similarly, make the necessary changes so that the user won't be able to move away from the TextBox for unit price if the data entered into the TextBox is not a valid number. Display a message using a MessageBox. Highlight the data so that the user can type over it.

Make sure the user is able to close the form by clicking the Exit button, even when the data entered into a TextBox is not valid.

Exercise 3-3

A retail clothing store offers a 10% discount if total sales is $50 or above and less than $100. If total sales is $100 or above, the store offers a 20% discount, and those who have a coupon get $10 off of total sales before applying the 20% discount.

Draw a flow chart that shows the logic for computing the subtotal (= total sales - all discounts).

Develop a new form named Exercise 3.3. You may develop the form within the existing project, Ch3_DecisionAndValidation, or a new project named Ch3_Exercises. Let the user enter the total sales into a TextBox on the form. Compute and display the total sales, coupon amount (if any), discount and subtotal, with the decimals aligned.

Hint: Use a Boolean variable for coupon. See Table 3-1 and Example 3 in this chapter. You may assign the value **true** or **false** to the Boolean variable within the program to test the code.

Exercise 3-4

A hotel chain offers special discounts for customers who have a gold or platinum status:
  Gold: 20% discount on reservations for 2 to 4 days and 30% discount for stays of 5 or more days
  Platinum: 30% discount on reservations for 2 to 4 days and 40% discount for stays of 5 or more days.

Develop a new form named Exercise 3.4. You may develop the form within the existing project, Ch3_DecisionAndValidation, or a new project named Ch3_Exercises. Let the user enter the number of days, customer status (Gold or Platinum) and room rate/day into three different TextBoxes. Compute and display the total cost (= rate/day * number of days), discount and net amount (= total cost - discount), with the decimals aligned.

Rentals Assignment 1

This is an expanded version of Exercise 2-3 from Chapter 2.

"Ace Auto Rentals" is a small business that rents automobiles. Ace would like you to develop an application that lets them compute the rental charges. Specifically, the system should do the following:

> Let the user enter the date rented, time rented, date returned, time returned, charge/day and a discount code. (If the study of Chapter 2 was postponed to a later time, you may enter the number of days rented instead of the date/time of rental/return.)
>
> Compute and display the days rented, the total rental cost (days rented x charge per day), discount and net amount.

Create a project named Assignment1, and develop a form named Rental to do the above tasks.

The specifications on the data and the computation and display of results are as follows:

> The discount code consists of an alphabetic letter followed by a digit. The first character of the discount code must be an "A," "B" or "C." The digit in the code is not used in this application.
>
> Number of days = Date/time returned - date/time rented.
>
> Any fractional day is considered as one full day (that is, if the item is returned later than the time rented, it is counted as a full day).
>
> Total charge = Days rented x charge/day.
>
> Discount is 20% of total charge for discount codes that start with "A," 10% for discount codes starting with "B" and 5% for codes starting with "C." There is no discount if there is no discount code entered into the field.
>
> The decimals of dollar amounts displayed in the ListBox must be aligned, even when they are of different sizes.

Validate the data as follows:

> Date Rented cannot be later than Date Returned.
>
> Make sure that a valid number is entered in the charge/day field. If the charge/day is not a valid number, do not let the user move away from the Textbox where the number is entered. Use an Error Provider to display an appropriate message, and let the user type over.
>
> Make sure that the first character of the discount code is an "A," "B" or "C."
>
> Use Try Catch to catch other errors.

Other general requiements are

> Use efficient code that is easy to understand.
>
> You must use meaningful and self-explanatory names for all variables and objects like forms, TextBoxes, etc., using the naming conventions followed in the textbook. Don't use variable names like tc (for total charge). You don't have to name Labels that are not used in the code.

Help Desk Assignment 1

This assignment develops software to support the services provided by the IT help desk in an organization. The help desk provides over-the-phone and on-site assistance with technical problems to employees from different departments at multiple locations. The department of the employee who requested the service is charged for the service.

The charge for service consists of a fixed charge and a variable charge. The fixed charge is $10, and the variable charge is the product of duration of service in minutes and charge/minute. The charge/minute varies with the area in which the service is provided, as specified later. The help desk would like to develop a system that maintains information on support provided to employees. The system should do the following:

> Let the user enter the department code for the employee, the start time and end time of the service (if the study of Chapter 2 was postponed to a later time, you may enter the number of minutes instead of the start time and end time), the fixed charge and the area of support.

> Compute and display the duration of service in minutes, charge/minute and total charge for the service.

The following are additional requirements of the input data and the computation and display of results:

> The department code consists of an alphabetical letter that represents the location, followed by two numeric digits that represent the department, as in G03.

> The duration of service = End time − Start time.
> Assume that each service ends the same day it is started.

> The valid entries for area of service are:
> > **computer system** (includes computer, printer, network, database and some applications), **ERP**, **BI** and **cloud**.

> The charge per minute is:
> > computer system: $1.00/min; ERP: $1.25/min; BI: $1.5/min; cloud: $1.25/min

> Total charge = fixed charge + (duration of service in minutes × charge/min)

> Use a ListBox to display results. The decimals of dollar amounts displayed in the ListBox must be aligned, even when they are of different sizes.

Validate data as follows:

> The start time cannot be later than the end time.

> Make sure that a valid number is entered in the fixed charge field. If the fixed charge is not a valid number, do not let the user move away from the Textbox where the number is entered. Use an Error Provider to display an appropriate message, and let the user type over.

> Make sure that the first character of the department code entered by the user is an "A," "F," "G" or "N."

> Use Try Catch to catch other errors.

Other general requirements are:

> Use efficient code that is easy to understand.

You must use meaningful and self-explanatory names for all variables and objects like forms, TextBoxes, and so on, using the naming conventions followed in the textbook. Don't use variable names like tc (for total charge). You don't have to name Labels that are not used in the code.

Create a project named HDassignment1, and develop a form named Helpdesk that meets the requirements specified above.

Chapter 4

Iteration Structure: Loops

All programs we discussed so far process each statement at most one time. A program, however, may need to process the same set of statements repeatedly. For example, searching a customer database to find a particular customer record may require reading thousands of similar records using the same set of statements.

Most programming languages, including C#, provide different types of loops to iteratively process a set of statements. In this chapter, you learn how to use different types of loops.

Learning Objectives

After studying this chapter, you should be able to:

- Create simple console applications.
- Use *while*, *do-while* and *for* loops to iteratively process a set of statements.
- Explain how *while*, *do-while* and *for* loops work.
- Increment/decrement values of variables inside a loop.
- Use the *break* and *continue* statements in a loop to deviate from the normal processing.

Topics

4.1	The while Loop	4.4	The do-while Loop
4.2	Console Applications	4.5	The for Loop
4.3	Incrementing Variables in a Loop	4.6	break and continue Statements

4.1 The while Loop

A commonly used form of loop is the *while* loop. This type of loop processes a set of statements while a Boolean expression is true. Figure 4-1 represents the structure of the while loop.

Figure 4-1: Structure of the while loop

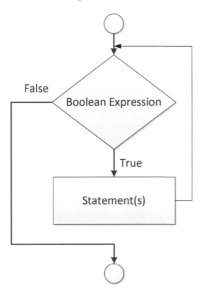

Let's look at an application that uses a while loop to check the password entered by a user when the user clicks the "Continue" button on a form, as shown in Figure 4-2. The while loop is used to check the password repeatedly and prompt the user for a valid password until the user enters the valid one. If the password is valid, a message tells the user to proceed.

Figure 4-2: The VerifyPassword form

Figure 4-3 shows the structure of the while loop. If the password is not equal to "easy" (which is the correct password), then the user is prompted to enter a new password, which is checked again. The process continues as long as the password entered is invalid.

Figure 4-3: Structure of loop to verify password

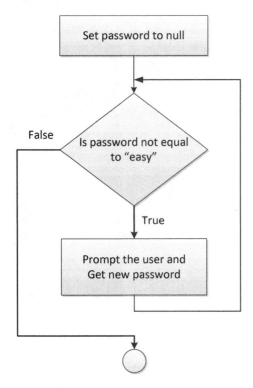

The general syntax of the code for a while statement is

    while (Boolean expression)
    {
        Statements
    }

The statements in the body of the while loop are processed repeatedly, while the test condition represented by the Boolean expression is true. The loop terminates if the test condition is false.

Next, you create the application. To simplify development, we will run it as a Console Application so that the user enters the password on the Console window, rather than into a TextBox as it typically is done.

4.2 Console Applications

In a Console Application, the user interacts with the application through the console rather than through the controls on a form. So, the development of the form is simpler; however, the user interface is limited.

To use the console, you need to change the ***Output type*** setting of the project using the following steps:

    Open the project. We will use the project named Ch4_Loops from Tutorial_Starts folder in the next tutorial.
    Open Properties window by double clicking Properties in Solution Explorer.
    Select *Applications*. Change *Output type* to *Console Application*.

The use of console is illustrated in Tutorial 1.

Tutorial 1: while Loop to Verify Password

This tutorial builds the application to verify the password using the while loop and illustrates the use of console.

Step 1-1: Open the project named Ch4_Loops from Tutorial_Starts folder. Open the form, VerifyPassword, shown in Figure 4.2. The form has a button named btnContinue with the caption "Continue."

Step 1-2: Open the Properties window by double clicking Properties in Solution Explorer. Change *Output type* to *Console Application*.

The code for while statement

The code that translates the flow diagram in Figure 4-3 is shown in Figure 4-4.

Figure 4-4: Code to verify password using the Console

```
19          private void btnContinue_Click(object sender, EventArgs e)
20          {
21              string password = null;
22              while (password != "easy")   // "!=" means not equal to
23              {
24                  // Display prompt in the Console.
25                  Console.WriteLine("Please enter password:");
26                  // Get password and store it in the variable password.
27                  password = Console.ReadLine();
28              }
29              MessageBox.Show("You may proceed now");
30              // Code to open the next form goes here
31          }
```

How does the program work? The first time the while statement is processed, the password is null. Thus, the Boolean expression (password! = "easy ") is true, which causes the statements in the while clause to be processed. Let's look at the two statements in the while clause.

Line 25: `Console.WriteLine ("Please enter password:") ;`

The **WriteLine** method of Console class displays a line of text in the console window and positions the screen cursor at the beginning of the next line so that the user can type some input data. Thus, line 25 displays the prompt "Please enter password: " on the console window, and the application waits for the user to enter an input data. The next statement in line 27 is processed only after the user enters the data and presses the *Enter* key.

Line 27: `password = Console.ReadLine ();`

The **ReadLine** method of Console class reads the data typed by the user into the current line and returns it as a string. Line 27 reads the password entered by the user and assigns it to the variable, password.

After the first iteration through the loop, the Boolean expression (password != "easy") in the header of the loop is evaluated using the password entered by the user.

If the password is not valid, the statements in the while clause are processed again to get a new password. This process is repeated as long as the password is invalid.

If the password is valid ("easy"), control passes to the statement immediately following the while loop (line 29) to displays the message, "You may proceed now." A real-world application typically may have code to open another form.

Step 1-3: Add the code from Figure 4-4 to btnContinue_Click event handler, and test the code.

Step 1-4: Put a break at Line 28. Run the form after entering an invalid password, and observe the value of the password when the program breaks at line 28. Enter a valid password and note how line 28 is skipped because the Boolean expression (password != "easy") is false.

4.3 Incrementing Variables in a Loop

The ability of a loop to process statements repeatedly makes it convenient to accumulate data in a variable, like adding a set of numbers or counting the number of records read from a file. We will look at examples using the while loop.

Using a Counter Inside the Loop to Limit the Number of Attempts

Let's add a counter within the loop to limit the number of attempts to enter the valid password. Figure 4-5 shows the revised code that includes the following steps:
    declaring a counter variable, count (line 21)
    incrementing the counter (line 25), and
    checking the number of attempts (lines 26–31) and terminating the program if the number of attempts exceeds 3.

Figure 4-5: while loop with a counter

```
19      private void btnContinue_Click(object sender, EventArgs e)
20      {
21          int count = 0;
22          string password = null;
23          while (password != "easy")
24          {
25              count = count + 1;
26              if (count > 3)
27              {
28                  MessageBox.Show("Incorrect password. Only 3 attempts allowed");
29                  Environment.Exit(0);    // Close the Console window & the form
30                  // Code 0 tells O.S. that it is a normal termination
31              }
32              Console.WriteLine("Please enter password"); // Display prompt
33              password = Console.ReadLine();  // Get password
34          }
35          MessageBox.Show(count + " Attempts. You may proceed now");
36          // Code to open the next form goes here
37      }
```

Chapter 4: Iteration Structure: Loops

Note that the counter variable, count, is initialized to zero at line 21 **outside** the loop, and its value is increased by one **within** the loop in line 25. Let's take a more detailed look at the statements.

Incrementing and Decrementing Variables

Line 25 uses the following statement to increase the value of count by 1:
    count = count + 1;

The ++ and -- operators

In place of **count = count + 1,** you may use the shorter alternate forms of this statement using the ++ or += operators:
    count ++;   // or, ++count
    count += 1;

To decrement a variable by 1, you replace the "+" sign by the "-" sign, as in
    count = count - 1;
    count --;   // or, --count
    count -= 1;

Line 26 checks whether count exceeded 3. If it did, a message is displayed, and the form and Console windows are closed using the Exit() method of Environment class.

If count is less than or equal to 3, lines 32 and 33 ask the user to enter another password as in the original code.

If the password is valid, control jumps out of the loop to display the message from line 35 that displays the value of count to show the number of attempts.

Step 1-5: Create a new form named VerifyPasswordLimittedAttempts, and add a button named btnContinue. Add the code from Figure 4-5 to the click event handler of the button. (You may copy the code from VerifyPasssword form and modify it.) Put breaks at lines 31 and 34.

    First, run the form with an invalid password two times, and then with the valid password.

    When the program breaks at line 34, observe the value of password and count. Note that lines 31 and 34 are skipped when the correct password is entered.

    Next, run the form with an invalid password three times, and verify that the Console and the form are closed.

    Remove the breaks.

Declaring Variables: Inside versus Outside the Loop

Note that variable count is declared outside the while loop. What happens if count is declared inside the loop?

Step 1-6: Move declaration of count to inside the loop, as shown in Figure 4-6.
    Put a break at the closing brace of the loop. Run the form with three or more incorrect passwords, and observe that the value of count is reset to 0 and incremented to 1 each time.

    To correct the problem, move declaration of count back to the original position at line 21.

Figure 4-6: Declaring counter inside the loop—the wrong way

```
19      private void btnContinue_Click(object sender, EventArgs e)
20      {
21          // int count = 0;    //moved to line 25
22          string password = null;
23          while (password != "easy")
24          {
25              int count = 0; //Wrong place. count gets reset to 0 every time
26              count = count + 1;
```

To increment a variable inside a loop, make sure that the variable is declared outside the loop.

Incrementing a Variable: Inside versus Outside the Loop

The loop in Figure 4-5 includes the statement, **count = count +1**, in line 25, to increment count by one in each iteration through the loop. What is the effect of placing this statement before the loop? Do Step 1-7 to find out.

Step 1-7: Increment count outside the loop before the loop begins, by moving the statement, **count = count + 1**, to the line immediately following line 21, as shown in Figure 4-7.

Put a break at the end of the loop on line 35. Run the form multiple times entering incorrect passwords, and observe that the value of count stays at one because count doesn't get incremented by the iterations through the loop.

Figure 4-7: Incrementing counter outside the loop—the wrong way.

```
19      private void btnContinue_Click(object sender, EventArgs e)
20      {
21          int count = 0;
22          count = count + 1;   // wrong place. count gets incremented only once
23          string password = null;
24          while (password != "easy")
25          {
26              // count = count + 1; moved to outside the loop
27              if (count > 3)
```

Next, we will look at the use of the while loop in a financial investment application where the value of accumulated investments determines the number of times the loop is processed.

Tutorial 2: while Loop: Financial Planning Application

This application uses a while loop in a financial planning application that helps you determine how many years it will take to reach a targeted amount of savings, as shown in Figure 4-8. This example also is used to show the application of do-while and for loops.

Figure 4-8: The FinancialPlanner form

In this example, yearly investment and yearly growth rate are used to forecast the future value of investments for each year, while investment value is less than the target of $500,000. It is assumed that investments are made at the beginning of each year. Figure 4-9 shows the flowchart of the while statement.

Figure 4-9: Flowchart of while loop in FinancialPlanner

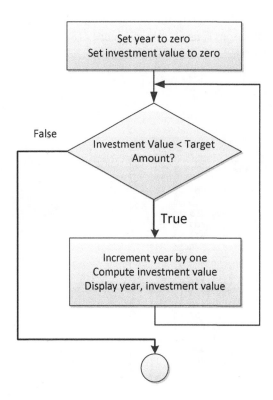

Step 2-1: Open the form named FinancialPlanner, shown in Figure 4.8, from the project Ch4_Loops in Tutorial_Starts folder. The form has the following controls:

>Three TextBoxes: txtInvestment, txtGrowthRate and txtTargetAmt
>Three Labels
>A button: btnHowLong
>A ListBox: lstDisplay

The complete code is shown in Figure 4-10.

The first three sections of the program consist of the following subtasks:

>Get input data from TextBoxes.
>Print headings.
>Set year and investment value to zero.

Figure 4-10: Code for the FinancialPlanner form

```csharp
20          private void btnHowLong_Click(object sender, EventArgs e)
21          {
22              // Get input data from Text Boxes:
23              double yearlyInvestment = double.Parse(txtInvestment.Text);
24              float YearlyGrowthPercent = float.Parse(txtGrowthRate.Text);
25              double targetAmt = double.Parse(txtTargetAmt.Text);
26
27              // Print heading
28              string formatCode = "{0,7}{1,20}";
29              lstDisplay.Items.Add(string.Format(formatCode, "years",
30                                                      "Investment Value"));
31              lstDisplay.Items.Add("");
32
33              // Set year and investment value to zero
34              int year = 0;
35              double investmentValue = 0;
36
37              // Use while loop to compute investment value for each year
38              while (investmentValue < targetAmt)
39              {
40                  year = year + 1;    // increment year by one
41                  investmentValue = (investmentValue + yearlyInvestment) *
42                                              (1+ YearlyGrowthPercent/100);
43                  lstDisplay.Items.Add(string.Format(formatCode, year,
44                                              investmentValue.ToString("C")));
45              }
```

Let's focus on the fourth section of the code that uses the while loop.

The while statement evaluates the Boolean expression (investmentValue < targetAmt). If it is true, the statements in the while clause are processed to increment the year and compute investment value; if not, control passes to the statement following the loop (the right bracket at the end of the method).

The first time the while statement is processed, investmentValue is zero, and the Boolean expression (0 < 500000) is true. So, the statements in the while clause are processed to do the following:

> increment year to 1,
> compute the investment value for year 1, using the expression in line 41, which yields the value 10,700, and
> display year and investment value.

The formula that computes investment value is discussed later.

As the flowchart in Figure 4-9 shows, after processing the last statement in the while clause, control automatically passes to the beginning of the loop, where the Boolean expression is evaluated using the investmentValue (10,700) computed in the first iteration, to start the second iteration.

In the second iteration, the Boolean expression
> (10700 < 500000)

is true. o, the statements in the while clause are processed the second time to increment year to 2, compute the new investment value (22,149) and display it.

Control again passes to the beginning of the loop for the third iteration, repeating the process.

The process continues through the twenty-second iteration because the Boolean expression in this iteration (480057 < 500000) is true. However, the investmentValue computed during the twenty-second iteration (524,361) exceeds 500,000, as shown in Table 4-1. Therefore, the Boolean expression (524361 < 500000) in the twenty-third iteration evaluates to false, causing the statements in the loop to be skipped and control to be passed to the statement following the loop (the left brace at the end of the method).

Table 4-1: Investment value in different iterations

Variable/expression	Value computed in iteration#					
	#1	#2	...	#21	#22	#23
(investmentValue<500000)	true	true	...	true	true	false
Year	1	2	...	25	26	
investmentValue	10700	22,149	...	480,057	524,361	

Step 2-2: Type in the code for the while loop from lines 38–45 in Figure 4-10.

Step 2-3: Change the project to a *Windows Application*. (Open the Properties window by double clicking Properties in Solution Explorer. Change *Output type* to *Windows Application*.) Run the form and test it.

Step 2-4: Change target amount to 50,000 (not 500,000). Put a break at the closing right brace of the while loop and another break at the closing right brace of the method.
Run the form, and observe the values of year and investmentValue for a few iterations.

After the fifth iteration, write down the values of year and investmentValue. Click Continue to complete the loop. When the program breaks at the last brace, observe that the values of year and investmentValue have not changed.

Infinite Loop

The statements in the body of a while loop are executed repeatedly while the test condition represented by the Boolean expression is true. The loop terminates when the value of the Boolean expression becomes false. However, logic errors in programming can cause the Boolean expression not to become false.

For example, if the variable investmentValue is initialized to zero within the loop, as shown below, its value may never reach the target amount, causing the test condition never to become false. This creates an "infinite loop" that continues to run until the program is interrupted.

```
double yearlyInvestment=0, YearlyGrowthPercent = 0, targetAmt=0;
int year = 0;
double investmentValue=0;
while (investmentValue < targetAmt)
{
    investmentValue = 0; // wrong place to initialize
    year = year + 1;
    investmentValue = (investmentValue + yearlyInvestment) *
                            (1 + YearlyGrowthPercent / 100);
    lstDisplay.Items.Add(string.Format(formatCode, year,
                            investmentValue.ToString("C")));
}
```

Pretest Loops

The while loop has a Boolean expression that is checked at the beginning of the loop. So, the while loop is called a pretest loop. Because it is a pretest loop, if the Boolean expression is false in the very first iteration, the statements in the loop will not be processed even once. For example, if investValue is initialized to a value greater than 500,000, the statements in the loop will not be processed.

Step 2-5: Change the initial value of investmentValue (in line 35) to 501,000. Run the form. Note that no output is displayed in the ListBox, indicating that the statements in the loop were not executed.

It's time to practice! Do Exercises 4-1 and 4-2.

Review Questions

4.1 Fix errors, if any, in the following program, which is intended to compute the total savings for 4 years if $750 is deposited each year, assuming that there is no interest added.
```
int yearlyDeposit = 750;
int totalSavings = 0;
while (year < 4)
{
        int year = 0;
        year = year + 1;
        totalSavings = totalSavings + yearlyDeposit;
}
```

4.2 What would be the output of the following program? (Pay special attention to the statements inside the loop.)
```
int yearlyDeposit = 750;
int totalSavings = 0;
int year = 0;
while (year < 3)
{
        year = year + 1;
        totalSavings = yearlyDeposit;
}
lstOutput.Items.Add(totalSavings);
```

4.3 What would be the output of the following program?
```
int year = 2016;
int population = 75000;
while (population < 77000)
{
        year = year + 1;
        population = population + 1000;
        lstOutput.Items.Add(year + " " + population.ToString());
}
```

4.4 How many times will the loop in the following program be executed?
```
int year = 3;
while (year < 3)
{
   year = year + 1;
}
```

4.4 The do-while Loop

A variation of the while loop is the do-while loop that checks the Boolean expression at the end of the loop, as shown below. So, do-while is a post-test loop.

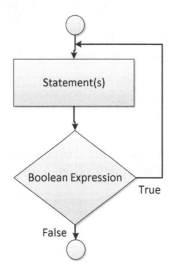

Figure 4-11 shows what the do-while version of the financial planning program looks like.

Figure 4-11: do-while version of FinancialPlanner

```
46
47                  do
48                  {
49                      year = year + 1;    // increment year by one
50                      investmentValue = (investmentValue + yearlyInvestment) *
51                                          (1 + YearlyGrowthPercent / 100);
52                      lstDisplay.Items.Add(string.Format(formatCode, year,
53                                          investmentValue.ToString("C")));
54                  }
55                  while (investmentValue < targetAmt);
```

The loop starts with the key word *do* and closes with the key word *while*, followed by the Boolean expression. It works the same way as the while loop, except that the Boolean expression is checked only at the end of the loop.

So, even if the investment value is initialized to a value greater than the target value, the statements in the loop will be processed the first time, and the iterations will continue as long as the computed investmentValue is less than the targetAmt.

Step 2-6: Comment out the while loop in lines 38-45. Add the code for the do-while loop from Figure 4-11. Make sure that investmentValue is initialized to 501,000, in line 35.
Run the form. Observe that the output is the same as when investmentValue was initialized to 0. Because the condition is checked at the end of the loop, the statements in the loop are processed the first time when investmentValue was set to a new computed value.

Review Question

4.5 How many times will the loop in the following program be executed?
```
int year = 3;
do
{
        year = year + 1;
}
while (year < 3)
```

4.5 The for Loop

The for loop is designed to perform a fixed number of iterations, determined by the value of a counter variable. The counter variable is initialized to a certain specified value and incremented/decremented at the end of each iteration by a specified value.

To help understand the for loop, we will look at a different version of the financial planning program. Instead of finding the number of years to reach a target amount, this version specifies the number of years and displays the investment value for the specified number of years, as shown in Figure 4-12.

The form, **FinPlannerForLoop,** has a TextBox named txtYears to enter the number of years, instead of the target amount.

Figure 4-12: FinPlannerForLoop form

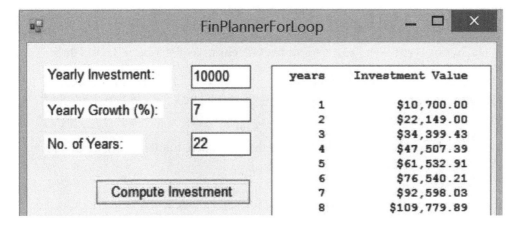

The code for the for loop version of FinancialPlanner is shown in Figure 4-13.

Figure 4-13: Code for FinPlannerForLoop form

```csharp
20          private void btnInvestment_Click(object sender, EventArgs e)
21          {
22              // Get input data from Text Boxes:
23              double yearlyInvestment = double.Parse(txtInvestment.Text);
24              float YearlyGrowthPercent = float.Parse(txtGrowthRate.Text);
25              int numberOfYears = int.Parse(txtYears.Text);
26
27              // Print heading
28              string formatCode = "{0,7}{1,20}";
29              lstDisplay.Items.Add(string.Format(formatCode, "years",
30                                                "Investment Value"));
31              lstDisplay.Items.Add("");
32
33              // Set investment value to zero
34              double investmentValue = 0;
35
36              // Use for loop to compute investment value for each year
37              int year;
38              for (year = 1; year <= numberOfYears; year = year + 1)
39              {
40                  investmentValue = (investmentValue + yearlyInvestment) *
41                                    (1 + YearlyGrowthPercent / 100);
42                  lstDisplay.Items.Add(string.Format(formatCode, year,
43                                       investmentValue.ToString("C")));
44              }
```

The for loop header in line 38, **for (year = 1; year <= numberOfYears; year = year + 1)**, has three expressions:

1. **Initialization expression** (year = 1) that initializes the counter variable, year. You may combine the declaration and initialization of the counter variable, as in
       for (int year = 1; year <= numberOfYears; year = year + 1);
   If you declare the variable in the loop header, the scope of the variable is limited to the loop. That is, you will not be able to access the variable outside the for loop.

2. **Boolean expression** (year <= numberOfYears). If it is true, the statements in the loop are executed; if not, control passes to the statement following the loop.

3. **Increment expression** (year = year + 1) that increments or decrements the counter variable. You may use *year++* or *year += 1* in place of *year = year + 1*
   A variable may be incremented by any amount. For example, year = year+5 would increment year by 5 after each iteration.

Note that the three expressions are separated by semicolons, not commas.

You also may decrement a variable. For example, you may start with year 5, decrement it by one each time, and process the loop as long as year is **greater than** or equal to 1. For example,

```
for (int year = 5; year >= 1; year = year - 1) ;
```

would give the following output:

```
years      Investment Value

  5         $10,700.00
  4         $22,149.00
  3         $34,399.43
  2         $47,507.39
  1         $61,532.91
```

Note that the Boolean expression has the ">=" operator.

Thus, the general syntax of the for loop is

> for (InitializationExpression, BooleanExpression, IncrementExpression)
> {
>   Statements
> }

How does the for loop shown in Figure 4-13 work? The first time the loop is processed, year is set to 1, and the Boolean expression (1 <= numberOfYears) is evaluated.

If the Boolean expression is true, the loop is processed to calculate and display the investment value. In addition, when an iteration is completed, the increment expression year=year+1 is executed to increment the value of year, and the Boolean expression is evaluated. This process continues until the Boolean expression is false.

If the Boolean expression is false, the processing of the loop ends, and control passes to the statement, if any, that follows the loop. If there are no statements after the loop, control passes to the end of the program. The equivalent flowchart is shown in Figure 4-14.

Note that the flowchart shown in Figure 4-14 is similar to that of the while loop that uses a counter variable. The for loop always has a counter variable whose value determines the termination of loop execution, whereas the while loop may use any Boolean expression to terminate the loop.

Figure 4-14: Flowchart of for loop

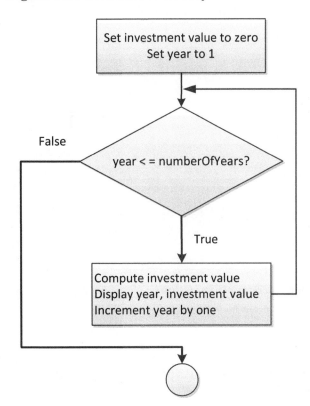

Tutorial 3: for Loop: Financial Planning Application

Let's build the FinPlannerForLoop form that uses the for loop to compute the investment value for a fixed number of years.

Step 3-1: Open the form named FinPlannerForLoop from the project, Ch4_Loops.
Add the code from lines 37-44 in Figure 4-13 to create the for loop.

Step 3-2: Put breaks at the header of the loop and at the beginning and ending braces.
Run the form, and observe the value of year the first time through the loop and how it changes between the end of the first iteration and the beginning of the second iteration.

Step 3-3: To help understand the termination of the loop, copy the statement that displays the output from lines 42 and 43 to the line right after the closing brace of the loop, as shown in Figure 4-15.
Run the form.
When the value of year become 23, the processing of the loop terminates because the Boolean expression (year<= numberOfYears) is false. So, investmentValue is not computed for year 23; it has the same value as for year 22, as shown below as shown in the following output:

```
21        $480,057.39
22        $524,361.41
23        $524,361.41
```

The variable, year, is available outside the loop because it is declared outside the loop.

Figure 4-15: Displaying output outside the loop

```
37                  int year;
38                  for (year = 1; year <= numberOfYears; year = year + 1)
39                  {
40                      investmentValue = (investmentValue + yearlyInvestment) *
41                                                  (1 + YearlyGrowthPercent / 100);
42                      lstDisplay.Items.Add(string.Format(formatCode, year,
43                                                  investmentValue.ToString("C")));
44                  }
45                  lstDisplay.Items.Add(string.Format(formatCode, year,
46                                                  investmentValue.ToString("C")));
```

Declaring Counter Variable Inside the Loop

Step 3-4: Modify the header of the loop as follows to declare the counter variable within the header:

```
37                  // int year;
38                  for (int year = 1; year <= numberOfYears; year = year + 1)
```
Run the form.

Error! The program would not run because the variable year is declared inside the loop, but it is used outside the loop in line 45 to display its value.
Delete the statement in lines 45 and 46 that uses the variable year outside the loop. The error disappears.

Again, the scope of a variable declared inside a loop is limited to the loop.

Incrementing Counter by a Value Other Than One

Step 3-5: Modify the header as follows to increment year by 5 in each iteration:
for (**int** year = 1; year <= numberOfYears; year = year + 5)
Run the form. Note how the value of year changes.
Warning! The computed investment values are incorrect except for the first year, because the formula assumes one year increments.

years	Investment Value
1	$10,700.00
6	$22,149.00
11	$34,399.43
16	$47,507.39
21	$61,532.91

Decrementing the Counter Variable

Step 3-6: Change the header of the loop as follows to decrement the counter variable:
for (int year = 5; year >= 1; year = year - 1)
Run the form. The value of year changes from 5 to 1, as follows:

```
years     Investment Value

  5          $10,700.00
  4          $22,149.00
  3          $34,399.43
  2          $47,507.39
  1          $61,532.91
```

Review Questions

4.6   What would be the output of the following program?

```
int population = 65000;
for (int year = 2016; year <= 2018; year = year + 1)
{
    population = population + 1000;
    lstOutput.Items.Add(year + " " + population.ToString());
}
```

4.7   What would be the output of the following program?
```
int population = 65000;
for (int yr = 2018; yr > 2016; yr = yr - 1)
{
    population = population - 1000;
    lstDisplay.Items.Add(yr + " " + population.ToString());
}
```

4.6 The break and continue Statements

The break and continue statements let you deviate from the normal processing of a loop.

The break Statement

You may jump out of a loop using the break statement.

Step 3-7: Modify the code by adding an if statement and a break statement, as shown in Figure 4-16, to break out of the loop, if investmentValue exceeds 100,000. Copy the display statement from line 42 to line 47 as shown, and make sure year is declared outside the loop.

Figure 4-16: if statement with break statement

```
37      int year;
38      for (year = 1; year <= numberOfYears; year = year + 1)
39      {
40          investmentValue = (investmentValue + yearlyInvestment) *
41                           (1 + YearlyGrowthPercent / 100);
42          lstDisplay.Items.Add(string.Format(formatCode, year,
43                              investmentValue.ToString("C")));
44          if (investmentValue > 100000)
45              break;
46      }
47      lstDisplay.Items.Add(string.Format(formatCode, year,
48                          investmentValue.ToString("C")));
```

Run the code. Execution of the loop terminates at the break statement when the investmentValue exceeds 100,000 in year 8, as shown by the output in Figure 4-17, although the Boolean expression in the header of the loop is true.

A break will process the remaining statements that follow the loop within the method. So, as shown in Figure 4-17, year and investmentValue are displayed by the statement in line 47, which is outside the loop.

Delete the if statement, including the break statement.

Figure 4-17: Effect of break in a for loop

```
years     Investment Value

  1          $10,700.00
  2          $22,149.00
  3          $34,399.43
  4          $47,507.39
  5          $61,532.91
  6          $76,540.21
  7          $92,598.03
  8         $109,779.89
  8         $109,779.89
```

The continue Statement

The continue statement causes the rest of the statements in a loop to be skipped, but continues processing the loop.

Step 3-8: Modify the code, as shown in Figure 4-18, to skip displaying the results if investmentValue exceeds 100,000.

Figure 4-18: Effect of continue statement in a loop

```
37    int year;
38    for (year = 1; year <= numberOfYears; year = year + 1)
39    {
40        investmentValue = (investmentValue + yearlyInvestment) *
41                          (1 + YearlyGrowthPercent / 100);
42        lstDisplay.Items.Add(string.Format(formatCode, year,
43                             investmentValue.ToString("C")));
44        if (investmentValue > 100000)
45            continue;
46    }
47    lstDisplay.Items.Add(string.Format(formatCode, year,
48                         investmentValue.ToString("C")));
```

Step 3-9: Run the code.

Figure 4-19 shows the output. The loop continues to be processed even after investmentValue exceeds 100,000, without displaying the year and investment value. How do we know? The final investmentValue is $524,361, and the year is 23. Note that the investment value exceeds 500,000 in year 22. But, that year is not displayed. The year is displayed only after year is incremented by 1 to 23, when control passes out of the loop.

Delete the if statement, including the continue statement.

Figure 4-19: Output with continue statement in the loop

```
years    Investment Value

  1       $10,700.00
  2       $22,149.00
  3       $34,399.43
  4       $47,507.39
  5       $61,532.91
  6       $76,540.21
  7       $92,598.03
 23      $524,361.41
```

The foreach Statement

A form of loop that is particularly useful to access each item from a collection of items is the **foreach** loop. Chapters 7, 8 and 9 discuss the foreach loop and how it is used to access items from arrays and collections.

It's time to practice! Do Exercises 4-3 and 4-4.

Review Question

4.8   What is the difference between the break and continue statements?

Exercises

Exercise 4-1
The price of laptops is expected to go down by 10% per year. Use a while loop to display the year and price when the price will be under a specified target amount. Allow the user to specify the current price and the target price using TextBoxes.

Exercise 4-2
Develop a form that allows the user to project the year when the population of your city exceeds a specified target number. Allow the user to specify the current year's population, the target number and the annual growth rate. Use a while statement to compute and display each year, starting with the current year, and the corresponding population.

Exercise 4-3
Develop a form to display the Fahrenheit temperature and corresponding Celsius temperature for Fahrenheit temperatures ranging from –40 to +40 degrees, at increments of 10. Use a for loop. The conversion formula is
   Celsius = (Fahrenheit - 32) * 5/9

Exercise 4-4
Develop a form that allows the user to project the population of your city for a specified number of years and growth rate. Allow the user to specify the current year's population, the number of years and the annual growth rate. Use a for loop to compute and display each year, starting with the current year and the corresponding population.

Chapter 5

Methods

A method is a program unit that performs a particular task. The methods you wrote in previous chapters were event handlers, which are connected to a certain event of a control, like the Click event of a button. The Click event handler, for example, is executed when the user clicks a button.

In this chapter, you will learn how to develop methods that are not event handlers—that is, methods not connected to any event; instead, they are called by other methods. The term method, unless qualified, would mean such methods that are not event handlers.

Learning Objectives

After studying this chapter, you should be able to:

- Describe a *method* and state the beneftis of using methods.
- Explain top-down design and develop a structure chart to show the design of a program.
- Create and call a *method* that receives values through parameters and returns one or more values.
- Describe the difference between passing by value and passing by reference; reference parameter and output parameter.
- Create and use an enumeration.

Topics

5.1	Introduction to Methods	5.5	Top-Down Design
5.2	Passing Values to Methods	5.6	Passing Arguments by Reference
5.3	Passing Arguments by Value	5.7	Enumerations: Limiting Parameter Values
5.4	Methods That Return a Value		

5.1 Introduction to Methods

What is the purpose of methods?

1. **Divide and conquer**: A method typically performs a subtask that is a part of a larger or more complex task. For example, processing an order might consist of the following subtasks:
    verify item numbers,
    check quantity on hand,
    compute order total.
   Rather than developing a single program to perform the entire task, breaking the task down to smaller subtasks that are performed by separate methods often helps make software development simpler and easier. This approach to software development is known as **top-down design**.

2. **Reuse**: A method, if written properly, can be reused in multiple projects/forms or called from multiple methods. A method to verify an item number, for example, could be used in multiple applications, like order processing and production scheduling.

Before we look at the process of top-down design, let's look at the structure of a method and how it is used.

Consider a form named MethodDemo, shown in Figure 5-1, which computes your optimal training heart rate using your age and resting heart rate. This form has an "About" button that uses a method to display a greeting and some information about the form in a MessageBox.

Figure 5-1: The MethodDemo form with "About" MessageBox

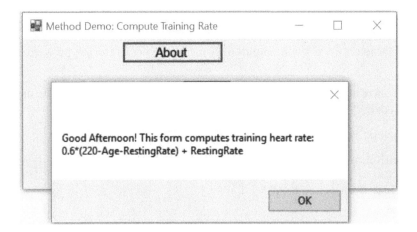

Example of a Method

Figure 5-2 shows the method named DisplayInfo() that displays the MessageBox with the greeting.

Figure 5-2: Example of a method

```
19
20      private void DisplayInfo()
21      {
22          string greeting;
23          if (DateTime.Now < DateTime.Parse("12:00 PM"))
24              greeting = "Good Morning!";
25          else if (DateTime.Now < DateTime.Parse("6:00 PM"))
26              greeting = "Good Afternoon!";
27          else
28              greeting = "Good Evening!";
29
30          MessageBox.Show(greeting + " This form computes training heart rate" +
31                          ": 0.6*(220-Age-RestingRate) + RestingRate");
32      }
```

Structure of a Method

DisplayInfo has the typical structure of a method:
        header (function definition): private void DisplayInfo(),
        body: consists of the statements enclosed in curly brackets.

The statements within the body of this method are not relevant in understanding the structure. For now, it is sufficient to understand that the statements in the body display one of three greetings based on the time of the day.

The first key word in the header, **private**, specifies that this method can be called (to execute it) only from within the class where it is declared—that is, only from the form, MethodDemo. By default, methods are private, so the word private can be omitted.

If you want to make the method callable from other forms, you specify **public** instead of private. The key word that specifies whether the method is private or public is termed an **access modifier**.

The second key word, **void,** specifies that this method does not return a value. For a method that returns a value, you specify the data type of the value in place of void. This key word is called the **return type** because it specifies the type of value returned by the method. A method whose return type is void is called a **void method**.

The pair of empty parentheses following the **method name** DisplayInfo indicates that no data can be passed to this method when it is called.

Thus, the general syntax of a method is

        AccessModifier ReturnType MethodName([parameterList])

        {

            Statements

        }

Naming a Method

Microsoft's recommendation is to use verbs or verb phrases to name methods, and use Pascal Casing—that is, the first letter of the method name and the first letter of subsequent words, if any, are capitalized, as in DisplayInfo. However, as you will learn later, it is a common practice to use nouns or noun phrases to name certain types of methods.

Calling a Method

The statement to call a method is just the name of the method followed by a pair of parentheses. In Figure 5-3, the single statement, **DisplayInfo();** within the btnAbout_Click event handler calls the method DisplayInfo.

Figure 5-3: Click event handler of btnAbout that calls DisplayInfo method

```
33
34   private void btnAbout_Click(object sender, EventArgs e)
35   {
36       DisplayInfo();
37       txtRestingRate.Focus();
38   }
```

When a method is called, the program jumps to that method, executes the statements in the body of the method and returns to the statement that follows the call statement. So, the statement **DisplayInfo()** causes the program to branch to the method DisplayInfo, display the message and return to the statement **txtScore.Focus().**

Tutorial 1 creates the form MethodDemo.

Tutorial 1: Methods: Compute Heart Rate

To explore the use of methods with and without parameters, let's build the MethodDemo form.

Step 1-1: Open the project named Ch5_Methods from the folder, Tutorial_Starts.

Step 1-2: Open the MethodDemo form.

Step 1-3: Open the code window to view the code for **DisplayInfo** method.

Add the two statements to the **btnAbout_Click** event handler, as shown in Figure 5-4.

Step 1-4: Run the form and test the code.

Step 1-5: To understand the path followed by the program, put breaks on lines 21, 36 and 37 by clicking on the left margin in the code window (see Figure 5-4).

Figure 5-4: Code to create DisplayInfo and to call it

```
19
20      private void DisplayInfo()
21      {
22          string greeting;
23          if (DateTime.Now < DateTime.Parse("12:00 PM"))
24              greeting = "Good Morning!";
25          else if (DateTime.Now < DateTime.Parse("6:00 PM"))
26              greeting = "Good Afternoon!";
27          else
28              greeting = "Good Evening!";
29
30          MessageBox.Show(greeting + " This form computes training heart rate" +
31                          ": 0.6*(220-Age-RestingRate) + RestingRate");
32      }
33
34      private void btnAbout_Click(object sender, EventArgs e)
35      {
36          DisplayInfo();
37          txtRestingRate.Focus();
38      }
```

Step 1-6: Run the form and click the About button.

The program breaks at the statement that calls the method DisplayInfo (line 36; your line numbers may be different).

Click the Continue button. The program jumps to DisplayInfo() method and breaks at the left curly brace (line 21).

Step 1-7: Click Continue to run the method and display the message.

Click "OK" to continue. The program returns to the called program (btnAbout_Click) and breaks at line 37.

Step 1-8: Click Continue. Execution stops.

Review Questions

5.1 What is a method?

5.2 What is an event handler?

5.3 True or false: A method typically performs a subtask that is a part of a larger or more complex task.

5.4 True or false: A method can be called from multiple methods.

5.5 Write the statement to call a method named, ComputeCost.

5.2 Passing Values to Methods

The previous method, DisplayInfo, did not need any input data to be provided by the calling program to display its output. However, it often is necessary to pass input data to a method from the calling program. To help understand passing data to methods, you will develop a method, ComputeTrainingRate(), to compute your training heart rate using your age and resting rate, and for now, display the rate in a MessageBox, as shown in Figure 5-5.

Figure 5-5: The MethodDemo form with Training Rate MessageBox

The age and resting rate are passed by the calling program to ComputeTrainingRate(). Figure 5-6 depicts the **structure chart** that shows the calling and called programs and the data transferred between them.

Figure 5-6: Structure chart of program to compute training rate

To receive the values of resting rate and age, the method ComputeTrainingRate has two **parameters**, restingRate and age, declared in the header, as shown in Figure 5-7.

Figure 5-7: ComputeTrainingRate with parameters

```
39
40      private void ComputeTrainingRate(float restingRate, float age)
41      {
42          float maximumRate, trainingRate;
43          // Compute maximum heart rate:
44          maximumRate = 220 - age;
45          // Compute training heart rate
46          trainingRate = 0.6F * (maximumRate - restingRate) + restingRate;
47
48          // Display Training heart rate:
49          MessageBox.Show("Your Training Heart Rate is: " + trainingRate.ToString("N0"));
50      }
```

The parameters restingRate and age get their values when a program calls this method. These parameters are used by the method to compute maximum heart rate and training heart rate.

The Scope of Parameters

Because parameters are declared in a method, they are available only within the method—that is, the scope of a parameter is the method in which the parameter is declared. Next, you create the method.

Step 1-9: Create the method, ComputeTrainingRate(), using the code from Figure 5-7.

Figure 5-8 shows how to call this method from the Click event handler of btnTrainingRate button to pass 72 to restingRate and 22 to age. This initial version of the event handler does not use the TextBoxes to get the resting rate and age; instead, the literals 72 and 22 are passed to the method.

Figure 5-8: Call ComputeTrainingRate

```
50
51          private void btnTrainingRate_Click(object sender, EventArgs e)
52          {
53              // Call ComputeTrainingRate. Pass 72 for restingRate and 22 for age.
54              ComputeTrainingRate(72, 22);
55          }
```

Step 1-10: Add the code from Figure 5-8 to the Click event handler of btnTrainingRate to call the method, ComputeTrainingRate().

Arguments and Parameters

The data items that are passed to a method by the calling program are called **arguments**. Thus, the values 72 and 22 in the call statement ComputeTrainingRate(72, 22) are arguments.

The variables declared in the header of a method to receive the arguments are called **parameters**. (As you will see later, parameters also may be used to pass values from the method to arguments in the calling program.) Thus, restingRate and age declared in the header of the method

```
private void ComputeTrainingRate(float restingRate, float age)
```

are parameters.

The arguments are passed to the parameters in the order in which they appear in the argument list. That is, the first argument is passed to the first parameter, the second argument to the second parameter and so on. When ComputeTrainingRate is called, the following assignments take place automatically:

```
            ComputeTrainingRate(72,    22);


     private void ComputeTrainingRate(float restingRate,   float age)
```
That is,
```
     restingRate = 72;
     age = 22;
```
The position of an argument in the argument list may be different from that of the corresponding parameter, if the name of the parameter is specified, as in

ComputeTrainingRate(age: 22, restingRate: 72);

The data type of the parameters and those of the arguments must be compatible—that is, they must obey the rules for assigning values to variables (discussed in Chaper 2).

Let's see how it works:

Step 1-11: Insert a break inside ComputeTrainingRate at the left brace (line 41) by clicking on the left margin.

Insert a second break inside the Click event handler at the call statement (line 54).

Run the form. Note that the TextBoxes and the ListBox on the form are not used in this version of the form.

Click the button to compute the training rate. The program breaks at the call statement. Click Continue. The program breaks at the left brace inside the method.

Move the pointer over the parameters restingRate and age to observe their values, 72 and 22, respectively.

Click Continue to display the training rate.

Step 1-12: Repeat Step 1-11 after changing the resting rate and age in the call statement.

Constants as Arguments

Instead of a literal, you may use a constant (or a variable, as discussed next) as an argument to pass a value to a method. For example, you may use a constant to represent the resting heart rate in the calling program btnTrainingRate_Click(0, as follows:

```
     const float restingRate = 72;
     ComputeTrainingRate(restingRate, 22);
```

Variables as Arguments

Rather than passing literals (72 and 22) or constants as arguments, variables may be used to pass data to a method to make it easy to change the values passed. Data are entered into two TextBoxes, txtRestingRate and txtAge, as shown in Figure 5-9. The data from the TextBoxes is stored in variables and passed to the method.

Figure 5-9: Variables as arguments

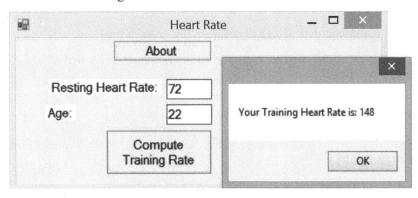

Step 1-13: Modify the code in the Click event handler, as shown in Figure 5-10, to get the resting rate and the age from TextBoxes, store them in variables and pass them as arguments to ComputeTraingRate().

Figure 5-10: Calling ComputeTrainingRate with variables as arguments

```
51
52    private void btnTrainingRate_Click(object sender, EventArgs e)
53    {
54        float restingRate, age;
55        // Get resting heart rate and age from Textboxes:
56        restingRate = float.Parse(txtRestingRate.Text);
57        age = float.Parse(txtAge.Text);
58
59        // Call ComputeTrainingRate
60        ComputeTrainingRate(restingRate, age);
61    }
```

Note that passing data through variables does not require any change in the called method ComputeTrainingRate(). It remains the same as in Figure 5-7, including the following header:

```
private void ComputeTrainingRate(float restingRate, float age)
```

When this method is called from line 59, the first variable in the argument list (restingRate) is assigned to the first variable in the parameter list (restingRate), and the second argument (age) to the second parameter (age).

```
ComputeTrainingRate(restingRate,    age);



private void ComputeTrainingRate(float restingRate, float age)
```

Because the value of restingHeartRate is 72 and that of age in the calling program is 22, in effect, the following assignment takes place when ComputeTrainingRate is called:

```
restingRate = 72;
age = 22;
```
Again, the variables in the argument list and variables in the parameter list must be compatible.

Step 1-14: Make sure that there is a break inside ComputeTrainingRate at the left brace of the method, ComputeTrainingRate (line 40). Run the form.

Step 1-15: Enter values for resting rate and age. Click the button btnTrainingRate.

When the program breaks at the curly left bracket within the method, verify that restingRate and age have the values you entered into the TextBoxes.

Names of arguments and corresponding parameters

In the current example, the names of the two parameters (restingRate and age) in the header of the called method,

```
private void ComputeTrainingRate(float restingRate, float age)
```

are the same as the names of the corresponding arguments in the calling statement,

```
ComputeTrainingRate(restingRate, age)
```

However, the calling program may use any name for the arguments. Thus, it would be valid for the calling program to represent the resting rate by the variable restingHeartRate and call the method using restingHeartRate as the argument name, as in Figure 5-11. Thus the calling program needs to know only the datatypes of the parameters in the called method. The developers of calling programs can use their own variable names to represent the values passed to a method.

Figure 5-11: Argument name different from the parameter name

```
51
52      private void btnTrainingRate_Click(object sender, EventArgs e)
53      {
54          float restingHeartRate, age;
55          // Get resting heart rate and age from Textboxes:
56          restingHeartRate = float.Parse(txtRestingRate.Text);
57          age = float.Parse(txtAge.Text);
58
59          // Call ComputeTrainingRate
60          ComputeTrainingRate(restingHeartRate, age);
61      }
```

The parameter names in the method remain the same:
```
private void ComputeTrainingRate(float restingRate, float age)
```

Step 1-16: Change restingRate to restingHeartRate, as shown in Figure 5-11. Put a break at the statement that calls ComputeTrainingRate (line 60).

Run the form. Enter values for resting rate and age. Click the btnTrainingRate button.

When the program breaks at the call statement, check the value of restingHeartRate.

When the program breaks at the curly left brace within the method, verify that restingRate within the method has the same value as restingHeartRate in the calling method.

Review Questions

5.6   What is an argument?

5.7   What is a parameter?

5.8   When a method has multiple parameters, how does C# determine which parameter is associated with an argument?

5.9   What is the scope of a parameter?

5.10 True or false: The name of an argument must be the same as the name of the corresponding parameter.

5.11 The following code shows a method named **Greetings**, and the Click event handler of a button, which calls the method:
```
private void btnGreeting_Click(object sender, EventArgs e)
{
    string word1 = "Hi", word2 = "There", word = "Hello";
    Greetings(word1);
    Greetings(word2);
}
private void Greetings(string word)
{
    lstDisplay.Items.Add(word);
}
```
What would be the output displayed in the List Box when you click the button?

5.3 Passing Arguments by Value

When a program calls ComputeTrainingRate(float restingRate, float age), the calling program, by default, passes the **value** of each argument to the corresponding parameter. This is called passing arguments by value, which is the default way of passing data to a method. An alternative is to pass values by reference, which is described in section 5.6.

When an argument is passed by value, if the value of the corresponding parameter is changed within the called method, that will not affect the value of the argument in the calling program—that is, if the value of age is changed within the ComputeTrainingRate method, that will not have any effect on the value of age in the calling program, btnTrainingRate_Click.

Step 1-17: Add a statement in the ComputeTrainingRate method at line 47 to change the value of age to 999, as shown in Figure 5-12, and put breaks at the left and right curly braces.

Figure 5-12: ComputeTrainaingRate that changes the value of age

```
40      private void ComputeTrainingRate(float restingRate, float age)
41      {
42          float maximumRate, trainingRate;
43          // Compute maximum heart rate:
44          maximumRate = 220 - age;
45          // Compute training heart rate
46          trainingRate = 0.6F * (maximumRate - restingRate) + restingRate;
47          age = 999;
48          // Display Training heart rate:
49          MessageBox.Show("Your Training Heart Rate is: " + trainingRate.ToString("N0"));
50      }
```

In addition, put breaks in the calling program at the call statement (line 60) and at the right (closing) brace, as shown in Figure 5-13.

Figure 5-13: Break in the calling program

```
52     private void btnTrainingRate_Click(object sender, EventArgs e)
53     {
54         float restingHeartRate, age;
55         // Get resting heart rate and age from Textboxes:
56         restingHeartRate = float.Parse(txtRestingRate.Text);
57         age = float.Parse(txtAge.Text);
58
59         // Call ComputeTrainingRate
60         ComputeTrainingRate(restingHeartRate, age);
61     }
```

Step 1-18: Run the form. Enter values for resting rate and age. Click the btnTrainingRate button.

Check the value of age when the program breaks at the call statement (line 60) and at the left brace within the method (line 41) to verify that it is the value you entered. Check it when the program breaks at the end of the method (line 50) to observe that age has the value 999.

Check age again when the program returns to the calling program (line 61) to observe that the value of age has not changed from its original value. Assigning the value 999 in the called method has no impact on the argument age in the calling program, because only the value of age from the calling program is passed to the parameter age in the method.

Delete the statement "age=999" from the method.

Default Arguments

You may assign a default value (argument) for one or more parameters of a method. For example, you may assign a default value (20) to age in the header of ComputeTrainingRate method:

```
private void ComputeTrainingRate(float restingRate, float age = 20)
```

The default value must be a literal or constant.

When you call a method that has a default value for one or more parameters, you have the option to pass no argument for age, as in the following call statement:

```
ComputeTrainingRate(restingRate)
```

In this case, age will have the default value 20.

Or, you may pass a value for age to override the default, as in

```
ComputeTrainingRate(restingRate, age)
```

If all parameters do not have a default value, the parameters with default value must be declared last.

Step 1-19: Modify ComputeTrainingRate to include a default argument for age, and modify the call statement so no argument is passed for age. Run the form and check the value of age within the method. Repeat by passing an argument for age with a value different from the default value.

Remove the default value. Remove the breaks.

It's time to practice! Do Exercise 5-1.

Cohesive Methods and Reusability

Consider the following method, ComputeTrainingRate, which you developed earlier:

```
39
40   private void ComputeTrainingRate(float restingRate, float age)
41   {
42       float maximumRate, trainingRate;
43       // Compute maximum heart rate:
44       maximumRate = 220 - age;
45       // Compute training heart rate
46       trainingRate = 0.6F * (maximumRate - restingRate) + restingRate;
47
48       // Display Training heart rate:
49       MessageBox.Show("Your Training Heart Rate is: " + trainingRate.ToString("N0"));
50   }
```

The above method computes the training rate and displays the result. Such a method that combines multiple tasks is not very reusable. For example, another program that needs to compute the training rate by calling this method may not want the result to be displayed in a Message Box.

Methods are more reusable when each method performs a single well-defined task, like compute training rate, and returns the results to the calling program so that the calling program can do whatever it wants with those results. Such methods are said to be **cohesive**.

Figure 5-14 shows the structure chart in which the training rate is returned by ComputeTrainingRate to the calling program.

Figure 5-14: ComputeTrainingRate that returns TrainingRate

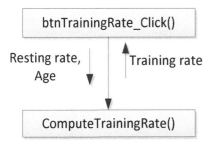

The next sections discuss how **to return values to the calling program**, which helps to create more cohesive methods.

Review Question

5.12   What is meant by passing an argument by value?

5.4 Methods That Return a Value

A program may often need values from the methods that it calls. The method itself can return values without the use of additional parameters. We refer to such a method as a **value returning method,** which can take the place of a simple variable to return a single value, or return multiple values using more advanced structures like a class, struct, tuple, array or List. Class and arrays are discussed in later chapters.

A less desirable alternative to using value returning methods to pass values to a calling program is to use reference parameters, which is discussed in the next section. An advantage of value returning methods is that the methods can be used as a part of an expression.

In general, we will use nouns or noun phrases as names of value returning methods; for example, *trainigRate* for a method that computes and returns the training rate, *indexOf* for a method that finds and returns the index of an item, *isValid* for a method that returns a Boolean indicating whether an id is valid and *custAddress* for a method that returns a structurte that contains the street and zipcode of a customer.

Next you will create a new form named **HeartRate** that computes the training heart rate using a method called TrainingRate(), as shown in Figure 5-15.

Figure 5-15: HeartRate form

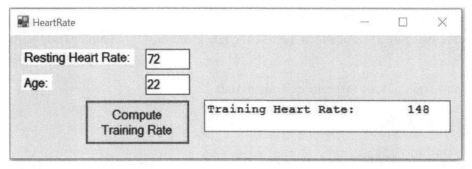

Figure 5-16 shows the TrainingRate() method, which is similar to the ComputeTrainigRate() method in the MethodDemo form, except that TrainingRate() returns the value of training heart rate to the calling program rather than displaying it. As discussed earlier, passing the results to the calling program, rather than displaying the results within a method, makes the method more reusable.

Figure 5-16: Value returning method, TrainingRate()

```
19
20      private float TrainingRate(float restingRate, float age)
21      {       // TrainingRate() computes and returns the training rate
22          float maximumRate = 220 - age;
23          // Compute training heart rate
24          return 0.6F * (maximumRate - restingRate) + restingRate;
25      }
```

Note that the header,
```
private float TrainingRate(float restingRate, float age)
```
shows no parameter to return the trainingRate. Then how is the training rate returned?

The method itself takes the place of a parameter and returns the value specified in the **return** statement,
```
return 0.6F * (maximumRate - restingRate) + restingRate;
```

Because the method returns a value, the type of value returned, called **return type**, is specified in the header using the key word **float** instead of void. Further, the name of the method (TrainingRate) indicates the value returned by the method, not what the method does.

Calling a Value Returning Method

This is how you call the method:
    float trainingRate = TrainingRate(restingRate, age);

The value returned by TrainingRate is assigned to the float type variable, trainingHeartRate. Figure 5-17 shows the Click event handler that calls the TrainingRate() method and displays the training rate.

Figure 5-17: Complete code for HeartRate form

```
19
20      private float TrainingRate(float restingRate, float age)
21      {       // TrainingRate() computes and returns the training rate
22          float maximumRate = 220 - age;
23          // Compute training heart rate
24          return 0.6F * (maximumRate - restingRate) + restingRate;
25      }
26
27      private void btnTrainingRate_Click(object sender, EventArgs e)
28      {
29          float restingRate, age, trainingtRate;
30          // Get resting rate and age from Textboxes:
31          restingRate = float.Parse(txtRestingRate.Text);
32          age = float.Parse(txtAge.Text);
33
34          // Call TrainingRate(), and store the returned value in trainingRate.
35          trainingtRate = TrainingRate(restingRate, age);
36          // Display training rate
37          lstResults.Items.Add(string.Format("{0,20}{1,10}", "Training Rate:",
38                                              trainingtRate.ToString("N0")));
39      }
```

Step 1-20: Open the project, Ch5_Methods, from the Tutorial_Starts folder. Open the form HeartRate.

Step 1-21: Create the method TrainingRate(), as shown in Figure 5-17.

Step 1-22: Add the statement to call TrainingRate (line 35, Figure 5-17).

Step 1-23: Run the form and test it.

Additional examples of value returning methods are provided in Chapter 6.

Returning Multiple Values from a Value Returning Method

The TrainingRate method returns only a single value to the calling program. A method may return multiple values using more advanced options like a tuple, class, struct, array or List. The next section discusses how to pass multiple values using a tuple.

It's time to practice! Do Exercise 5-2.

Review Question

5.13 The following code shows a method named **ComputeSum** and the Click event handler of a button, which calls the method and displays the result in a ListBox named lstDisplay:

```
private void btnExamScore_Click(object sender, EventArgs e)
{
    Int score1 = 90;
    Int score2 = 80;
    int totalScore = ComputeSum(score1, score2);
    lstDisplay.Items.Add(totalScore);
}

private int ComputeSum(int examScore1, int examScore2)
{
    Return (score1 + examScore2);
}
```

Note that the argument, totalScore, is passed by value.

What would be the output displayed in the ListBox?

5.5 Top-Down Design

Top-down design involves breaking down a larger problem into smaller subtasks, each performed by a separate method, to make software development simpler and easier.

Though the task of computing the training heart rate is fairly simple, for the purpose of illustrating top-down design, let's look at the subtasks of this problem. Figure 5-18 shows the code that you developed in HearRate form, with a couple of additional statements to display the input data (resting rate and age), in addition to the result (training rate). This event handler calls the TrainingRate method to compute the training heart rate.

Figure 5-18: Revised Click event handler

```
private void btnTrainingRate_Click(object sender, EventArgs e)
{
    restingRate, age, trainingHeartRate;

    // Get personal data (resting heart rate and age) from Textboxes:
    restingRate = float.Parse(txtRestingRate.Text);
    age = float.Parse(txtAge.Text);

    // Compute training heart rate
    trainingRate = TrainingRate(restingHeartRate, age);

    // Display training heart rate
    lstResults.Items.Add(string.Format("{0,20}{1,10}", "Resting Heart Rate:",
        restingRate));
    lstResults.Items.Add(string.Format("{0,20}{1,10}", "Age:", age));
    lstResults.Items.Add(string.Format("{0,20}{1,10}", "Training Heart Rate:",
        trainingHeartRate.ToString("N0")));
}
```

The task performed in this program can be broken down to three subtasks:

1. Get personal data.
2. Compute training rate.
3. Display results.

You already developed a method for the second subtask (compute the training rate) in the HeartRate form. Now, you will develop two additional methods corresponding to the subtasks (get personal data and display results) and call all three methods from btnTrainingRate_Click event handler, as represented in Figure 5-19.

Figure 5-19: Top-down design for computing training rate

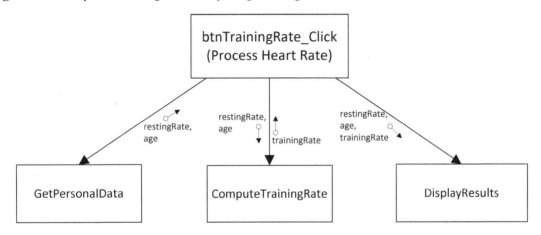

Step 1-24: Create a new form named HeartRateTopDown. Copy the controls from the HeartRate form. The user interface for HeartRateTopDown should appear as follows:

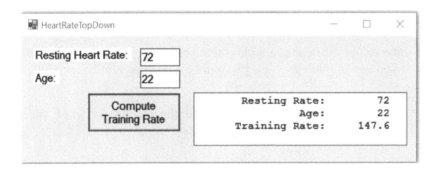

Method to Get Personal Data Using a Tuple

Corresponding to the task *Get Personal Data*, you will develop a method named **PersonalData** that returns multiple values (resting rate and age). As stated earlier, you can return multiple values using a structure like tuple, class, struct, array or List. We will use a tuple because it is relatively simpler to group and access simple data types using this structure. Tuple is particularly suited when the use of the data is limited. However, when the data is to be used widely, other sturctures like struct and class, which are more self-documenting, should be used.

Tuples

A tuple combines multiple related values into a single group. Each value is stored as an element of the tuple. The tuple itself need not have a name. A tuple is represented by a pair of parentheses that contains multiple values separated by commas. Consider the following statement that shows a simple way to create a tuple containing the elements last name, heart rate and age.

```
(string lastName, float restingRate, float age) = ("Hart", 72, 35);
```

The left-hand side of the equal sign defines a tuple with its first, second and third elements named *lastName*, *heartrate* and *age*, respectively. The tuple also specfies the data type of each element. Note that the tuple itself is not given a name. The right-hand side specifies a tuple that contains the values to be assigned to the three elements in the tuple on the left side.

When the statement is executed, the tuple on the right side is split (deconstructed) to its three component values, and the values are assigned to the corresponding elements of the tuple on the left. The lastName, for example, will be assigned the value "Hart," and the heart rate will be assigned 72.

Once you create the tuple, you can access the values of the elements by using their names. For example, you can display the last name in a ListBox as follows:

```
lstResults.Items.Add("Last Name: " + lastName);
```

To store the values in a **named tuple**, you assign the values to a variable of type *var*:

```
var personalData = (lastName, restingRate, age);
```

C# infers the data type of the variable personalData based on the data type of what is assigned to it. Thus, after execution of the statement, personalData will be a tuple that has three elements of type (string, float, float) with the names lastName, restingRate and age, respectively. You access the elements using their names, as in **personalData.restingRate** and **personalData.age.** The PersonalData method that is discussed next illustrates the use of named tuples.

You need **C# version 7.1** or greater to access an element by its inferred name as in this example. You may select the C# version in Visual Studio as follows:
- Right-click the project name in Solution Explorer
- Select Properties -> Build -> Advanced
- Select *Language Version*

The advantage of combining three values into a single tuple is that you can pass the entire tuple from a method to a calling program. If you use simple variables, you will be able to pass only one value from a method.

The tuple is built on the structure, ValueTuple. So, you need to install the package *System.ValueTuple* on versions of .Net earlier than 4.7, as specified in Step 1-25.

PersonalData() method

Next, you develop a method called PersonalData() that gets the resting heart rate and age from TextBoxes and returns those two values using a tuple. Figure 5-20 shows the method.

Figure 5-20: The PersonalData() method using a tuple

```
19      private (float restingRate, float age) PersonalData()
20      {   // The return type of PersonalData() is a tuple with two elements
21
22          // Get resting rate and age from Textboxes:
23          float restingHeartRate = float.Parse(txtRestingRate.Text);
24          float age = float.Parse(txtAge.Text);
25
26          // Return the tuple that contains the values of restingHeartRate and age
27          return (restingHeartRate,age);
28      }
```

Two key parts of the above method are the header and the return statement.

Header: `private (float restingRate, float age) PersonalData()`

The parentheses that contain the data types and names of multiple values specify that the method PersonalData() returns a tuple with two float type elements named restingRate and age.

return statement: `return (restingHeartRate,age);`

The parentheses specify the values returned by the tuple. Realize that the element names of the tuple returned by the method are determined by the names specified in the header. The variable names used within the method have no effect on the names of the elements. So, it doesn't matter what the names of internal variables are.

Calling PersonalData()

The PersonalData() method returns a tuple that has two elements of type (float, float) with the names restingRate and age. To access the values, you store the returned tuple in a variable of type *var*. Figure 5-21 shows the code within the Click event handler of the button to invoke the method, store the returned tuple in the variable personalData and get the values from the variable.

Figure 5-21: Calling PersonalData() within the Click event handler

```
private void btnTrainingRate_Click(object sender, EventArgs e)
{
    // Call PersonalData() and store the returned tuple in a variable
    var personalData = PersonalData();
    // Get values of resting rate and age from
    float restingRate = personalData.restingRate;
    float age = personalData.age;
    // Call TrainingRate()
    …
    // Call DisplayResults()
}
```

After the tuple is stored in the variable, it is easy to access the elements using their names and store individual values in simple variables:

```
float restingRate = personalData.restingRate;
float age = personalData.age
```

Step 1-25: If you are using a version of .Net earlier than 4.7, install the System.ValueTuple package using NuGet:
Right click the project Ch5_Methods in the Solution Explorer.
Select **Manage NuGet Packages.** You will see the NuGet Ch5_Methods window.
Select **Browse**.
Type "**System.ValueTuple**" in the search box. The name of the package will appear in the bottom left pane.
Select the package and click **Install**. Click OK.

Step 1-26: Create the method PersonalData() and call it from the btnTrainingRate_Click() event handler:
Use the code from Figure 5-20 to create PersonalData().
Add the statements within btnTrainingRate_Click event to call GetPersonalData (Figure 5-21).
Put a break in the closing brace of the event handler.
Run the form, enter values for resting rate and age, and click the button.
When the program breaks at the brace, make sure that the two elements of the tuple (restingRate and age) have the values you entered.

Method to Compute Training Rate

You already developed the method to compute training rate in the HeartRate form. Figure 5-22 shows the method.

Figure 5-22: TrainingRate() method

```
29
30    private float TrainingRate(float restingRate, float age)
31    {      // TrainingRate() computes and returns the training rate
32        float maximumRate = 220 - age;
33        // Compute training heart rate
34        return 0.6F * (maximumRate - restingRate) + restingRate;
35    }
```

Calling TrainingRate() method

Calling TrainingRate() involves passing the resting rate and age to the method as shown in Figure 5-23.

Figure 5-23: Calling PersonalData() and TrainingRate() within the Click event handler

```
private void btnTrainingRate_Click(object sender, EventArgs e)
{
    // Call PersonalData() and store the returned tuple in a variable
    var personalData = PersonalData();
    // Get values of resting rate and age from the variable
    float restingRate = personalData.restingRate;
    float age = personalData.age;

    // Call TrainingRate() and store the returned value in trainingRate
    float trainingRate = TrainingRate(restingRate, age);
    // Call DisplayResults()
}
```

Step 1-27: Create the method trainingRate and call it.

Use the code from Figure 5-22 to create trainingRate() and insert the code immediately after the method PersonalData(). You may copy the code from the HeartRate form.

Add the statements within btnTrainingRate_Click event handler to call trainingRate (see Figure 5-23).

Run the form, enter values for resting rate and age, and click the button.

When the program breaks at the brace, make sure that that the variable trainingRate has the computed value.

Method to Display Results

Data need to be passed from the calling program to DisplayResults() method, but DisplayResults doesn't return any values. Figure 5-24 shows the method that has the input parameters restingRate, age and trainingRate.

Figure 5-24: The DisplayResults() method

```
37      private void DisplayResults(float restingRate, float age, float trainingRate)
38      {
39          // Display Results
40          lstResults.Items.Add(string.Format("{0,20}{1,10}", "Resting Rate:", restingRate));
41          lstResults.Items.Add(string.Format("{0,20}{1,10}", "Age:", age));
42          lstResults.Items.Add(string.Format("{0,20}{1,10}", "Training Rate:", trainingRate));
43      }
```

Calling DisplayResults method

Because DisplayResults doesn't return any values, the call statement involves only passing the values of the two elements of the tuple (restingRate and age) and the value of trainingRate from the Click event handler to the method.

```
        DisplayResults(restingRate, age, trainingRate);
```

The complete code for the form HeartRateTopDown is shown in Figure 2-25.

Step 1-28: Add the code from Figure 5-24 to create the DisplayResults() method immediately after the TrainingRate() method. Call the method from the Click event handler (see Figure 5-24).

Figure 5-25: Complete code for HeartRateTopDown

```
19      private (float, float) PersonalData()
20      {   // The return type of PersonalData() is a tuple of type (float, float).
21
22          // Get resting rate and age from Textboxes:
23          float restingHeartRate = float.Parse(txtRestingRate.Text);
24          float age = float.Parse(txtAge.Text);
25
26          // Return the tuple that contains the values of restingHeartRate and age
27          return (restingHeartRate, age);
28      }
```

```csharp
        private float TrainingRate(float restingRate, float age)
        {       // TrainingRate() computes and returns the training rate
            float maximumRate = 220 - age;
            // Compute training heart rate
            return 0.6F * (maximumRate - restingRate) + restingRate;
        }

        private void DisplayResults(float restingRate, float age, float trainingRate)
        {
            // Display Results
            lstResults.Items.Add(string.Format("{0,20}{1,10}", "Resting Rate:", restingRate));
            lstResults.Items.Add(string.Format("{0,20}{1,10}", "Age:", age));
            lstResults.Items.Add(string.Format("{0,20}{1,10}", "Training Rate:", trainingRate));
        }

        private void btnTrainingRate_Click(object sender, EventArgs e)
        {
            // Call PersonalData(), which returns a tuple conataining (float, float).
            // store the returned tuple in the tuple (restingRate, age)
            float restingRate, age;
            (restingRate, age) = PersonalData();

            // Call TrainingRate(), and store the returned value in trainingRate.
            float trainingRate = TrainingRate(restingRate, age);

            // Call DisplayResults
            DisplayResults(restingRate, age, trainingRate);
        }
```

Getting the Big Picture

An advantage of the top-down design is that the code in the calling program, btnTrainingRate_Click, primarily consists of three statements to call the three methods, as shown in Figure 5-24. Thus, it is easier to get a high-level view of the application by looking at the code in the calling program. In addition, the methods corresponding to the subtasks are more reusable, compared to having all code in a single program.

It's time to practice!

Modify TrainingRate, DisplayResults and the call statements as follows:

TrainingRate should return both **maximumRate** and trainingRate to the calling program. Pass the restingRate, age, maximumRate and trainingRate to DisplayResults from within the Click event handler, as shown in the following structure chart. DisplayResults must display all four values.

Chapter 5: Methods 161

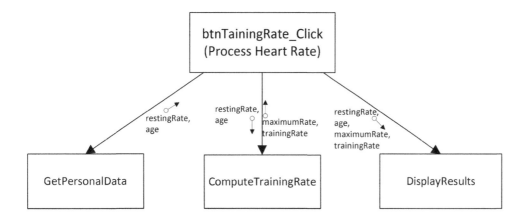

An Undesirable Design Using Noncohesive Methods

In the top-down design shown in Figure 5-19, the task of computing the training heart rate is broken down to three subtasks represented by the methods PersonalData, TrainingRate and DisplayResults. These methods are called in sequence by a main program (btnTrainingRate_Click). Input data from PersonalData() and the result from TrainingRate() is passed to the calling program through the methods. As discussed earlier in this chapter, such a design makes the methods cohesive and reusable. An alternate undesirable design that involves less reusable and noncohesive methods is shown in Figure 5-26.

Figure 5-26: Alternate undesirable design involving noncohesive methods

In this design, the main program (the Click event procedure, btnTrainingRate_Click) calls the method GetPersonalData, as in the top-down design.

However, the method GetPersonalData gets personal data, and instead of passing the data to the main program, it calls ComputeTrainingRate, passing restingRate and age to the method.

ComputeTrainingRate computes the training rate and then calls DisplayResults passing restingRate, age and trainingRate to the method.

DisplayResults displays the results as in the top-down design.

The above design suffers from several problems:

1. The methods GetPersonalData and ComputeTrainingRate are not cohesive. For example, ComputeTrainingRate doesn't do a single well-defined task. It computes training rate and also displays results by calling DisplayResults. Similarly, GetPersonalData gets restingRate and age, and also computes results by calling ComputeTrainingRate.
2. The methods GetPersonalData and ComputeTrainingRate are less reusable. For example, ComputeTrainingRate does more than computing the results, and it doesn't return the result to the calling program.
3. This design makes it necessary to pass certain data to a method where it is not needed. For example, if the personal data includes the person's name, the only way it can be passed to DisplayResults is by first passing it to ComputeTrainingRate, where it is not needed. Passing data to a method where is not needed makes the maintenance of code difficult.

Review Questions

5.14 What is top-down design?

5.15 Describe two benefits of top-down design.

5.6 Passing Arguments by Reference

Earlier sections described how to use a value returning method to pass data back from the method to the calling program. Passing multiple values requires that the values are combined together into a structure like tuple, class or struct.

An alternative is to pass each value using a reference parameter. When you use this approach, because the method itself doesn't return any values, the method is given a name that indicates what it does, like VerifyCustomerId, rather than what values are passed using the parameters, like a flag named IsValid. So, it may be less clear what values are passed back to the calling program. However, it is important to understand this approach, because it has been commonly used in programming, especially by those who find it easier to use simple variables as parameters.

In general, this approach of passing arguments by reference allows the called method to change the value of arguments in the calling program. This allows the results computed in a method to be made available (or returned) to the calling program.

When a variable is passed by reference, the memory location (address or reference) of the variable is passed to the parameter so that the argument and the parameter access data from the same memory location, as represented below:

So, if the value of a parameter is changed within a method, in effect, it is changing the value of the corresponding argument.

Similarly, the value assigned to an argument before calling a method becomes the value of the parameter when the method is called. That is, in effect, values can be transferred from the calling program to the called method, and also from the called method to the calling program.

You will create an alternative version of the HeartRateTopDown form called HeartRateTopDownAlt that uses reference parameters to pass data to the calling program. The user interface is the same as in the HeartRateTopDown form:

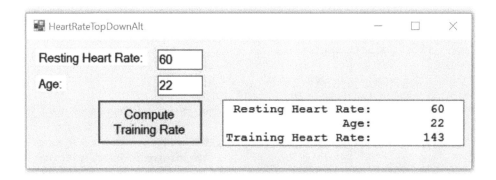

Step 1-29: Create a new form named HeartRateTopDownAlt. Copy the controls from the form HeartRateTopDown.

First you will create a method to compute training rate, which uses reference parameters to pass the training rate to the calling program, as shown in Figure 5-26. Because it is not a value returning method, it is named **ComputeTrainingRate**, to indicate what the method does.

Figure 5-26: Passing arguments by reference

```
25
26    private void ComputeTrainingRateAlt(float restingRate, float age, ref float trainingRate)
27    {
28        float maximumRate;    //trainingRate is not declared here; it is a parameter
29        // Compute maximum heart rate:
30        maximumRate = 220 - age;
31        // Compute training heart rate
32        trainingRate = 0.6F * (maximumRate - restingRate) + restingRate;
33    }
```

In the header (line 26), the key word **ref** specifies that trainingRate is a **reference parameter**. Note that because trainingRate is declared as a parameter, it is not declared within the method.

Line 32 assigns the computed value to the ref parameter trainingRate, which shares the value with the calling program. Note that there is no *return* statement.

Line 60 in Figure 5-27 shows the statement to call ComputeTrainingRate() that includes trainingHeartRate as an argument.

5.6 Passing Arguments by Reference

Figure 5-27: Calling ComputeTrainingRate

```
50
51      private void btnTrainingRate_Click(object sender, EventArgs e)
52      {
53          float restingHeartRate, age;
54          // Get resting rate and age from Textboxes:
55          restingHeartRate = float.Parse(txtRestingRate.Text);
56          age = float.Parse(txtAge.Text);
57
58          // Call ComputeTrainingRate
59          float trainingHeartRate = 0;
60          ComputeTrainingRate(restingHeartRate, age, ref trainingHeartRate);
61          // Display training rate
62          lstResults.Items.Add(string.Format("{0,20}{1,10}", "Training Heart Rate:",
63                                              trainingHeartRate.ToString("N0")));
64      }
```

The key word "ref" in the calling statement in line 60 specifies that the parameter is reference type. Note that the name of the argument (trainingHeartRate) is different than the name of the parameter (trainingRate).

An argument passed by reference must be declared in the calling program, and it must be set to a value. The argument trainingHeartRate is declared and it is set to zero.

What happens when ComputeTrainingRate is called?

> First, the reference to trainingHeartRate is passed to the parameter trainingRate. That means trainingRate will take the value of the argument trainingHeartRate.

Thus, passing by reference makes sharing between the calling program and the called program like a two-way street, as shown, whereas passing by value is like a one-way street.

Step 1-30: Create the ComputeTrainingRate() method shown in Figure 5-26.

Put breaks at the beginning and ending braces of the method.

Add the btnTrainingRate_Click event handler shown in Figure 5-27. Your line numbers will be different because your program is not complete

Run the form. At the first break inside ComptueTrainingRate, observe that the parameter trainingRate has the value 0 (same as the value of the argument restingHeartRate). At the second break at the end of the method, restingRate has the computed value.

When control returns to the calling program, note that the argument restingHeartRate has the value that was assigned to the parameter restingRate, because both variables share the same memory location.

Returning Multiple Values from a Method Using Parameters

The ComputeTrainingRate method returns only a single value to the calling method. A method may use multiple reference parameters, each parameter returning a single value or a set of values using objects and different data structures like struct and tuple. The GetPersonalData method provides an example of a method that returns two values.

The method named GetPersonalData gets restingRate and age from TextBoxes and passes them to the calling program using reference parameters, as shown in Figure 5-28.

Figure 5-28: GetPersonalData method

```
private void GetPersonalData(ref float restingRate, ref float age)
{
    // Get resting rate and age from Textboxes:
    restingRate = float.Parse(txtRestingRate.Text);
    age = float.Parse(txtAge.Text);
}
```

Calling GetPersonalData: To call GetPersonalData, the calling program would use reference arguments that must be assigned an initial value before calling the method. Figure 5-29 shows the click event handler that calls GetPersonalData.

Figure 5-29: Event handler that calls all three methods

```
private void btnTrainingRate_Click(object sender, EventArgs e)
{
    // Call GetPersonalData
    float restingHeartRate = 0, age = 0;
    GetPersonalData(ref restingHeartRate, ref age);

    // Call ComputeTrainingRate
    float trainingHeartRate = 0;
    ComputeTrainingRate(restingHeartRate, age, ref trainingHeartRate);

    // Call DisplayResults
        ...
}
```

Note that the resting heart rate is represented by the variable restingHeartRate in the calling program, but the corresponding parameter in GetPersonalData is called restingRate.

Step 1-31: Create the method GetPersonalData, as shown in Figure 5-28.

Modify the Click event handler to call GetPersonalData, as shown in Figure 5-29.

Put a break in the closing brace of the Click event handler.

Run the program, enter values for resting rate and age, and click the button.

When the program breaks at the closing brace, make sure that restingHeartRate and age have the values you entered.

DisplayResults method

DisplayResults method is the same as in HeartRateTopDown form. The complete code for the form is shown in Figure 5-30.

Step 1-32: Copy DisplayResults method from HeartRateTopDown form, and call it from the Click event handler. Test the code.

Figure 5-30: Complete code for HeartRateTopDownAlt form

```csharp
19      private void GetPersonalData(ref float restingRate, ref float age)
20      {
21          // Get resting rate and age from Textboxes:
22          restingRate = float.Parse(txtRestingRate.Text);
23          age = float.Parse(txtAge.Text);
24      }
25
26      private void ComputeTrainingRate(float restingRate, float age, ref float trainingRate)
27      {
28          float maximumRate; //trainingRate is not declared here; it is a parameter
29          // Compute maximum heart rate:
30          maximumRate = 220 - age;
31          // Compute training heart rate
32          trainingRate = 0.6F * (maximumRate - restingRate) + restingRate;
33      }
34
35      private void DisplayResults(float restingRate, float age, float trainingRate)
36      {
37          // Display Results
38          lstResults.Items.Add(string.Format("{0,20}{1,10}", "Resting Rate:", restingRate));
39          lstResults.Items.Add(string.Format("{0,20}{1,10}", "Age:", age));
40          lstResults.Items.Add(string.Format("{0,20}{1,10}", "Training Rate:",
41                                              trainingRate.ToString("N0")));
42      }
43
44      private void btnTrainingRate_Click(object sender, EventArgs e)
45      {
46          // Call GetPersonalData
47          float restingHeartRate = 0, age = 0;
48          GetPersonalData(ref restingHeartRate, ref age);
49
50          // Call ComputeTrainingRate
51          float trainingHeartRate = 0;
52          ComputeTrainingRate(restingHeartRate, age, ref trainingHeartRate);
53
54          // Call DisplayResults
55          DisplayResults(restingHeartRate, age, trainingHeartRate);
56      }
```

It's time to practice! Do Exercise 5-3.

The out Parameter

When the key word "ref" is used to specify passing by reference, the parameter is called a reference parameter. Another method to pass an argument by reference is to use the key word "out" in place of "ref," as shown in the following code, to specify that the parameter is an **output parameter**.

```
ComputeTrainingRate(restingHeartRate,    age,    out trainingHeartRate);

private void ComputeTrainingRate(float restingRate, float age, out float trainingRate)
```

An output parameter does not require the corresponding argument to be assigned a value before calling the method. In addition, an output parameter must be assigned a value inside the method. Except for these differences, the output parameter works like a reference parameter.

You may use the out parameter when the calling program must access the value of a parameter from the called method, but the called method doesn't have to access the value of the corresponding argument from the calling program, as in the above example.

Why Pass by Value?

If pass by reference can pass data both ways, then why use pass by value, which can pass only from calling program to called method? What is wrong with using pass by reference, even when the method does not have to pass a value to the calling program?

Consider data items like resting rate and age, which are needed in a method, but their values should not be changed by the method. If such arguments are passed by reference, the called method can inadvertently change their values, which, in turn, affects other methods that might use those arguments.

For example, the developer of ComputeTrainingRate might decide to reset the values of restingRate and age to zero after using them to compute the trainingRate. If the calling program passes restingRate and age to other methods, expecting the variables to retain their original values, then those methods will not work correctly. The advantage of passing by value is that it prevents any unintended damage to the rest of the methods, due to one method changing the value of a variable.

Review Questions

5.16   What is meant by passing an argument by reference?

5.17   The following code shows a method named **ComputeSum,** and the Click event handler of a button that calls the method:
```
private void btnExamScore_Click(object sender, EventArgs e)
{
        int exam1 =150, exam2 =100, total = 0;
        ComputeSum(exam1, exam2, ref total);
        lstDisplay.Items.Add(exam1 + "   " + exam2 + "   " + total);
}

private void ComputeSum(int exam1, int exam2, ref int sum)
{
        sum = exam1 + exam2;
        exam1 = 0;
        exam2 = 0;
}
```

What would be the output displayed in the ListBox when you click the button?

5.18 The following code shows a method named **ComputeSum**, and the Click event handler of a button that calls the method (note that the parameters are declared differently from the previous question):

```
private void btnExamScore_Click(object sender, EventArgs e)
{
    int exam1 = 10, exam2 = 15, total = 0;
    ComputeSum(ref exam1, ref exam2, ref total);
    lstDisplay.Items.Add(exam1 + " " + exam2 + " " + total);
}
private void ComputeSum(ref int test1, ref int test2, ref int sum)
{
    sum = test1 + test2;
    test1 = 0;
    test2 = 0;
}
```

What would be the output displayed in the ListBox when you run the Click event handler?

5.7 Enumerations: Limiting Parameter Values

An enumeration is a set of named integer constants. For example, Figure 5-31 shows an enumeration named CheckState with three named constants, **Checked, Indeterminate and Unchecked**. The statement in Figure 5-31 assigns the selected value (0, 1 or 2, represented by the three named constants) to the CheckState property of the CheckBox.

Figure 5-31: Enumeration example

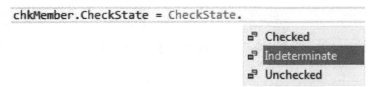

Each name represents an integer. By default, the first name ("Checked") has the value 0; each of the following names has a value one greater than that of the previous name. So, in effect, the statement assigns a number (0, 1 or 2) to the CheckState property.

An enumeration provides a convenient way to select from a limited set of numbers. In addition, an enumeration uses meaningful names in place of numbers, which helps minimize errors in making a selection.

The syntax to create an enumeration is
enum enumerationName [: optional integer type] {list of named constants separated by commas};

For example,
```
enum CheckState { Checked, Indeterminate, Unchecked };
```
creates the enumeration CheckState, which is of enumerated type, where the named constants represent integer values 0, 1 and 2, respectively.

An enumeration cannot be declared inside a method. It can be declared inside the class, inside the namespace or outside the namespace.

Though the named constants represent integer values, you need to cast them to integer type before assigning to an integer variable, as in
```
int checkState = (int)CheckState.UnChecked;
```

You may assign values other than the default set of values that starts with zero. Here is an example:
```
enum size { Small = 10, Medium = 20, Large = 30 }
```

To better understand the use of enumeration, you will create an application that computes hotel room rates based on the discount type selected from a ComboBox, as shown in Figure 5-32.

Figure 5-32: DiscountTypeEnum form

The discount types are "No Discount," "AAA," "Federal Government" and "Senior."

To make it easy to work with the above discount types within the program, we will represent them by the set of integers 0, 1, 2 and 3, respectively, and give these numbers more meaningful names: **NoDiscount, AAA, Govt and Senior**.

The following statement would create the set of named constants (an enumeration) named discountTypes that represents the set of numbers 0, 1, 2 and 3:
```
enum discountTypes { NoDiscount, AAA, Govt, Senior};
```

After creating an enumeration, you can create variables of that type. You create a variable named selectedDiscType of type, discountTypes, as follows:
```
discountTypes selectedDiscType;
```

The coding for the form shown in Figure 5-27 includes the following steps:

1. Declare an enumeration named discountTypes and a variable named selectedDiscType of the type discountTypes, both at the class level.

2. Add discount types ("No Discount," "AAA," "Federal Government" and "Senior") to the ComboBox, in the Load event handler of the form.

3. When the user selects a discount code from the ComboBox, assign the corresponding value (0, 1, 2 or 3, represented by NoDiscount, AAA, Govt and Senior) to the variable selectedDiscType. This is done within the SelectedIndexChanged event handler of the ComboBox.

4. Create a method that determines the room rate based on the value of selectedDiscType. This method is called to compute the room rate.

Tutorial 2: Enumeration: Enumeration Type Parameter

This tutorial illustrates the use of enumeration by building the form, DiscountTypeEnum.

Step 2.1: Open the project Ch5_Methods. Open the form named DiscountTypeEnum. Open the code window.

Step 2.2: Add the code to declare the enumeration and the variable at the class level, as shown in Figure 5-33.

Figure 5-33: Declare enumeration and a variable of enumerated type

```csharp
19          enum discountTypes { NoDiscount, AAA, Govt, Senior};
20          discountTypes selectedDiscType;
```

Step 2.3: Insert the code to add discount types to the ComboBox, within the Load event handler of the form, as shown in Figure 5-34.

Figure 5-34: Add items to ComboBox

```csharp
22          private void DiscountTypeEnum_Load(object sender, EventArgs e)
23          {
24              string[] discTypes = { "No Discount", "AAA", "Federal Government", "Senior" };
25              for (int index = 0; index <= 3; index++)
26                  cboDiscountType.Items.Add(discTypes[index]);
27          }
```

Step 2.4: Based on the discount type selected by the user, assign a value to the variable, selectedDiscType, within the SelectedIndexChanged event of the ComboBox. See Figure 5-35.

Figure 5-35: SelectedIndexChanged event handler

```csharp
28
29          private void cboDiscountType_SelectedIndexChanged(object sender, EventArgs e)
30          {
31              switch(cboDiscountType.SelectedItem.ToString())
32              {
33                  case "No Discount":
34                      selectedDiscType = discountTypes.NoDiscount;
35                      break;
36                  case "AAA":
37                      selectedDiscType = discountTypes.AAA;
38                      break;
39                  case "Federal Government":
40                      selectedDiscType = discountTypes.Govt;
41                      break;
42                  case "Senior":
43                      selectedDiscType = discountTypes.Senior;
44                      break;
45              }
46          }
```

Note that if the room rates are to be used just within this form, you could assign a value to the room rate within this switch statement without using an enumeration.

Here, we store the selected discount type in a class level variable of enumerated type so that it can be accessed from other classes (by making it public) where it can be used potentially for other purposes.

Next, we create a method that accepts an argument of discountTypes type and returns the room rate.

Step 2-5: Create a method that determines the room rate based on the value of selectedDiscType. See code in Figure 5-36.

Figure 5-36: Method with an enumerated type parameter

```
47
48      private float RoomRate(discountTypes discountType)
49      {
50          float rate=0;
51          switch (discountType)
52          {
53              case discountTypes.NoDiscount:
54                  rate = 125;
55                  break;
56              case discountTypes.AAA:
57                  rate = 110;
58                  break;
59              case discountTypes.Govt:
60                  rate = 120;
61                  break;
62              case discountTypes.Senior:
63                  rate = 105;
64                  break;
65          }
66          return rate;
67      }
```

Note that the parameter discountType is of enumerated type discountTypes, and it will not accept any value other than the enumerated constants, thus providing some data validation that is not provided by a string type parameter.

Step 2-6: Call the method ComputeRate() from the Click event handler of btnComputeRate, and display the rate in the ListBox. See Figure 5-37.

Figure 5-37: btnComputeRate Click event handler

```
68
69     private void btnComputeRate_Click(object sender, EventArgs e)
70     {
71         float roomRate = RoomRate(selectedDiscType);
72
73         lstDisplay.Items.Clear();
74         lstDisplay.Items.Add("Rate for disc. type, " + selectedDiscType
75                             + ":  " + roomRate.ToString("C"));
76     }
```

Line 71 calls RoomRate method and passes the selectedDiscType, which is a variable of discountTypes type. The return value is stored in the variable roomRate. Passing a value like an integer constant 1 or 2 will result in an error unless it is converted to the enumerated type.

Step 2-7: Run the form and test your code.

Review Questions

5.19   What is an enumeration?

5.20   What are the benefits of using an enumeration?

Exercises

Exercise 5-1

Create a form named GradeWithMethod_1 that lets you to do the following:

> Enter a score (0–100) into a TextBox named txtScore, and
> compute the letter grade by clicking a button.

Use the scale 90, 80, 70 for A, B and C, respectively. The letter grade must be computed using a separate method named ComputeGrade, which displays the letter grade using a MessageBox. Call this method from the Click event of the button passing the score as a parameter to the method.

Exercise 5-2

This is a modified version of the form specified in Exercise 5-1. You may copy the controls and code to a new form named GradeWithMethod_2 and modify the code.

Create a form named GradeWithMethod_2 that lets you do the following:

> Enter a score (0–100) into a TextBox named txtScore, and
> compute the letter grade by clicking a button.

Use the scale 90, 80, 70 for A, B and C, respectively. The letter grade must be computed using a separate method named Grade. Call this method from the Click event of the button, passing the score as a parameter to the method. Grade must return the letter grade to the calling program **without using a parameter**. The calling program displays the data in a Label named lblLetterGrade.

Exercise 5-3

This is a modified version of the form specified in Exercise 5-1. You may copy the controls and code to a new form named GradeWithMethod_3 and modify the code.

Create a form named GradeWithMethod_3 that lets you do the following:
>Enter a score (0–100) into a TextBox named txtScore, and
compute the letter grade by clicking a button.

Use the scale 90, 80, 70 for A, B and C, respectively. The letter grade must be computed using a separate method named ComputeGrade. Call this method from the Click event of the button passing the score as a parameter to the method. **ComputeGrade program must return the letter grade to the calling program through a parameter.** The calling program displays the data in a Label named lblLetterGrade.

Exercise 5-4

Create a form named GradeTopDown that lets you to do the following:
>Enter a score (0–100) and a name into two different TextBoxes, and
compute the letter grade by clicking a button.

Use the scale 90, 80, 70 for A, B and C, respectively. The structure chart of the task to process the grade is as follows:

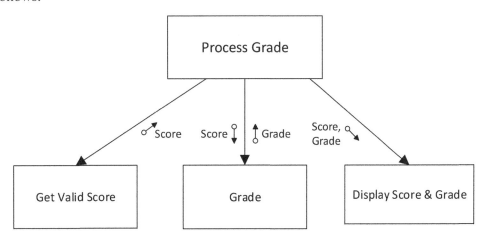

The Click event handler of the button must call three separate methods, corresponding to the three subtasks, described below:

1. A method named ValidScore that gets the name and score from the TextBoxes, validates the score and passes the name and valid score to the Click event handler.
2. A second method named Grade that computes the letter grade. Pass the score from the calling program to Grade. The method must return the letter grade **without using a parameter**. (This method is the same as in Exercise 5-2.)
3. A third method that displays the name and letter grade. Pass the name and letter grade as parameters to this method.

Chapter 6

Graphical User Interface Controls

This chapter introduces you to graphical user interface (GUI) controls that help minimize data entry errors and make it easier for users to interact with a program.

Learning Objectives

After studying this chapter, you should be able to:

- Effectively use ScrollBars, CheckBoxes, RadioButtons, ComboBoxes and ListBoxes to enhance user interface.
- Set up GUI controls on a form to let the user provide input data.
- Get the value(s) selected by the user from a control.
- Use appropriate events to take actions based on user selections.

Topics

6.1	Working with ScrollBars	6.5	Getting User Input from Controls
6.2	Working with CheckBoxes	6.6	Working with ComboBoxes
6.3	Working with RadioButtons	6.7	Working with ListBoxes
6.4	Validating Input		

To understand the use of GUI controls, consider an application that lets you calculate the total price for tickets to a show in a theater. The user interface, shown in Figure 6-1, uses a ScrollBar to specify the number of tickets, a set of RadioButtons to select the section of the theater and a CheckBox to specify whether the customer is a member.

Figure 6-1: User interface for ticket sales application

Next, we look at how to work with each of these controls.

6.1 Working with ScrollBars

The ScrollBar controls, HScrollBar and VScrollBar, let you select a number between two limits determined by the values of the **Minimum** and **Maximum** properties. The two ScrollBars are essentially the same, except that the VScrooBar lets you scroll horizontally, and the VScrollBar lets you scroll vertically. In this section, you will use the HScrollBar.

To select a number, you may move the slider by clicking the arrows at either end by clicking the inside area or by dragging the slider. The number selected by the user is given by the **Value** property. However, the selected number is not displayed by the control. Typically, a separate Label is used to display the selected value. Table 6-1 shows the important properties of HScrollBar.

Table 6-1: Properties of HScrollBar

Property name	Description
Value	The integer value of the ScrollBar, selected by moving the slider.
Minimum	Lower limit of the range of values. Default is zero. User may specify a different limit.
Maximum	Upper limit of the range of values. Default is 100. User may specify a different limit.
SmallChange	Amount by which the value changes when the user clicks on one of the two arrows. Default is 1.
LargeChange	Amount by which the value changes when the user clicks the ScrollBar area on either side of the slider. Default is 1. If LargeChange is greater than 1, you must increase the Maximum by adding (LargeChange - 1) to Maximum. For example, if you want a range of 0 to 20, and LargeChange is 4, then you must set the Maximum to 23 (20 + (4-1)).

Tutorial 1: Theater Tickets Sale Application

In this tutorial, you will create the TheaterTickets form, shown in Figure 6-1, to illustrate the use of GUI controls.

Step 1-1: Create a new project named Ch6_GUIcontrols. Rename the form to TheaterTickets.

Add an HScrollBar (found in the *All Windows Forms* group in the Toolbox) to the form, and name it **hsbNumOfTickets**.

Set Maximum property to 10. It is assumed that purchases of more than 10 tickets are processed differently at group rates in a different form.

Change the value of the Value property from 0, and observe how it affects the position of the slider. Make sure you change it back to 0.

Step 1-2: Run the form. Move the slider to change the value of the ScrollBar by clicking the arrows, clicking the ScrollBar area, or dragging the slider. But you will not see the value.

How do you view the value of the ScrollBar?

Events of ScrollBars

There are two important events of the ScrollBar that can be used to display the value in a Label: the **ValueChanged** event and the **Scroll** event.

ValueChanged event

This event is triggered when the value of a ScrollBar is changed by moving the slider manually, and also when the value is changed at runtime by executing code. The following code uses the ValueChanged event handler to display the value of the ScrollBar in a Label whenever the value changes at runtime:

```
private void hsbNumOfTickets_ValueChanged(object sender, EventArgs e)
{
        lblNumOfTickets.Text = hsbNumOfTickets.Value.ToString();
}
```

The ValueChanged event handler allows you to display the value in the Label when the user moves the slider and also when the ScrollBar is reset to zero at runtime using code by clicking the clear button. So, you will use this event handler to display the value of the ScrollBar.

Scroll event

This event is triggered only when the value of a ScrollBar is changed by moving the slider manually; not when the value is changed at runtime by code, like resetting the ScrollBar using code. You create the scroll event handler by double clicking the ScrollBar in design view.

Creating and Deleting Event Handlers

How do you create an event handler for an event like **ValueChanged**, which is **not** the default event handler that is created when you double click the ScrollBar. (**Scroll** is the default event handler.)
To create an event handler that is not the default, follow these steps:

Select the object, which is the ScrollBar in this example.
In the properties window, click on the **events** button to display all events for the object, as shown in Figure 6-2.
Double click the event (ValueChanged) to open the event handler.

Figure 6-2: Events of the ScrollBar

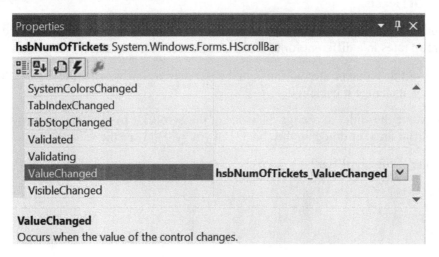

If you double clicked the ScrollBar and created the Scroll Event handler that you don't need, you may delete it using the following steps, as described in Chapter 1:

Display the events of the object (hsbScrollBar) in the Properties window, which should look as follows:

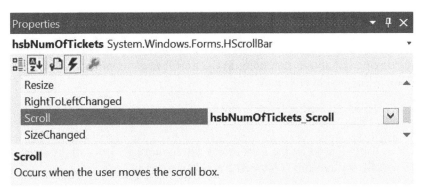

Select the event (Scroll).

Click the dropdown arrow and select the event handler (hsbNumOfTickets_Scroll) you want to delete. (You need to do it even if the event handler already is displayed.)

Right click the event handler and select **Delete**.

Step 1-3: For practice, double click the ScrollBar to create the Scroll event handler. Then, delete the event handler.

Next, create the ValueChanged event handler for the ScrollBar. Enter the following statement to display the ScrollBar's value in the Label:

```
lblNumOfTickets.Text = hsbNumOfTickets.Value.ToString();
```

Step 1-4: Put a break within the Value Changed event handler, and run the form.
Move the slider, and observe that the ValueChanged event is triggered.
Remove the break.
Move the slider from the left end to the right end, and note the range of values determined by the default values of Minimum and Maximum properties.
Change the value of LargeChange to 4.
Click inside the ScrollBar, and observe that the slider moves by 4 and the largest value is only 7, though the value of Maximum property is 10.
To get the full range of 0 to 10, set Maximum to 13 (10+4-1), and check the largest value. It should be 10.
Set LargeChange back to 1.

Step 1-5: Create the Click event handler of Clear button, and enter the statement to reset the ScrollBar value to 0, as follows:
```
private void cmdClear_Click(object sender, EventArgs e)
{
    hsbNumOfTickets.Value = 0;
}
```

Step 1-6: Run the form. Move the slider to select a value greater than zero. Click the Clear button. The slider moves to zero position because the ScrollBar's value is set to zero. This, in turn, triggers the ValueChanged event and also sets the Label's value to zero.

Note that if the statement,
```
lblNumOfTickets.Text = hsbNumOfTickets.Value.ToString();
```
were in the Scroll event handler, resetting the ScrollBar value would not have reset the value displayed in the Label. You may test it by copying the above statement to the Scroll event handler and commenting out the statement inside the ValueChanged event handler.

Review Questions

6.1   A form uses a ScrollBar, hsbUnits, to let the user select whole numbers only in the range 0 to 10. What property of the ScrollBar should be changed from its default value?

6.2   Write the statement to display the value of the ScrollBar, hsbUnits, in a Label named lblUnits.

6.3   The value of a ScrollBar named hsbUnits is to be automatically displayed in a Label named lblUnits whenever the value of the ScrollBar is changed (either by clicking inside the ScrollBar, or by executing code, for example, when the ScrollBar is reset by clicking a "Clear" button). The code required to display the value should be entered in which event handler?

6.2 Working with CheckBoxes

A CheckBox is used to let the user select a true or false Boolean value. For example, we use a CheckBox in this application for the user to specify whether or not a customer is a member. Two commonly used properties of CheckBox are shown in Table 6-2.

Table 6-2: Properties of CheckBox

Property	Settings of the property	Description
CheckState	Checked	Checking the CheckBox at runtime sets CheckState to Checked.
	UnChecked	Unchecking the CheckBox at runtime sets CheckState to UnChecked. This is the default state.
	Indeterminate	The CheckState can be set to Indeterminate in the design window, or by executing code at runtime. Setting this value checks the CheckBox. When the user checks (or unchecks) the CheckBox, the CheckState changes to Checked (or UnChecked). You may set the CheckBox to Indeterminate state initially, so that the code can check whether the user forgot to make a selection.
Checked	True	Indicates that the CheckBox is in the Checked or Indeterminate state.
	False	Indicates that the CheckBox is in the UnChecked state. This is the default.

Step 1-7: Add a CheckBox named chkMember to the form. Change the value of CheckState property, and observe the effect on the CheckBox and on the value of Checked property.

How do you make sure that the user made a selection of member status (member or nonmember) at runtime? To help verify, you set the CheckState property to Indeterminate state within the Load event of the form, and within the Click event handler of Clear button using the code from Figure 6-3. If the CheckState property remains as Indeterminate when the user clicks the compute button, that means the user did not make a selection. You will later add the code to verify the CheckState status.

Figure 6-3: Set CheckState property to Indeterminate

```
private void cmdClear_Click(object sender, EventArgs e)
{
    hsbNumOfTickets.Value = 0;
    chkMember.CheckState = CheckState.Indeterminate;
}

private void TheaterTickets_Load(object sender, EventArgs e)
{
    chkMember.CheckState = CheckState.Indeterminate;
}
```

Step 1-8: Run the form. Note that when the form loads, the state is Indeterminate.
Click the CheckBox to change the CheckState to Checked, and then to UnChecked (you cannot set the state to Indeterminate at runtime by clicking the CheckBox).
Click the Clear button, and make sure the state changes to Indeterminate.

Events of CheckBoxes

Two important events of the CheckBox are the **CheckStateChanged** event and the **CheckedChanged** event.

> The **CheckStateChanged** event is triggered when the value of CheckState property changes.
> The **CheckedChanged** event is triggered when the value of Checked property changes.

You may use these event handlers to take actions when the Checked or CheckState properties change.

Review Questions

6.4 What is the value of Checked property of a CheckBox when the CheckState property is set to Indeterminate?

6.5 The CheckState property of a CheckBox is set to ____ when the Checked property is set to True?

6.3 Working with RadioButtons

RadioButtons are used as a group to let the user make a single selection from a group of limited options, like selecting a student's class in college or an employee's title. In this example, we use a set of RadioButtons to let the user select a section of the theater for seating:

The **Checked** property of the RadioButton is set to True if it is checked and False if it is not checked. At any time, only one RadioButton in a group can be checked.

You may create multiple groups by adding RadioButtons inside a **GroupBox** control on the form. If no GroupBox is used, all RadioButtons on a form are treated as part of a single group, even if they are physically in multiple groups. The GroupBox also lets the user move all RadioButtons that are within it, as a group.

RadioButtons versus CheckBoxes

RadioButtons are used when the number of options to choose from is limited (typically under 10) and also static (i.e., the number of items in the list does not change).

When not to use RadioButtons

When the selection is true/false or yes/no, the standard practice is to use a CheckBox, **not** two RadioButtons. For example, the following use of RadioButton is **not** recommended:

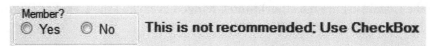

Step 1-9: Add a GroupBox to the form. Change its Text property to Section#.

Add five RadioButtons inside the GroupBox, as shown:

Change the value of Text property of the first four RadioButtons to 1, 2, 3 and 4, and that of the fifth button to zero.

Name the first four RadioButtons, **rdbSec1, rdbSec2, rdbSec3** and **rdbSec4**, respectively. Name the fifth, rdbDefault, and change its **Visible** property to False.

The first four represent four sections of the theater. The fifth button will be hidden from the user and is used as the default. It serves two purposes:

1. You can reset (uncheck) all four buttons by checking the fifth button using code.
2. If the fifth button is checked, that indicates that the user did not make a selection.

Next, change the Checked property of any one of the RadioButtons to True.

Chapter 6: Graphical User Interface Controls

The button changes to checked status. Now, change the Checked property of another button to True. Note that its status changes to checked, and the previously checked button becomes unchecked. Change the Checked property back to False.

Step 1-10: To help verify that the user checked a RadioButton, add the following code to the Load event of the form, and to the Click event handler of Clear button:

```
rdbDefault.Checked = true;
```

How does it help? If the Default RadioButton is checked when the user clicks the button to compute cost, that means the user did not check one of the first four buttons.

Note that the above statement assigns the Boolean value **true** to the Checked property, which has the same effect as setting the Checked property to **True** in the Desgin window of the form.

Figure 6-4 shows the modified code.

Figure 6-4: Resetting RadioButtons

```csharp
private void cmdClear_Click(object sender, EventArgs e)
{
    hsbNumOfTickets.Value = 0;
    chkMember.CheckState = CheckState.Indeterminate;
    rdbDefault.Checked = true;
}
private void TheaterTickets_Load(object sender, EventArgs e)
{
    chkMember.CheckState = CheckState.Indeterminate;
    rdbDefault.Checked = true;
}
```

Step 1-11: Run the form. Click any RadioButton, and click Clear to test the clear button.

The code to compute the total price of tickets using the data from the controls has three major parts:

1. Verify that the user specified the number of tickets, selected a section of the theater and specified the membership status.
2. Get the number of tickets from the ScrollBar, get the unit price based on the section selected, and get the discount based on membership status.
3. Compute the total cost.

The next section discusses the code to verify user input.

Review Questions

6.6 Write the statement that would make the RadioButton, rdbDefault, selected.

6.7 A form has two GroupBoxes, each containing a set of RadioButtons. What is the maximum number of RadioButtons that can be selected by the user from the entire form at runtime?

6.4 Validating Input

The code required to verify that the user provided all input is shown in Figure 6-5.

Figure 6-5: Verify that the user provided all input data

```
41          // Verify that the user selected number of tickets:
42          if (hsbNumOfTickets.Value == 0)
43          {
44              MessageBox.Show("Please select number of tickets");
45              return;
46          }
47          // Verify that the user selected a section:
48          if (rdbDefault.Checked == true)
49          {
50              MessageBox.Show("Please select the section");
51              return;
52          }
53          // Verify that the user selected membership status:
54          if (chkMember.CheckState == CheckState.Indeterminate)
55          {
56              MessageBox.Show("plese specify membership status");
57              return;
58          }
```

Line 41 checks whether the value of the ScrollBar is 0—that is, the user did not select a value.

Line 47 checks whether the Boolean value of the Checked property of the Default button is true (as asssinged in the Load event or Click event handler of the Clear button)—that is, none of the other buttons are checked.

Line 53 checks whether the CheckState property of the CheckBox is Indeterminate—that is, the user did not specify membership status.

The return statement within each if statement would prevent processing of the remaining code, if any, in the event handler.

Step 1-12: Add the code shown in Figure 6-5 to the Click event handler of btnComputePrice. Test the code by clicking the Compute button with missing data.

6.5 Getting User Input from Controls

The previous section verified that the user provided all inputs. Next, you get the inputs from controls: number of tickets, discount and unit price.

You get the **number of tickets** from the Value property of the ScrollBar, as follows:
```
int numOfTickets = hsbNumOfTickets.Value;
```

To compute the **discount**, you create a method named Discount and invoke (call) it. Members get a $20 discount per ticket. Figure 6-6 shows the method, and the statement that invokes it.

Figure 6-6: Code to compute discount

```
float discount = Discount();
float Discount()
{
    switch (chkMember.CheckState)
    {
        case CheckState.Checked:    // or, case 0
            return 20;
            break;
        default:
            return 0;
            break;
    }
}
```

The switch statement is used to compute discount if the CheckState property of the CheckBox has the value *Checked*.

To compute **unit price**, we create a method called UnitPrice and invoke it. Unit price for sections 1, 2, 3, and 4 are $70, $60, $52, and $45, respectively. Figure 6-7 shows the method, and the statement that invokes it.

Figure 6-7: Code to compute unit price

```
float unitPrice = UnitPrice();
    float UnitPrice()
    {
        if (rdbSec1.Checked)
            return 70;
        else if (rdbSec2.Checked)
            return 60;
        else if (rdbSec3.Checked)
            return 52;
        else if (rdbSec4.Checked)
            return 45;
        else
            return 0;
    }
```

Figure 6-8 shows the complete code that includes additional code to compute and display the amount due (lines 68, 69). Note that amtDue is declared at the class level so that it can be accessed from the Click event handler of btnComputePrice, and also from another event handler to be developed later.

Figure 6-8: Complete code to compute price

```csharp
float amtDue;
private void hsbNumOfTickets_ValueChanged(object sender, EventArgs e)
{
    lblNumOfTickets.Text = hsbNumOfTickets.Value.ToString();
}

private void btnClear_Click(object sender, EventArgs e)
{
    hsbNumOfTickets.Value = 0;
    chkMember.CheckState = CheckState.Indeterminate;
    rdbDefault.Checked = true;
    lblAmtDue.Text = "";
}

private void TheaterTickets_Load(object sender, EventArgs e)
{
    chkMember.CheckState = CheckState.Indeterminate;
    rdbDefault.Checked = true;
}

private void btnComputePrice_Click(object sender, EventArgs e)
{
    // Verify that the user selected number of tickets
    if (hsbNumOfTickets.Value == 0)
    {
        MessageBox.Show("Please specify number of tickets");
        return;
    }
    // Verify that the user selected a section
    if (chkMember.CheckState == CheckState.Indeterminate)
    {
        MessageBox.Show("Please specify membership");
        return;
    }
    // Verify that the user selected membership status
    if (rdbDefault.Checked == true)
    {
        MessageBox.Show("Please specify Section");
        return;
    }
    // Get number of tickets
    int numOfTickets = hsbNumOfTickets.Value;

    // Compute discount
    float discount = Discount();

    // Compute unit price
    float unitPrice = UnitPrice();

    amtDue = numOfTickets * (unitPrice - discount);
    lblAmtDue.Text = amtDue.ToString("$#,###.00");
}
```

```
72      float Discount()
73      {
74          switch (chkMember.CheckState)
75          {
76              case CheckState.Checked:    // or, case 0
77                  return 20;
78                  break;
79              default:
80                  return 0;
81                  break;
82          }
83      }
84
85      float UnitPrice()
86      {
87          if (rdbSec1.Checked)
88              return 70;
89          else if (rdbSec2.Checked)
90              return 60;
91          else if (rdbSec3.Checked)
92              return 52;
93          else if (rdbSec4.Checked)
94              return 45;
95          else
96              return 0;
97      }
```

Step 1-13: Copy the rest of the code shown in Figure 6-8 from Ch6_GUIcontrols/Ch6_ Code.txt to the Click event handler of ComputePrice.

Step 1-14: Run the form. Test the code using the input data from Figure 6-9 and verify the output.
Clear the data, and run it with another set of valid selections.
Manually compute the result, and verify the output from the code.
Run it when one or more of the three items is not selected. Make sure you get the proper message.

Figure 6-9: The TheaterTickets form with sample data

6.6 Working with ComboBoxes

A ComboBox lets the user select an item from a group of items. Consider a modified version of the TheaterTicket form that includes a ComboBox to select the name of the show, as shown in Figure 6-10. In the modified form, the name of the show and the total cost are written to a ListBox. (Typically, in a real-world application, the data would be written to a file or database.) It is assumed that the price of a ticket depends only on the section, not on the show.

Figure 6-10: TheaterTicket form with ComboBox

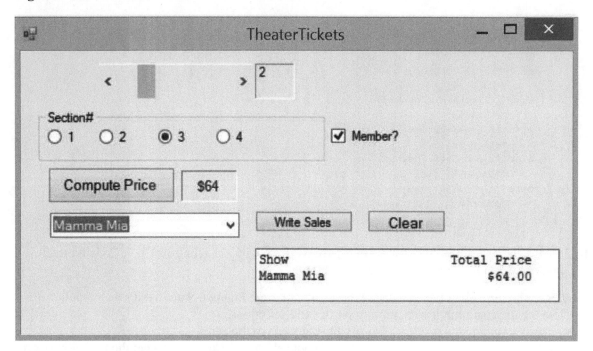

A Set of RadioButtons versus a Single ComboBox

A ComboBox, as opposed to a set of RadioButtons, is recommended when the list is large and/or the list is dynamic. The list is dynamic when the content of the list or the number of items in the list is likely to change. In the current example, we use a ComboBox because the list of shows is likely to change over time. Similarly, to let the user select a month from a list of the 12 months, for example, a ComboBox generally is recommended over a set of RadioButtons because the list is large.

Properties and Events of ComboBoxes

Table 6-3 shows commonly used properties and events of the ComboBox and ListBox.

Table 6-3: Commonly used properties and events of ComboBox and ListBox

Property	Description
DropDownStyle	Determines the style and whether user can type in an item that is not in the list
Items	The collection of all items in the list
SelectedIndex	The index of the selected item
SelectedItem	The selected item (of object type)
SelectedItems	The collection of selected items in a **ListBox**
SelectionMode	Determines whether more than one item can be selected from a **ListBox**
Sorted	If set to True, the items in the list are sorted
Text	The text displayed in the ComboBox
Events	
SelectedIndexChanged	Raised when the user selects an item from the list

The Items collection

The Items collection is an important property of the ComboBoxes and ListBoxes. It has several properties and methods, as shown in Table 6-4.

Table 6-4: Properties and methods of the Items Collection

Property	Description
Count	Number of items in the list
Item (index)	The item at the specified index
Method	
Add (object)	Adds the specified item to the list
Clear()	Removes all items from the list
Insert (index, object)	Inserts the specified item at the specified index
Remove (index)	Removes the item at the specified index
Remove (object)	Removes the specified item

Adding items to a ComboBox

You may add items to the Items collection in Design view using the GUI interface, or at runtime using code. Adding items at runtime makes future maintenance of the list easier.

To add items to the ComboBox in Design view, open the Items collection window by clicking on the ellipses on the right side of the Items property, as depicted in Figure 6-11.

Figure 6-11: Items collection of ComboBox

Next, type the items into the **String Collection Editor** window, and click OK. Figure 6-12 shows the window.

Figure 6-12: Window for adding items to a ComboBox

Sort property of ComboBox

You may sort the items in the collection by setting the Sort property to True. Sorting would make it easy for the user to select items from the list.

Step 1-15: Add a ComboBox named cboShows to the form (see Figure 6-10), and set its Text property to "Select a Show." Set the Sort property to True.
Open the Items collection window, type in the list of shows from Figure 6-12, and click OK.
Run the form and display the list. Note that the list is sorted. Stop the program.
Now, delete the items from the window so you can add items using code at runtime.

Adding Items to ComboBox at Runtime Using Code

To add items to the list in the ComboBox, use the Add method of the Items collection. You may add one item at a time, using a separate statement for each item, as shown in the following example:
```
cboShow.Items.Add("Mamma Mia");
cboShow.Items.Add("Rock of Ages");
…
```
However, it is more convenient to store the items in an **array** and use the **foreach** loop to add each item to the ComboBox. Unlike the simple variables you used in previous chapters, an array can store multiple data items of the same type. The foreach loop is a type of loop that can be conveniently used to access each data item stored in an array. Arrays and foreach loops are discussed in more detail in Chapter 7.

For example, the names of the shows can be stored in an array named **shows** as follows:
```
string[] shows = {"Mamma Mia","Rock of Ages", "Les Miserables",
        "Motown The Musical","The Lion King", "The Phantom of the Opera"};
```

The following code gets each show from the array, stores it in the variable **show**, and adds it to the ComboBox:
```
foreach (string show in shows)
      cboShow.Items.Add(show);
```
The foreach loop loops through each data item from the array and adds the item to the ComboBox.

Step 1-16: Insert the code from Figure 6-13 to add the names of shows to the ComboBox. The code should be added to the Load event of the form so that the ComboBox will show the list when the form is displayed.

Figure 6-13: Code to add items to the ComboBox

```
34      private void TheaterTickets_Load(object sender, EventArgs e)
35      {
36          chkMember.CheckState = CheckState.Indeterminate;
37          rdbDefault.Checked = true;
38          string[] shows = { "Mamma Mia", "Rock of Ages", "Les Miserables",
39                  "Motown", "The Lion King", "The Phantom of the Opera" };
40          foreach (string show in shows)
41              cboShow.Items.Add(show);
42      }
```

Step 1-17: Run the form, and make sure the list is displayed in the sorted order when you click the arrow on the ComboBox.

ComboBox styles

You can choose one of three different styles for a ComboBox by changing the value of the DropDownStyle property shown in Figure 6-14.

Figure 6-14: DropDownStyle property

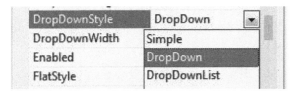

DropDown style

This is the default value of the DropDownStyle property. Setting the DropDownStyle property to DropDown style provides a dropdown list when the user clicks the arrow, as shown in Figure 6-12. The user also may type in an item that is not in the list.

Figure 6-15: The DropDown style

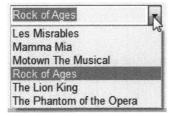

DropDownList style

This style is similar to DropDown style, except that the user cannot type in an item.

Simple style

With this style, the ComboBox does not have an arrow to provide a dropdown list. A ScrollBar is provided to view items, if any, that cannot be displayed in the available space. Figure 6-16 shows what Simple style looks like.

Figure 6-16: Simple style of ComboBox

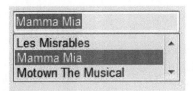

Step 1-18: Change the DropDownStyle property of the ComboBox to DropDownList, run the form and observe the difference. Repeat for Simple style.

Getting the selected item

There are three major properties related to the item Selected in a ComboBox: SelectedIndex, SelectedItem and Text property.

SelectedIndex

SelectedIndex is the index of the selected item. If the first item is selected, the SelectedIndex would be 0; if the second item is selected, it would be 1, and so on.

If no item is selected, the SelectedIndex has the value -1. You may reset the ComboBox so that no item is selected by setting the SelectedIndex to -1, as follows:

```
cboShow.SelectedIndex = -1;
```

It is important to note that the index of an item in the original list that was added to the ComboBox may not be the same as the index of the same item in the list displayed in a sorted ComboBox at runtime. For example, the index of Les Miserables in the original list was 2, whereas in the sorted list that is displayed in the ComboBox, it is 0. As items are added or deleted from the list, the index of the item changes. **Therefore, it is not reliable to use the SelectedIndex to identify the item selected by the user.**

SelectedItem

SelectedItem gives the item selected by the user. If no item is selected, its value is null.

Text property

The value of Text property is whatever is displayed in the TextBox part of the ComboBox, which could be an item selected by the user, the prompt specified in the Text property (e.g., "Select a Show"), or any text typed into the TextBox part of the ComboBox.

Step 1-19: Add the code from Figure 6-17 to the Click event handler of Clear button to reset the ComboBox.

Chapter 6: Graphical User Interface Controls   191

Figure 6-17: Code to clear the ComboBox

```
24
25      private void btnClear_Click(object sender, EventArgs e)
26      {
27          hsbNumOfTickets.Value = 0;
28          chkMember.CheckState = CheckState.Indeterminate;
29          rdbDefault.Checked = true;
30          lblAmtDue.Text = "";
31          cboShow.SelectedIndex = -1;
32      }
```

Step 1-20: Add a button named btnWriteSales that writes the name of the selected show and the computed price to a ListBox. Add a ListBox named lstDisplay. Figure 6-18 shows the lower part of the form where button and ListBox are added.

Figure 6-18: TheTheaterTickets form with added Write button and ListBox

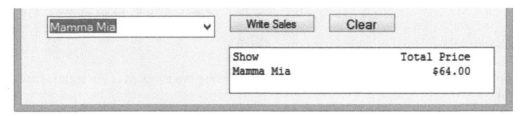

Step 1-21: Add the necessary code from Figure 6-19 to the Click event handler of btnWriteSales, to do the following tasks:
Verify that the user computed the price.
Verify that the user selected a Show.
Write the name of the selected show and the price to the ListBox.

Figure 6-19: Code to write the name of the show and price

```
105     private void btnWriteSales_Click(object sender, EventArgs e)
106     {
107         if (cboShow.SelectedIndex == -1)
108         {
109             MessageBox.Show("Please select a show");
110             return;
111         }
112         if (amtDue == 0)
113         {
114             MessageBox.Show("Please compute amount");
115             return;
116         }
117         string fmtStr = "{0,-25}{1,12:C}";
118         lstDisplay.Items.Clear();
119         lstDisplay.Items.Add(String.Format(fmtStr, "Show", "Total Price"));
120         lstDisplay.Items.Add(String.Format(fmtStr, cboShow.SelectedItem, amtDue));
121     }
```

The two key statements related to the ComboBox are lines 107 and 120:

Line 107 checks whether the value of SelectedIndex is -1—that is, the user did not select a show. **Line 120** displays the name of the show, given by the SelectedItem property, and the price given by amtDue.

SelectedText Property

ComboBox also has a property called SelectedText. Like the SelectedItem property, the SelectedText property gives the name of the item selected by the user. However, if you select an item in a ComboBox and move focus away from the ComboBox by clicking a button, for example, the value of the SelectedText property in the Click event handler of the button will be an empty string. This is because the selection is cleared when the focus is moved away. Similarly, the SelectedText will be an empty string also within the SelectedIndexChanged and SelectedValueChanged event handlers (discussed later in this section) because at the time of these events, the selected value has not yet been set. So, the SelectedText property is not suitable to get the selected item from an event handler.

Accessing Items from the Items Collection

Any item in the collection of items in a ComboBox may be accessed using its index number. For example,
  cboShow.Items[0] gives the first item in the list, and
  cboShow.Items[1] gives the second item in the list.
To access every item in a ComboBox, you may use a loop to vary the index over the entire range, as shown below:

```
for (int index=0; index < cboShow.Items.Count; index++)
    lstDisplay.Items.Add (cboShow.Items[index]);
```

In the above code, cboShow.**Items.Count** property gives the number of items in the Items collection. Because the index starts at zero, the index of the last item is one less than the count of items.

An alternate method that does not require the count of items is to use the foreach loop. The following code would display each show from the Items collection of the ComboBox, cboShow:

```
foreach (string show in cboShow.Items)
    lstDisplay.Items.Add(show);
```

SelectedIndexChanged event

When the user selects an item from a ComboBox, the action triggers the SelectedIndexChanged event. You can use this event to take additional actions. For example, when the user selects a product from a ComboBox, the SelectedIndexChanged event handler can look up the price for that product.

Consider another revision to the TheaterTicket form where the name of the show and the price are displayed automatically in the ListBox, when the user selects a show, without having to click the Write Sales button. To do this, we add the code to the SelectedIndexChanged event handler of the ComboBox, shown in Figure 6-20.

Figure 6-20: SelectedIndexChanged event handler of ComboBox

```
123    private void cboShow_SelectedIndexChanged(object sender, EventArgs e)
124    {
125        if (amtDue == 0)
126        {
127            MessageBox.Show("Please compute amount");
128            return;
129        }
130        string fmtStr = "{0,-25}{1,-12:C}";
131        lstDisplay.Items.Clear();
132        lstDisplay.Items.Add(String.Format(fmtStr, "Show", "Total Price"));
133        lstDisplay.Items.Add(String.Format(fmtStr, cboShow.SelectedItem, amtDue));
134    }
```

This code is the same as in the Click event handler of the Write Sales button (Figure 6-19), except that this code does not check whether the user selected a show. Note that with this approach, you do not need the Write Sales button and code in the Click event handler of the button.

Step 1-22: Double click the ComboBox to create the SelectedIndexChanged event handler of the ComboBox. Add the code from Figure 6-20 to the event handler and test it.

Quite often, items are added to ComboBoxes by reading data from a text file or database. Reading from text file is discussed in Chapter 8, and getting data from a database is discussed in Chapter 12.

The following code shows the complete revised code that includes selecting a show using the ComboBox and displaying it. The code in the Click event handler of btnWrite is not shown, because the code in the SelectedIndexChanged event handler of the ComboBox does the same function.

```csharp
        float amtDue;
        private void hsbNumOfTickets_ValueChanged(object sender, EventArgs e)
        {
            lblNumOfTickets.Text = hsbNumOfTickets.Value.ToString();
        }

        private void btnClear_Click(object sender, EventArgs e)
        {
            hsbNumOfTickets.Value = 0;
            chkMember.CheckState = CheckState.Indeterminate;
            rdbDefault.Checked = true;
            lblAmtDue.Text = "";
            cboShow.SelectedIndex = -1;
        }

        private void TheaterTickets_Load(object sender, EventArgs e)
        {
            chkMember.CheckState = CheckState.Indeterminate;
            rdbDefault.Checked = true;
            string[] shows = { "Mamma Mia", "Rock of Ages", "Les Miserables",
                    "Motown", "The Lion King", "The Phantom of the Opera" };
            foreach (string show in shows)
                cboShow.Items.Add(show);
        }

        private void btnComputePrice_Click(object sender, EventArgs e)
        {
            // Verify that the user selected number of tickets
            if (hsbNumOfTickets.Value == 0)
            {
                MessageBox.Show("Please specify number of tickets");
                return;
            }
            // Verify that the user selected a section
            if (chkMember.CheckState == CheckState.Indeterminate)
            {
                MessageBox.Show("Please specify membership");
                return;
            }
            // Verify that the user selected membership status
            if (rdbDefault.Checked == true)
            {
                MessageBox.Show("Please specify Section");
                return;
            }
            // Get number of tickets
            int numOfTickets = hsbNumOfTickets.Value;

            // Compute discount
            float discount = Discount();

            // Get unit price
            float unitPrice = UnitPrice();

            amtDue = numOfTickets * (unitPrice - discount);
            lblAmtDue.Text = amtDue.ToString("$#,###.00");
        }
```

```
 76
 77   float Discount()
 78   {
 79       switch (chkMember.CheckState)
 80       {
 81           case CheckState.Checked:    // or, case 0
 82               return 20;
 83               break;
 84           default:
 85               return 0;
 86               break;
 87       }
 88   }
 89
 90   float UnitPrice()
 91   {
 92       if (rdbSec1.Checked)
 93           return 70;
 94       else if (rdbSec2.Checked)
 95           return 60;
 96       else if (rdbSec3.Checked)
 97           return 52;
 98       else if (rdbSec4.Checked)
 99           return 45;
100       else
101           return 0;
102   }
121
122   private void cboShow_SelectedIndexChanged(object sender, EventArgs e)
123   {
124       if (amtDue == 0)
125       {
126           MessageBox.Show("Please compute amount");
127           return;
128       }
129       string fmtStr = "{0,-25}{1,-12:C}";
130       lstDisplay.Items.Clear();
131       lstDisplay.Items.Add(String.Format(fmtStr, "Show", "Total Price"));
132       lstDisplay.Items.Add(String.Format(fmtStr, cboShow.SelectedItem, amtDue));
133   }
```

Review Questions

6.8   A comboBox, cboNames, shown below, is used to display a set of student names, and let the user select a name.

Which property of the ComboBox represents the currently selected name if the user selected an item, and would be empty if the user had not selected an item?

6.9 A comboBox, cboNames, shown below, is used to display a set of student names, and let the user select a name.

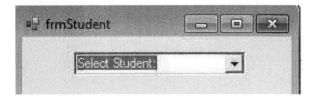

Write the statement to reset the ComboBox so that no student is selected.

6.10 A sign-up form for the activities of a club requires the user to specify whether the user is a member of the club. The prompt for the user on the form is
    Are you a member of the club?

A control (or controls) is to be placed to the right of the above prompt to let the user specify the membership status. Based on standard practice, which one of the following control(s) would you choose?

- A CheckBox
- Two CheckBoxes (one with a "Yes" label and the other with a "No" label)
- A RadioButton
- Two RadioButtons (one with a "Yes" label and the other with a "No" label)
- A TextBox

6.7 Working with ListBoxes

An important difference between a ComboBox and a ListBox is that the ListBox allows the user to select more than one item from the list.

Table 6-3 presented the commonly used properties and events of ComboBoxes and ListBoxes, which included the SelectionMode property that determines whether more than one item can be selected from a ListBox. Properties and methods of the Items collection, which is an important property of ListBoxes, are presented in Table 6-4.

A Set of CheckBoxes versus a Single ListBox

When do you use a set of ListBoxes rather than a ListBox to select multiple items from a list? Again, the guideline is to use a ListBox if the list is large and/or dynamic. A list is dynamic when the content of the list or the number of items in the list is likely to change. Thus, a ListBox is preferable to select multiple ingredients for a pizza, if the list of ingredients is dynamic.

SelectionMode Property

To enable selection of multiple items, the SelectionMode property shown in Figure 6-21 needs to be set to MultiSimple or MultiExtended. The two settings are similar, except that the MultiExtended setting makes it easier to select a set of consecutive items using the shift key.

Figure 6-21: SelectionMode property of ListBox

You already used ListBoxes to display output by adding text to the Items collection using the Add method, as in

```
lstDisplay.Items.Add ("Total Price: " + lblAmtDue.Text);
```

Similar to the ComboBox, the Add method adds a new item to the ListBox.

You also may add a list of items to the ListBox by storing the items in an array and then using a loop to add the items to the ListBox. For example, the following code stores a list of pizza toppings in an array named toppings, and adds them to a ListBox named lstToppings using a foreach loop.

```
String[] toppings = {"Pepperoni","Sausage", "Onion", "Green Pepper",
                     "Mushrooms", "Xtra Cheese", "Ham", "Anchovies"};
foreach (string topping in toppings)   // Add toppings to ListBox
       lstToppings.Items.Add(topping);
```

To better understand the use of ListBoxes, let's look at a pizza order form that lets the user select the size of pizza and the toppings, compute the total price, and display the total price and the toppings selected, as shown in Figure 6-22.

Figure 6-22: The PizzaOrder form

The list of toppings is added to the ListBox named lstToppings using code in the Load event of the form, shown in Figure 6-23.

Figure 6-23: Code to add toppings to the ListBox

```
19              // This form uses a ListBox to let users select multiple items from a list:
20              private void PizzaOrder_Load(object sender, EventArgs e)
21              {
22                  string[] toppings = {"Pepperoni","Sausage", "Onion", "Green Pepper",
23                                       "Mushrooms", "Xtra Cheese", "Ham", "Anchovies"};
24                  foreach (string topping in toppings)  // Add toppings to ListBox
25                      lstToppings.Items.Add(topping);
26
27                  rdbDefault.Checked = true;
28              }
```

In **line 27**, rdbDefault is a hidden RadioButton that makes it easier to reset the other RadioButtons and to check whether the user selected a size.

MultiColumn property

If there is not enough space to display the entire list of items in a ListBox, a vertical ScrollBar will be provided. If you want to avoid a vertical ScrollBar, you may set the MultiColumn property of the ListBox to True so that the items are displayed in multiple columns, as shown in Figure 6-24.

Figure 6-24: Display when MultiColumn property is set to True

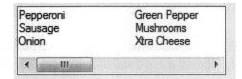

Step 1-23: Open the form named PizzaOrder. This form has the following controls. See Figure 6-22 for captions.

> RadioButtons named rdbSmall, rdbMedium, rdbLarge, rdbDefault
> ListBoxes named lstToppings and lstDisplay
> Buttons named btnCompute, btnClear

Step 1-24: Set the **SelectionMode** property of lstToppings to **MultiSimple**.

Step 1-25: Open the load event handler of the form, which already has the code from Figure 6-23, to add the toppings to the ListBox. Review the code.

> Run the form, and verify that toppings are displayed in the ListBox lstToppings and that you can select multiple items.

Getting the Selected Items

For ComboBoxes, the SelectedItem property gives the item selected by the user. How do we get the selected items from a ListBox when there are multiple items?

SelectedItems collection

The **SelectedItems** (note the "s" at the end that makes it plural) is the collection of all items selected by the user. You use a loop to get each item from the collection. The following code gets each selected item from the ListBox and displays it in another ListBox named lstDisplay:

```
foreach (string topping in lstToppings.SelectedItems)
      lstDisplay.Items.Add("    " + topping);
```

You may access any selected item from the SelectedItems collection by specifying the index. Here are some examples:
      lstToppings.SelectedItems[0] gives the first selected item (Sausage)
      lstToppings.SelectedItems[1] gives the second selected item (Green Pepper), etc.

Instead of using a foreach loop, you may use a for loop to vary the index using a for loop to access each selected item, and display it, as in the following code:

```
for (int index = 0; index < lstToppings.SelectedItems.Count; index++)
      lstDisplay.Items.Add("    " + lstToppings.SelectedItems[index]);
```

In the above code, **SelectedItems.Count** gives the number of items selected.

Items verses SelectedItems

Note that **Items** is the collection of all items in a ListBox, whereas **SelectedItems** is the collection of only selected items. Consider a ListBox with items selected, as follows:

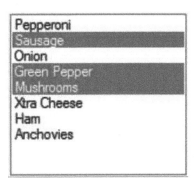

In this ListBox, Items[0] is the first item in the list (Pepperoni), but, SelectedItems[0] is the first selected item (Sausage).

Now, let's look at the code to compute the price based on the size selected and the number of toppings. Figure 6-25 shows the code, except for the statements to compute the total price and display selected items, which are left for you to complete.

Line 33 checks whether the user selected a size:
      if (rdbDefault.Checked == true),
If the default RadioButton is selected, that means none of the other RadioButtons that represent the size are selected.

Lines 39–41 should invoke the UnitPrice method, shown in Figure 6-25, which returns a tuple that has two elements of type (float, float) named basePrice and costPerTopping.

Step 1-26: Add the missing code in lines 39–41 to call UnitPrice and compute the total price (base price + cost per toppings × number of toppings).

Hint: Count property of SelectedItems collection gives the number of selected items.

Add the missing code in lines 55–56 to display the selected toppings in the ListBox, lstDisplay. You may use a foreach loop to access each item in the SelectedItems collection.

Figure 6-25: Code to compute price based on size and number of toppings

```
29
30      private void btnCompute_Click(object sender, EventArgs e)
31      {
32          // Verify that the user selected size:
33          if (rdbDefault.Checked == true)
34          {
35              MessageBox.Show("Please select size");
36              return;
37          }
38          // Invoke UnitPrice to get base price and cost per topping
39
40
41
42
43          // Compute and display price
44          // (insert code here to compute the price: base price + cost of toppings)
45          float price = 0;
46          lstDisplay.Items.Clear();
47          lstDisplay.Items.Add ("Price:   " + price.ToString("C2"));
48          // Display Toppings:
49          lstDisplay.Items.Add("");
50          lstDisplay.Items.Add ("Toppings");
51          if (lstToppings.SelectedItems.Count == 0)  // if no item is selected
52              lstDisplay.Items.Add("    None");
53          else
54          {
55              // (insert the code to display selected items)
56
57          }
58      }
59
60      private (float basePrice, float costPerTopping) UnitPrice()
61      {
62          float basePrice = 0, costPerTopping = 0;
63          if (rdbSmall.Checked)
64          {
65              basePrice = 6.95F;
66              costPerTopping = 1.5F;
67          }
68          else if (rdbMedium.Checked)
69          {
70              basePrice = 8.95F;
71              costPerTopping = 2.0F;
72          }
73          else if (rdbLarge.Checked)
74          {
75              basePrice = 11.95F;
76              costPerTopping = 2.5F;
77          }
78          return (basePrice, costPerTopping);
79      }
```

Step 1-27: View the code in the Click event handler of Clear button, shown in Figure 6-26.

Figure 6-26: Code to reset controls

```
80
81   private void btnClear_Click(object sender, EventArgs e)
82   {
83       rdbDefault.Checked = true;
84       lstToppings.SelectedIndex = -1;
85       lstDisplay.Items.Clear();
86   }
```

Step 1-28: Run the form. Test the code by clicking the Compute button when size and/or toppings are not selected. Test it with proper selection of size and toppings.

ListView control

The ListView control is another commonly used control, which allows multiple data items to be displayed in a row. Because arrays make it easier to add data to this control, we will discuss this control in Chapter 8.

It's time to practice! Do Exercise 6-1.

6.8 Windows Presentation Foundation (WPF)

If you want to build applications with rich graphical user interfaces, Windows Presentation Foundation (WPF) is a Microsoft technology that can help you. The WPF features include extensive support for 2-D and 3-D graphics, animation, media and high quality text. Further, the layout system in WPF allows creating dynamic layouts that adjust to changes in window size and display settings. Additional components of WPF applications include

1. XAML (Extensible Application Markup Language): The user interface is created using XAML, which is a Microsoft XML-based markup language. Just like in a common Windows Form application, you can drag WPF controls from the Toolbox onto the Design widow, which will generate the corresponding XAML tags that specify the controls. You also may use XAML to create controls. XAML allows you to do extensive customization and nesting of controls. WPF controls also have properties and events that are not available in Windows Form controls.
2. Code-behind: Code-behind represents C#, Visual Basic and other .Net language programs that are attached to specific events of WPF controls, like the Click event of a button. This code is similar to the code in Windows Forms that you are familiar with.

To create a WPF project, select
   File, New, Project, C#, Windows, WPF Application

For tutorials and additional information on creating WPF applications, please see
   https://msdn.microsoft.com/en-us/library/aa970268(v=vs.110).aspx

Review Questions

6.11 You want to develop a form that allows the user to select one section of an introductory IS course from a list of all sections. The number of sections varies from semester to semester within the range 3 to 8. The prompt for the user on the form would be

> Please select a section.

Based on the standard practice, which one of the following controls would you use to let the user select a section: a ListBox, a set of CheckBoxes, a set of RadioButtons, or a ComboBox?

6.12 It is desired to develop a form that allows the user to select one or more days from a list of all seven days of a week. The prompt for the user on the form would be

> Please select one or more days:

Based on industry standards, which one of the following controls would you use to let the user select one or more days: a ListBox, a set of CheckBoxes, a set of RadioButtos, or a ComboBox?

6.13 A form displays a list of sports in a ListBox named lstSports. A user selects Basketball and Soccer from the list at runtime, as shown below:

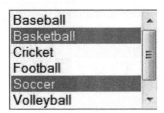

What would be the value of lstSports.SelectedItems(0)?

6.14 A form displays a list of sports in a ListBox named lstSports. A user selects Basketball and Soccer from the list at runtime, as shown below:

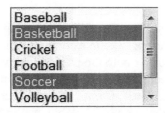

What would be the value of lstSports.Items(2)?

Exercises

Exercise 6-1

Develop a form, Icecream.cs, as shown below:

The form must let the user do the following:

Specify the number of scoops (max three) using a ScrollBar, and display the specified number in a Label named lblScoops.
Select the type of sundae from a list of sundaes (Ice Cream, Custard, Yogurt, etc.) displayed in a ComboBox sorted alphabetically. Because the items in this list may change over time, do not assume that, for example, the third item will remain as Yogurt.
Use a CheckBox to specify whether the user wants nuts for topping.
Compute the amount due.
Clear all data so that the user can make another set of selections. Exit the form.

The amount due is given by
Cost of Sundae (Ice cream - $3.50; Custard - $3:75; Yogurt - $4:00) * Number of Scoops + Cost of Nuts (0.50)

Do not use defaults for the items to be specified by the user. Defaults make it difficult to check whether the user forgot to select an item.

Follow the guidelines to make programs easy to understand, maintain, and reuse.

Exercise 6-2

A charitable organization uses volunteers to put on 10 to 15 different types of fund-raising events per year. Examples of the types of events are brat fry, walkathon, bake sale and golfing. A section of the volunteer signup form collects the following information:

> Name
> Types of events a person would like to volunteer for (a person may select one or more types)
> Number of hours available per week (The options are 5-10, 11-20, 21-30, 31-40.)
> Whether he/she is a new volunteer

Develop a form named Exercise 6.2 that collects the above information. Select appropriate controls based on the guidelines discussed in your textbook. Write the data entered by the user into a ListBox. If a volunteer is new, display the word "New"; if not, display the word "Experienced." A sample of the display is shown below:

Name: Sam Adamson

Events:
                Walkathon
                Golfing

Hours/week: 11-20

Status: New

Do not specify default values for any control. Defaults make it difficult to check whether the user forgot to select an item.

Chapter 7

Arrays

The variables we used so far can store only one value at a time. So, if you want to store 30 test scores in memory and do computations with them simultaneously, you would have to declare 30 different variables.

A single array allows you to store multiple data items of the same type. For example, a single one-dimensional array can store all 30 scores at the same time. Because one-dimensional (1-D) arrays are the simplest type of arrays, we will discuss them first.

Learning Objectives

After studying this chapter, you should be able to:

- Describe the difference between value types and reference types.
- Create one-dimensional (1-D) arrays.
- Store a set of data in a 1-D array.
- Access data from 1-D arrays using the *for* loop and *for each* loop.
- Copy a 1-D array.
- Look up a value from a 1-D array.

Topics

7.1	Introduction to One-Dimensional Arrays	7.4	Accessing Elements of an Array Using foreach Loop
7.2	Assigning a Value to an Element of an Array	7.5	Copying an Array
		7.6	Looking Up Values in an Array
7.3	Accessing Elements of an Array Using the Index		

7.1 Introduction to One-Dimensional Arrays

An array can be thought of as a container of data with multiple storage locations or slots, one location for each data item. An array called testScore that stores 10 test scores may be represented as follows:

testScore

85	92	87	67	83	91	73	98	83	84
0	1	2	3	4	5	6	7	8	9

Element number:
(subscript or index)

Each storage location of the array is called an element. The elements are numbered consecutively, starting with zero. An individual element of an array is accessed using the array name, followed by the element number (subscript or index) in brackets. The subscript must be an integer. For example, testScore[0] refers to the first element of the array, that is, the number 85,

testScore[1] refers to the second element of the array—that is, the number 92, and

testScore[9] refers to the tenth element of the array, the number 84.

Different arrays can hold data items of different types, including numbers, strings, arrays and objects. However, within an array, all elements must be of the same type. Thus, **an array is a collection of data items of the same type.**

An array, like testScore, which can store one set of data, is called a **one-dimensional** (1-D) array. A two-dimensional (2-D) array, discussed later in Chapter 8, can store multiple sets of data, all of the same type.

To help explain how to create an array, we will discuss another important property of arrays.

Value Types versus Reference Types

The integer, real and Boolean types we discussed in earlier chapters are called **value types** because each variable of these types holds the actual value assigned to that variable. Thus, the memory location allocated to a value type variable holds the actual value assigned to that variable.

By contrast, arrays (and other objects) are called **reference types** because each variable of these types holds only a reference to the data—that is, the memory location or address of the data, not the actual data. A variable that references an object, like an array, is called a **reference variable**. Let's first look at the value types through some examples.

Consider two value type variables, previousSales and currentSales, declared below:

    int previousSales, currentSales;

The effect of declaring the variables would be to store the default value zero in the memory locations for the two variables, as follows:

0	0
previousSales	currentSales

Assigning a value for currentSales using the statement, **currentSales = 245**; stores the value in the memory location for currentSales, as represented below:

0	245
previousSales	currentSales

When currentSales is assigned to previousSales using the statement

    previousSales = currentSales;

the value from currentSales is copied to the memory location for previouSales, so that **both locations store the same data,** as shown:

245	245
previousSales	currentSales

As discussed in Chapter 1, if you later change the value of currentSales using a statement like

```
currentSales = 300;
```

it will have no effect on the value of previousSales. The locations for the two variables will appear as follows:

```
   ┌─────┐        ┌─────┐
   │ 245 │        │ 300 │
   └─────┘        └─────┘
 previousSales  currentSales
```

Now, let's look at how arrays differ.

Creating Arrays

Creating and using an array, which is a reference type, involves two steps:

1. Create the array object in memory to hold a set of data items, and
2. Create a variable, called reference variable, which references (holds the address of) the array, as shown in Figure 7.1

Figure 7-1: Array and reference variable

testScore

(location of the array)	→	85	92	87	67	83	91	73	98	83	84
	indexes:	0	1	2	3	4	5	6	7	8	9

Here is a statement that creates an array called testScore:

```
int[] testScore = new int[10];
```

The right hand side, **new int[10]**, creates an array in memory. Let's look at each part:

The key word **new** creates a new object (in memory) of the type specified by **int[10]**, and returns the memory location of the new array.

The expression **int[10]** specifies that the object is an array of int type, and that the **array size** is 10—that is, the array has 10 elements whose indexes ranges from 0 to 9. When a new array is created, each element has the default value zero.

The left-hand side, **int[] testScore**, declares a reference variable that we can use to access the array. The pair of square brackets after the variable type is what distinguishes it from a value type variable and specifies that it is a variable to reference an array.

The = operator stores a reference to the array (memory location of the array) in the reference variable testScores.

Figure 7-2 represents the effect of executing the statement, int[] testScore = new int[10], to create the array testScore and assign the reference to testScore.

Figure 7-2: Array and reference variable

You may split the statement, int[] testScore = new int[10], to two separate statements:

```
int[] testScore;
testScore = new int[10];
```

The first statement, **int[] testScore,** declares testScore as a variable to reference an int type array. The second statement, **testScore = new int[10]**, creates an int type array of size 10 using the key word new. The = operator assigns the memory location of the new array to testScores.

The general syntax to create an array

The general syntax of the statement to create an array is

Type[] VariableName = new *Type(ArraySize)*

where *ArraySize* is the number of elements of the array, *Type* is the type of data to be stored in the array and *VariableName* is any valid variable name.

Alternatively, you may use two different statements:

Type[] VariableName

VariableName = new *Type(ArraySize)*

Tutorial 1: Working with 1-D Arrays: Test Scores Application

In this tutorial, you will work with simple one-dimensional arrays.

Step 1-1: Open the project, Ch7_Arrays, from the folder Tutorial_Starts.

Step 1-2: Open the form, Array_1D_Intro, shown below.

Step 1-3: Open the Click event handler of btnDemo.

Step 1-4: Add the statement to create an array testScores within the Click event handler of btnDemo, as shown in Figure 7-3, and put a break at the closing brace.

Figure 7-3: Code to create an array

```
19      private void btnDemo_Click(object sender, EventArgs e)
20      {
21          // Create array testScore with default values:
22          int[] testScore = new int[10];
```

Step 1-5: Run the project, and click the Demo button.

Step 1-6: When the program breaks, display the Locals window (Debug, Windows, Locals). Expand testScore to display the values of the 10 elements of the array corresponding to index 0 to 9, as shown in Figure 7-4.

Figure 7-4: Locals window displaying the array

Locals	
Name	Value
⊞ ● this	{Ch7_Arrays.Array_1D_Intro, Text: I-Dime
⊞ ● sender	{Text = "Demo Array"}
⊞ ● e	{X = 43 Y = 21 Button = Left}
⊟ ● testScore	{int[10]}
● [0]	0
● [1]	0
● [2]	0
● [3]	0
● [4]	0
● [5]	0
● [6]	0
● [7]	0
● [8]	0
● [9]	0

Creating Arrays with Initial Values Other Than Zero: Array Initializer

An array may be created with initial values other than the default of zeros, using **array initializer**, which is a comma-separated list of values enclosed in braces, as in

```
int[] testScore = { 85, 92, 87, 67, 83, 91, 73, 98, 83, 84 };
```

This way of creating an array does not require the key word new, or the type and the array size specified on the right side of the = sign. The C# compiler infers the array size from the number of data items in the list and creates the array just like when the key word is used.

Alternately, you may split it to two statements as follows:

```
int[] testScore;
testScore = { 85, 92, 87, 67, 83, 91, 73, 98, 83, 84 };
```

Step 1-7: Comment out the statement int[] testScore = new int[10], and add the statement that creates the array using an array initializer instead of the key word new, as in Figure 7-5. Put a break at the closing brace of the event handler.

Figure 7-5: Use of the array initializer

```
19      private void btnDemo_Click(object sender, EventArgs e)
20      {
21          // Create array testScore with default values:
22          // int[] testScore = new int[10];
23
24          // Create array testScore using array initializer
25          int[] testScore = { 85, 92, 87, 67, 83, 91, 73, 98, 83, 84 };
```

Step 1-8: Run the project, and click the Demo button.

When the program breaks, display the Locals window (Debug, Windows, Locals). Expand testScore to display the values of the 10 elements of the array, as shown in Figure 7-6.

Figure 7-6: Locals window showing the effect of the array initializer

testScore	{int[10]}
[0]	85
[1]	92
[2]	87
[3]	67
[4]	83
[5]	91
[6]	73
[7]	98
[8]	83
[9]	84

Review Questions

7.1 How does a reference variable differ from a value type variable?

7.2 Consider the following code that computes and displays the total sales in a ListBox named lstDisplay:

```
int sales=2, totalSales = 10;
totalSales = totalSales + sales;
sales = 3;
lstDisplay.Items.Add(totalSales);
```

What would be the value of totalSales displayed in the ListBox?

7.3 Write a single statement to create an array that can hold a maximum of twenty student names referenced by a variable named student.

7.4 Write a statement to create a variable named student that can reference a string type array. Write a second statement to create an array that can hold a maximum of twenty student names referenced by student. 7.5 Write a single statement to create an array consisting of the abbreviated names of the seven days of a week, referenced by a variable named weekDay.

7.2 Assigning a Value to an Element of an Array

Array initializer is a simple way to assign values to the elements of an array when the array is created. However, in applications where data is read from a source, like a file or database, typically data is read from its source and assigned to the elements of an array using a loop, as discussed later.

To help understand this concept, first we look at how to assign a value to an individual element of an array by specifying the index of the element. As an example, the statement

```
testScore[2] = 99;
```

would assign the value 99 to the third element of the array testScore to replace the value 87.

You also may use a variable as the index, as in

```
int index = 2;
testScore[index] = 99; // you may use any valid variable name for the index.
```

This method of specifying the index using a variable is used within a loop to assign values to every element of an array.

Step 1-9: Add the code from Figure 7-7 to the Click event handler to change the third test score to 99, and the first test score to 89.

Figure 7-7: Changing the values of array elements

```
19      private void btnDemo_Click(object sender, EventArgs e)
20      {
21          // Create array testScore with default values:
22          // int[] testScore = new int[10];
23
24          // Create array testScore using array initializer
25          int[] testScore = { 85, 92, 87, 67, 83, 91, 73, 98, 83, 84 };
26
27          // Change the first test score to 89, and the third test Score to 99:
28          testScore[0] = 89;
29          testScore[2] = 99;
```

Step 1-10: Run the project, and Click the Demo button.

Step 1-11: When the program breaks, display the Locals window (Debug, Windows, Locals). Expand testScore to display the values of the 10 elements of the array. The updated values of the array are shown in Figure 7-8.

Figure 7-8: Updated testScore

Step 1-12: Add the necessary statement to change the last test score of the array to 79. Consider this your challenge—figure it out! Run the project, and verify that the score is changed to 79. Delete the added statement.

Chapter 8 shows you how to assign values to every element of an array by reading data items from a file using a loop.

Review Question

7.6   Write the statement to change the second element ("Mon") in the following array to "Monday":

```
string[] weekDay = {"Sun", "Mon", "Tue", "Wed", "Thu", "Fri", "Sat"};
```

7.3 Accessing Elements of an Array Using the Index

As discussed earlier, an element of an array is accessed by specifying the corresponding index. For example, the following code displays the third element of testScore in a ListBox named lstDisplay:

```
lstDisplay.Items.Add(testScore[2]);
```
Alternatively, you may use a variable to represent the index:

```
int index = 2;
lstDisplay.Items.Add(testScore[index]);
```

Step 1-13: Add the code from lines 31–33 in Figure 7-9 to display the third element of testScore.

Figure 7-9: Displaying an element of testScore

```
19      private void btnDemo_Click(object sender, EventArgs e)
20      {
21          // Create array testScore with default values:
22          // int[] testScore = new int[10];
23
24          // Create array testScore using array initializer
25          int[] testScore = { 85, 92, 87, 67, 83, 91, 73, 98, 83, 84 };
26
27          // Change the first test score to 89, and the third test Score to 99:
28          testScore[0] = 89;
29          testScore[2] = 99;
30
31          // Display 3rd element of the array
32          int index = 2;
33          lstDisplay.Items.Add(testScore[index]);
```

Step 1-14: Run the form to observe the output. You should see the score 87.

It's time to practice! Do Exercise 7-1.

Accessing Elements of an Array Using *for* Loops

Next, we will develop the code to display each element of the array, along with the value of its index, in the ListBox lstDisplay, as shown in Figure 7-10.

Figure 7-10: Using for Loop to display each element of an array

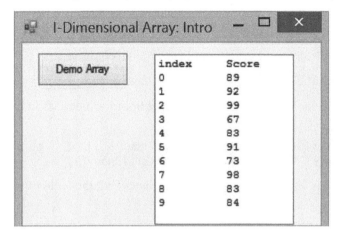

We use a **for loop** to change the value of the index of the element from 0 to 9. Note that the upper limit of the index (9) is equal to array size (10) minus 1.

Because an array is an object, it has different properties, including the **Length** property that gives the array size. Thus,

```
testScore.Length
```

gives the array size (10). Note that the highest value of the index (9) is one less than the value of Length property (testScore.Length - 1).

The following code shows the **for** loop that changes the index from 0 to the upper limit, (testScores.Length - 1), and increments the index by 1 in each iteration.

```
for (int index = 0; index <= testScore.Length-1; index = index + 1)
    lstDisplay.Items.Add(index + "        " + testScore[index]);
```

Thus, first time through the loop, the index has the value 0, and the ListBox will display the value of testScore[0], and increment the index by 1.

The second time through the loop, the index has the value 1, and the ListBox will display the value of testScore[1].

The process continues while the value of the index is less than or equal to the upper limit of (testScore.Length - 1), which is equal to 9.

Step 1-15: Delete lines 31–33 in Figure 7-9, which displays a single element. Add code that displays every element of the array, as shown in Figure 7-11.

Figure 7-11: Use of for loop to display every element of an array

```
31            // Display each element of the array using a for loop
32            lstDisplay.Items.Add("index" + "    " + "Score");
33            for (int index = 0; index <= testScore.Length - 1; index = index + 1)
34                lstDisplay.Items.Add(index + "        " + testScore[index]);
```

Step 1-16: Run the form after removing the break, and observe the output.

Step 1-17: Put a break at line 34, and observe the value of the index during each iteration, and the value of testScore.Length property.

Watch Out for Out of Range Indexes!

A common source of error when using arrays is using a value for index that is outside the valid range. The statement

```
lstDisplay.Items.Add(testScore[10]);      // invalid index
```

results in an error because 10 is outside the valid range of index for testScore, which is 0–9.

It is easy to make this error by mistakenly using the array size as the upper limit for index, as in the following statement:

```
for (int index = 0; index <= testScore.Length; index = index + 1)
    lstDisplay.Items.Add(index + "          " + testScore[index]);
```

In the above for loop, testScore.Length (10) is used as the upper limit for index, but the valid range of the index for testScore is 0–9.

Step 1-18: In line 33, change the upper limit for index to **testScore.Length** (from testScore.Length - 1). Put a break at line 34. Run the form, and iterate through the loop. Observe that when the value of index reaches 10, you get an error in line 34.

Step 1-19: Change line 33 back to the original form to specify (**testScore.Length - 1**) as the upper limit.

Using for Loop to Compute the Average Score

Let's modify the for loop that displays the scores, as shown in Figure 7-12, so that it computes the sum of all scores to compute the average.

Figure 7-12: Using for loop to compute the average

```
31          // Use a for loop to display each array element and compute the average score:
32          int totalScore = 0, currentScore;   //totalScore must be initialized to zero
33          lstDisplay.Items.Add("index" + "     " + "Score");
34          for (int index = 0; index <= testScore.Length - 1; index = index + 1)
35          {
36              currentScore = testScore[index];
37              lstDisplay.Items.Add(index + "          " + currentScore);
38              totalScore = totalScore + currentScore;
39          }
40          float averageScore = (float)totalScore / testScore.Length;
41          lstDisplay.Items.Add("Average:   " + averageScore.ToString("N"));
```

For each iteration through the loop, the first statement within the loop (line 36) stores the value of the current element in the variable currentScore so that it can be used in two different statements in lines 37 and 38 without having to access the array two times.

Step 1-20: Modify the Click event handler of the demo button so that it computes and displays the average score, as shown in Figure 7-12.

Step 1-21: Put a break on line 39, run the project, and observe the values of currentScore and totalScore in the first few iterations of the loop.

Review Questions

7.7 Write a statement to display the second element ("Mon") of the following array in a ListBox named lstDisplay:

```
string[] weekDay = {"Sun", "Mon", "Tue", "Wed", "Thu", "Fri", "Sat"};
```

7.8 Use a **for** loop to display all elements of the following array in a ListBox named lstDisplay:

```
string[] weekDay = {"Sun", "Mon", "Tue", "Wed", "Thu", "Fri", "Sat"};
```

7.9 Identify errors, if any, that would cause a runtime error in the second statement:

```
string[] weekDay = {"Sun", "Mon", "Tue", "Wed", "Thu", "Fri", "Sat"};
lstDisplay.Items.Add(testScore[7]);
```

7.4 Accessing Elements of an Array Using the foreach Loop

An alternative to using the for loop to access each element of an array is to use the foreach loop. In general, using the foreach loop is a convenient way to access individual items from any collection of items. This is how you would use a foreach loop to display each test score:

```
foreach (int score in testScore)
    lstDisplay.Items.Add(score);
```

In each iteration, the foreach loop would automatically copy the next element of the array into the **iteration variable** score, starting with the first element. Thus, the first time through the loop, score is assigned the first element (89) and it is displayed in the ListBox; the second time, score is assigned the second element (92); and the process continues until all scores are displayed.

Compare the foreach loop to the for loop introduced earlier:

```
for (int index = 0; index <= testScore.Length-1; index = index + 1)
    lstDisplay.Items.Add(index + "          " + testScore[index]);
```

Note that when using the **foreach** loop, there is no need to initialize a loop variable, increment it and specify the upper limit because it is not necessary to specify the index to access an element. Because no loop variable is used in the foreach loop, the output from the foreach loop, shown in Figure 7-13, does not include the value of the index.

Figure 7-13: Displaying each score using foreach loop

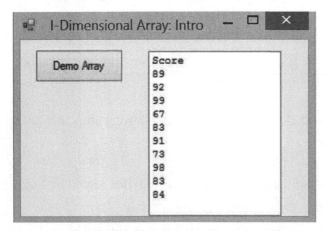

for versus foreach

Compared to the foreach loop, the for loop, in general, provides more flexibility in working with the elements. Within a for loop, you also may assign a value to an element of the array, and you can conveniently work with multiple arrays of the same size.

If the application requires accessing the elements of a single array sequentially without having to modify them, then it is mostly a matter of personal preference.

Now, let's modify the foreach loop, as shown in Figure 7-14, to compute the total score, and then compute the average.

Figure 7-14: Using foreach loop to compute the average

```
int totalScore = 0;
lstDisplay.Items.Add("Score");
foreach (int score in testScore)
{
        lstDisplay.Items.Add(score);
        totalScore = totalScore + score;
}
float averageScore = (float)totalScore / testScore.Length;
lstDisplay.Items.Add("Average:   " + averageScore.ToString("N"));
```

General syntax of foreach loop

    Foreach (Type VariableName in ArrayName)
        {
            Statements
        }

Step 1-22: Comment out the code that uses the for loop to compute the average (lines 31–41 in Figure 7-12) and add the code from Figure 7-14, which uses a foreach loop.

Step 1-23: Run the project and test the code.

Should Array Names Be Singular or Plural?

Should the name of an array be singular like testScore or plural like testScores? Because an array holds multiple data items, the plural form might make more intuitive sense. When you use a foreach loop to access each element of the array, as in the following code, it is convenient to use the plural form (testScores) for the array and the singular form (testScore) for the variable representing an individual score:

```
foreach (int testScore in testScores)
    lstDisplay.Items.Add(testScore);
```

However, when you refer to an individual score using the index, the singular form of the noun makes more sense. For example, testScore[1] might be a clearer representation of a single score than testScores[1]. In addition, the singular form of the noun makes more sense in certain advanced uses of arrays. In this book, we use both forms.

Review Question

7.10 Use a **foreach** loop to display all elements of the following array, to a ListBox named lstDisplay:

```
string[] weekDay = {"Sun", "Mon", "Tue", "Wed", "Thu", "Fri", "Sat"};
```

7.5 Copying an Array

Because arrays are reference types, you cannot create a separate copy of an array by assigning the reference variable to another reference variable. For example,
```
int[] temp1 = testScore;
```
will not create a copy of the array testScore. Instead, both temp and testScore will refer to the same array, as shown.

testScore   temp1

☐ → | 89 | 92 | 99 | 67 | 83 | 91 | 73 | 98 | 83 | 84 | ← ☐

That is, there is only one array, but two variables refer to it. That means if you change an element of testScore, in effect, you also are changing that element for temp1.

Step: 7-24: Add the following code to the end of the Click event handler of Demo button, and insert breaks at line 56 and 57:

```
53
54          // Assign testScore to another reference variable - doesn't copy the array
55          int[] temp1 = testScore;
56          testScore[9] = 69;
57      }
```

Step 1-25: Run the project. When the program breaks at line 56 (before testScore[9] is changed to 69), expand testScore and temp in Locals window (Debug, Windows, Locals) and observe that temp has the same values as testScore.

Step 1-26: Click Run to continue execution. When the program breaks at line 57, expand testScore and temp in Locals window (Debug, Windows, Locals) and observe that temp[9] has same values as testScore[9], as shown in Figure 7-15, though it looks like you changed the value of only testScore.

Figure 7-15: Changed value of temp1

temp1	{int[10]}
[0]	89
[1]	92
[2]	99
[3]	67
[4]	83
[5]	91
[6]	73
[7]	98
[8]	83
[9]	69

You may change any element of the array using temp1 or testScore. For example, temp1[8] would change the ninth element of the array that also is referenced by testScore.

Copying an Array Using Array.Copy Method

To create a separate copy of an array, you may use the **Array.Copy** method. The general syntax is

Array.Copy(source array, starting index in the source array, target array, starting index in target array, number of elements to copy)

Here is how you would copy the array referenced by testScore to another array referenced by temp2:

```
int[] temp2 = new int[10];
Array.Copy(testScore, 0, temp2, 0, 10);
```

If the entire array is to be copied, you may use the Length property in place of 10:

```
int[] temp2 = new int[testScore.Length];
Array.Copy(testScore, 0, temp2, 0, testScore.Length);
```

You may copy the last four elements of testScore (starting at index 6) to a new array, temp3, as follows:

```
int[] temp3 = new int[4];
Array.Copy(testScore, 6, temp3, 0, 4);
```

Step 1-27: Add the code from Figure 7-16 to copy the entire array testScore to another array temp2, and to copy the last four elements of testScore to temp3.

Figure 7-16: Using Array.Copy to copy an array

```
57
58      //Copy the entire array testScore using Array.Copy method to temp2:
59      int[] temp2 = new int[10];
60      Array.Copy(testScore, 0, temp2, 0, 10);
61      // copy the last four elements of testScore to temp3:
62      int[] temp3 = new int[4];
63      Array.Copy(testScore, 6, temp3, 0, 4);
64
```

Step 1-28: Run the project. When the program breaks at end of the method (line 64), display the Locals window and observe temp2 that should look like testScore, and temp3 that should display the values shown in Figure 7-17.

Figure 7-17: Using Array.Copy to copy a subset of the elements of an array

temp3	{int[4]}
[0]	73
[1]	98
[2]	83
[3]	69

Because the Array.Copy method makes a copy of the array referenced by testScore and references the copy by temp2, as shown in Figure 7-18, changes in one array do not affect the other array.

Figure 7-18: temp2 created by copying testScore

testScore

→ | 89 | 92 | 99 | 67 | 83 | 91 | 73 | 98 | 83 | 69 |

temp2

→ | 89 | 92 | 99 | 67 | 83 | 91 | 73 | 98 | 83 | 69 |

For example, executing the statement

```
testScore[9] = 99
```

changes the last element of testScore, but it does not affect temp2 or temp3.

However, temp1 is changed because temp1 was assigned the reference to testScore using the following statement; thus, both temp1 and testScore point to the same array:

```
int[] temp1 = testScore;
```

Step 1-29: Add the following code to change the last element of testScore to 99.

```
64
65              testScore[9] = 99;
66          }
```

Step 1-30: Run the project. When the program breaks at line 66, display Locals window and observe that the last element of testScore and temp1 is 99, but the last element of temp2 and temp3 is unchanged.

Copying an Array Using a for Loop

You also may use a for loop to copy the elements of an array to another array. Here is the code that copies the elements of testScore to an array named temp4:

```
int[] temp4 = new int[10];   // or, int[] temp4 = new int[testScore.Length]
for (int index = 0; index <= testScore.Length - 1; index = index + 1)
    temp4[index] = testScore[index];
```

Here, the index is varied from 0 to the upper limit to access each element. So, the statement within the loop is equivalent to 10 different statements:

```
temp4[0] = testScore[0];
temp4[1] = testScore[1];
temp4[2] = testScore[2];
etc.
```

Step 1-31: Add the following statements to copy the elements of testScores to temp4:

```
66
67          // copy testScore to another array using for loop:
68          int[] temp4 = new int[10];   // or, int[] temp4 = new int[testScore.Length]
69          for (int index = 0; index <= testScore.Length - 1; index = index + 1)
70              temp4[index] = testScore[index];
71      }
```

Step 1-32: Run the project. When the program breaks, verify that temp4 has the same values as testScore.

Again, because testScore and temp4 are two separate arrays, the two arrays can be changed independently without affecting each other.

We will discuss one-dimensional arrays further, and introduce two-dimensional arrays, in the next chapter that discusses accessing data from files and storing in arrays.

Review Questions

7.11   The general syntax of the Array.Copy method is

   Array.Copy(source array, starting index in the source array, target array,
         starting index in target array, number of elements to copy)

   Write a statement to copy all days, except the weekends (Sat, Sun), from the following array to another array named workDay.

   ```
   string[] weekDay = {"Sun", "Mon", "Tue", "Wed", "Thu", "Fri", "Sat"};
   ```

7.12   Consider the following code that uses a ListBox named lstDisplay to display an element of an array:

   ```
   string[] weekDay = {"Sun", "Mon", "Tue", "Wed", "Thu", "Fri", "Sat"};
   string[] DaysOfWeek = weekDay;
   weekDay[1] = "Monday";
   lstDisplay.Items.Add(DaysOfWeek[1]);
   ```

   What would be the value displayed in the ListBox?

7.13 Consider the following code that uses a ListBox named lstDisplay to display an element of an array:

```
string[] weekDay = {"Sun", "Mon", "Tue", "Wed", "Thu", "Fri", "Sat"};
string[] DaysOfWeek = new string[7];
Array.Copy(weekDay, 0, DaysOfWeek, 0, 7);
weekDay[1] = "Monday";
lstDisplay.Items.Add(DaysOfWeek [1]);
```

What would be the value displayed in the ListBox?

7.6 Looking Up Values in an Array

One-dimensional arrays can be used conveniently to lookup values, like finding the price for a specified item or looking up the phone number of a person.

Consider a string array of names of months and another int array that stores the units sold for a product in each month, created as shown in Figure 7-19. It should be noted that in real-world applications, such data often are read from a database table or file, rather than hard-coded.

Figure 7-19: Code to create arrays, months and unitsSold

```
string[] months = { "January", "February", "March", "April", "May", "June",
    "July", "August", "September", "October", "November", "December" };

int[] unitsSold = { 572, 645, 693, 564, 580, 756, 960, 756, 663, 678, 745, 960};
```

To find the units sold in a specified month, first find the index of the month from the months array. Then, find the data item stored in unitSold array at that index position. This method works only if the **months and units sold are stored in the same sequence** (that is, the first element of unitsSold corresponds to the first month in months array, the second element in unitsSold corresponds to second element in months array, etc.).

For example, to find the units sold in March, you find the index of March, which is 2, and then find the data item at the index position 2 in unitsSold array, which is 693. An easy way to find the index of a specific data item in an array is to use the Array.IndexOf method.

Array.IndexOf Method

The Array.IndexOf method returns the index of a specified data item in a 1-D array.

This method has the syntax

```
Array.IndexOf(ArrayName, SearchValue)
```

as in this example:

```
Array.IndexOf(months, "March").
```

ArrayName is the name of a 1-D array, and SearchValue is the value you are trying to find in an array. The method returns the index of the first element that matches the specified value. If no value is found, the method returns -1. Because months and units sold are stored in the same sequence, you can use the index of the month to find the unitsSold.

Tutorial 2: Lookup Units Sold Using 1-D Arrays

This tutorial creates a form named LookUpUsingArray that allows the user to look up the sales in a specified month, as shown in Figure 7-20.

Figure 7-20: LookUpUsingArray form

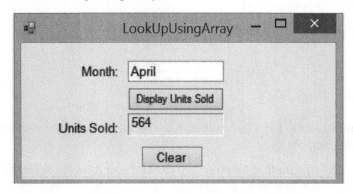

Step 2-1: Open the form named LookupUsingArray within the project Ch7_Arrays.

Open the code window. The two statements in lines 19–21 create two arrays, months and unitsSold at the class level, as shown in Figure 7-21. These arrays are created at the class level, rather than within the Click event handler of btnDisplayUnitsSold, so that they don't have to be created repeatedly each time the user clicks the button to find the sales.

Step 2-2: Add the code shown in lines 22–37 in Figure 7-21 to the Click event handler of btnDisplayUnitsSold to display units sold in a specified month.

Figure 7-21: Code to find the units sold

```
19    string[] months = { "January", "February", "March", "April", "May", "June", "July",
20                        "August", "September", "October", "November", "December" };
21    int[] unitsSold = { 572, 645, 693, 564, 580, 756, 859, 756, 663, 678, 745, 958 };
22
23    private void btnDisplayUnitsSold_Click(object sender, EventArgs e)
24    {
25        string monthName = txtMonthName.Text;
26        int monthIndex, unitsSoldInMonth;
27
28        // Find the index of the month. Use the index to find units sold
29        monthIndex = Array.IndexOf(months, monthName);
30        if (monthIndex >= 0)   // if month is not found, index would be -1
31            unitsSoldInMonth = unitsSold[monthIndex];   // find units sold
32        else
33        {
34            MessageBox.Show("Month not found");
35            return;
36        }
37        lblUnitsSold.Text = unitsSoldInMonth.ToString();
```

Line 29 finds the index of the specified month from the months array using Array.IndexOf method, described earlier.

Line 31 finds the units sold from unitsSold array, using the index of the specified month.

Step 2-3: Run the form and test the code.

Looking Up Values in an Array Using a Loop

An alternative to using the Array.IndexOf method to find the index of a data item is to use a loop to compare each element of the array to the data item whose index is to be found. For example, to find the index of "March" in the months array, compare each month in the months array to "March" until a match is found or all elements are compared. Figure 7-22 shows the code that uses a for loop.

Figure 7-22: Look up sales using a for loop

```
38
39    // Alternate method using a loop:
40    // Compare each month in the array to the specified month
41    int index = 0;
42    for (index = 0; index < months.Length; index++)
43    {
44        if (months[index] == monthName)
45        {
46            unitsSoldInMonth = unitsSold[index];
47            lblUnitsSold.Text = unitsSoldInMonth.ToString();
48            break; // if month is found jump out of the loop
49        }
50    }
51    // Diplay a message, if month was not found
52    if (index >= months.Length)
53    {
54        MessageBox.Show("Month not found");
55        return;
56    }
```

Lines 42 and 44: The for loop in line 42 varies the value of index from 0 to 11 in increments of 1, so that the statement in line 44,

    `if (months[index] == monthName)`

would be equivalent to 12 different statements:

    `if (months[0] == monthName)`
    `if (months[1] == monthName)`
    ...
    `if (months[11] == monthName).`

Line 48: If the specified month matches a month in the array, the *break* statement in line 48 transfers control out of the loop.

Line 52: If the specified month doesn't match any month in the array, the value of index would be equal to the size of the array (12), given by the Length property. Line 52 checks whether index is greater than or equal to the array size.

Step 2-4: Comment out the statements in lines 29-37, and add the code from lines 38-55 in Figure 7-22. Run the form and test the code.

It's time to practice! Do Exercises 7-2 and 7-3.

Exercises

Exercise 7-1

Open the project Ch7_Arrays from the folder Tutorial_Starts. Open the form named Exercise 7_1. Add the necessary code to the Click event handler of the btnDisplayMonth to display the name of the month for the month number typed in by the user. For example, if the user types in the number 3, the code should display "March," as follows:

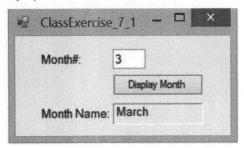

Hint: Create a string array with the twelve month names (January through December) assigned to it as the initial values. To avoid creating the array each time the button is clicked, create the array at the class level outside the Click event handler. Use the array within the Click event handler to find and display the month name.

Exercise 7-2

Part A: This exercise finds and displays the month(s) with the highest sale. Add a button named btnTopMonths to the LookUpUsingArray form that you used in Tutorial 2 . Add the necessary code to the Click event handler of the button to find and display the highest sale, and the month(s) with the highest sales, in a ListBox. You may use the following method:

First, find the highest sale by looping through the unitSold array, as follows:

> Set highest sale to the first element of unitsSold (int highestSale = unitsSold[0];)
> Set highestSalePosition to zero.
> Use a loop (that initializes and increments the index) to compare each array element to the highest sale. If the array element is greater than highest sale, update highest sale and highestSalePosition, using the following logic:
>
>> if (array element > highestSale)
>> set highestSale to the current element of unitsSold
>> set highestSalePosition to the current value *of index*

Next, display the highestSale and the month at the highestSalePosition.

Part B: This exercise is similar to Part A, except that it finds and displays the month(s) with the lowest sale. Add a button named btnBottomMonths to the LookUpUsingArray form that you used in Tutorial 2. Add the necessary code to the Click event handler of the button to find and display the lowest sale, and the month(s) with the lowest sales, in a ListBox. You may use the following method:

First, find the lowest sale by looping through the unitSold array, as follows:
    Set lowest sale to the first element of unitsSold (int lowestSale = unitsSold[0];)
    Set lowestSalePosition to zero.
    Use a loop (that initializes and increments the index) to compare each array element to the lowest sale. If the array element is less than lowest sale, update lowest sale and lowestSalePosition, using the following logic:

        if (array element < lowestSale)
        set lowestSale to the current element of unitsSold
        set lowestSalePosition to the current value of index

Next, display the lowestSale and the month at the lowestSalePosition.

Exercise 7-3

The TheaterTicketsWithArray form in Tutorial_Starts/Ch7_Arrays project performs essentially the same function as TheaterTickets form from Ch6_GUIcontrols. Rather than using a button to compute the cost, TheaterTicketsWithArray form combines computing cost and displaying cost into the SelectedIndexChanged event handler of the ComboBox. In addition, this form doesn't use a separate method to compute the discount.

Modify this form so that the amount due includes a markup for the selected show. When the user selects a show from the ComboBox, add the markup for that show to the amount due and display the amount. You may use the following approach:

Add the necessary statement to the form to create a second array named markups that contains the markup amount (a positive or negative amount) for each show.

Add the code to the SelectedIndexChanged event handler of the ComboBox to find the markup amount for the selected show, and add the amount to the amount due. Display the amount in the ListBox. Make sure that the reference variable for the array, shows, is declared at the class level so that it can be used in the Load event of the form and in the SelectedIndexChanged event handler of the ComboBox.

Chapter 8

Sequential Files and Arrays

Data stored in variables and arrays are lost when the program terminates. **Sequential Access Files** provide a relatively easy way to store data for longer terms in devices like hard disks, magnetic tapes and flash drives. This chapter discusses writing data into sequential files and reading from files. Further, we look at reading data from files into one-dimensional and two-dimensional arrays.

Learning Objectives

After studying this chapter, you should be able to:

- Describe sequential access, direct access, text and binary files.
- Read data from Text files into arrays.
- Write data to Text files.
- Use the SaveFile Dialog control to let users select a file.
- Pass data to a method using an array.
- Create a two-dimensional (2-D) array.
- Access and lookup data from a 2-D array.

Topics

8.1	Introduction to Text Files	8.5	Writing to Text Files
8.2	Splitting a Row: Split Method	8.6	SaveFileDialog Control
8.3	Reading Data from Files into Arrays	8.7	Passing Arrays to Methods
8.4	Additional Methods of Arrays	8.8	Two-Dimensional (2-D) Arrays

8.1 Introduction to Text Files

In a sequential access file, data are accessed sequentially, starting with the first set of data items, followed by successive sets. To find a particular data item, you need to go over all data that precedes it, like playing a song on a cassette tape. By contrast, data may be accessed directly from a **direct access** (**random access**) file, similar to playing a song on a CD.

Sequential files are of two types: **text files** and **binary files**. In a text file, all data, including numbers and dates, are stored as text. For example, the number 9.95 is stored as a combination of four characters (three digits and a decimal) that forms a string. Hence, text files can be created and edited using a text editor like Notepad. Binary files, discussed in **Appendix A**, store data in its original form without converting it to text. The number 9.95 is stored as single number using the standard internal representation of numbers.

Reading Lines from Text Files

Typically, text files store each record in a separate line, with the data items separated by delimiters like commas, tabs, or the pipe character (|). Because text files are sequential access files, data are read sequentially from the beginning to the end of the file.

Example 1: Reading and Displaying Exam Scores

Here is an example of a text file that contains the name and two exam scores of students:
>Mike Burns, 75, 92
>Cathy Arnold, 86, 95
>Wesley Mathews, 82, 90
>Jerry Luke, 70, 89
>Gina Rodriguez, 77, 88

You can create such a file in a text editor or using a program. In either case, an invisible new line character is inserted automatically at the end of each line.

Let's see how you read the above data from a text file, named **Exams.txt**. You use the .txt extension, which is used commonly to indicate that the file contains text data.

A variety of classes are available in the **System.IO** namespace of Framework Class Library to work with files and directories. You use the **StreamReader** class to read data from files. The data in the text file is viewed as a stream of data—hence the name StreamReader.

To read from a text file, you create a StreamReader object using the StreamReader class. The statement to create the StreamReader object has the following syntax:
```
StreamReader VariableName = new StreamReader(@"Path and FileName");
```

The new key word creates the StreamReader object to read the specified file, and returns a reference to the object. The reference (address) of the object is stored in the variable specified on the left side.

Specifying the Path for a File

If no path is specified, the default path, **ProjectDirectory\Bin\DeBug**, is assumed. You do not need to use the prefix @ if no path is specified. If the file is not found, an exception is thrown.

For example, a StreamReader object to read from E:\Csharp\Exams.txt is created as follows:
```
StreamReader examsReader = new StreamReader(@"E:\Csharp\Exams.txt");
```
If the file is in the default directory, Bin\Debug, you skip the path:
```
StreamReader examsReader = new StreamReader("Exams.txt");
```

If the file is stored in a folder above the default folder (Bin\Debug), you may use the code "..\\" for each level. Here are examples:
>"..\\Exams.txt"—the file Exams.txt is in the Bin folder that is one level above the default folder, Debug.9.5

>"..\\..\\Exams.txt"—the file is in the project folder that is two levels above the default folder, Debug.

Using Directive for System.IO

Because StreamReader class is in the System.IO namespace, you must add the following using directive to the form:
```
using System.IO;
```

ReadLine method

The StreamReader provides a method called **ReadLine** to read the line at the current read position and then advance the position to the next line. This method is used within a loop to sequentially read each line of data. This is how you would use a loop to read each line and display the line in a ListBox named lstDisplay:

```
string currentLine;
while (examsReader.EndOfStream == false)  // or, !examsReader.EndOfStream
{
        // Read current line and store it in a string variable
        currentLine = examsReader.ReadLine();
        lstDisplay.Items.Add(currentLine); // Display the line
}
examsReader.Close();
```

EndOfStream property

The EndOfStream property of StreamReader has the value true when the current read position is at the end of the file. The header of the while loop in the above code,
```
while (examsReader.EndOfStream == false),
```
uses the **EndOfStream** property to terminate the loop when EndOfStream is true. That is, the loop continues while EndOfStream is false.

The statement
```
currentLine = examsReader.ReadLine();
```
reads the current line and stores the entire line as a single string in the string variable, currentLine. At the end of the first iteration through the loop, the string variable currentLine would have the following value:
"Mike Burns, 75, 92"

Thus, the loop reads each line, stores it in the string variable currentLine, and then displays it to give the output shown in Figure 8-1.

Figure 8-1: Data from the Exams.txt file

```
Mike Burns, 75,   92
Cathy Arnold, 86, 95
Wesley Mathews, 82, 90
Jerry Luke, 70, 89
Gina Rodriguez, 77, 88
```

Tutorial 1: Reading Text Files

This tutorial reads and works with the name and exam scores from the file Exams.txt.

Step 1-1: Open the project Ch8_FilesAndArrays from the Projects_Starts folder. Open the form ReadTextFiles. Figure 8.2 shows the form.

Figure 8-2: The ReadTextFiles form

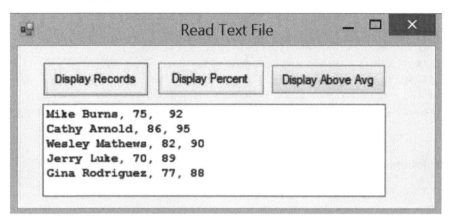

Step 1-2: Open the file Exams.txt from the Bin\Debug folder in **Ch8_FilesAndArrays** project directory, and view the records. Close the file.

Step 1-3: Make sure that the System.IO namespace is added, as shown in Figure 8-3.

Figure 8-3: The using directives with System.IO namespace

```
1  using System;
2  using System.Collections.Generic;
3  using System.ComponentModel;
4  using System.Data;
5  using System.Drawing;
6  using System.Linq;
7  using System.Text;
8  using System.Threading.Tasks;
9  using System.Windows.Forms;
10 using System.IO;    //Add this statement
11 namespace Ch8_FilesAndArrays
```

Step 1-4: Add the code from Figure 8-4 to the Click event handler of btnDisplayRecords. Please note that, in line 32, the name of the ListBox, lstDisplay, starts with the letter "l," not the number 1, though the font makes it look like number 1.

Figure 8-4: Code to read and display each line from Exams.txt

```csharp
20      private void btnDisplayRecords_Click(object sender, EventArgs e)
21      {
22          string currentLine;
23          // Create StreamReader object to read Exams.txt
24          StreamReader examsReader = new StreamReader("Exams.txt");
25
26          lstDisplay.Items.Clear();
27          while (examsReader.EndOfStream == false)
28                  // or, while (!examsReader.EndOfStream)
29          {
30              // Read the line at the current read position
31              currentLine = examsReader.ReadLine();
32              lstDisplay.Items.Add(currentLine); // Display current line
33          }
34          examsReader.Close();
35      }
```

Step 1-5: Make sure the Program.cs file specifies ReadTextFile as the startup form.

Put a break at line 33, and run the project. Click the Display Records button.

When the program breaks at the end of the loop (line 33), display the Locals window (Debug, Windows, Locals), and observe the value of currentLine for each iteration. Observe the value of EndOfStream property of examsReader after the fourth iteration (false), and after the fifth iteration, when it has the value, true, as shown in Figure 8-5.

Figure 8-5: The EndOfStream Property

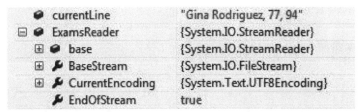

Review Questions

8.1   What namespace contains the classes that work with text files?

8.2   Describe what happens when the ReadLine method of a StreamReader is executed.

8.2 Splitting a Row: Split Method

The ReadLine method reads an entire row as a single string. How do you get the name and the two scores from the string so that you can compute the percent score for each student and display it, as shown in Figure 8-6?

Figure 8-6: The revised ReadTextFiles form with percent scores

```
Read Text File
Display Records    Display Percent    Display Above Avg
Name              Exam1    Exam2    Percent
Mike Burns         75       92       83.5
Cathy Arnold       86       95       90.5
Wesley Mathews     82       90       86.0
Jerry Luke         70       89       79.5
Gina Rodriguez     77       88       82.5
```

To get individual data items from a line, you use the **Split** method of the StreamReader that splits a string and **returns a string array** containing individual data items. This method uses a specified delimiter character (a character that separates the different data items in a row) to split the string. The syntax is

> VariableName = StringVariableName.Split('delimiter character');

For example,

```
string [] fields = currentLine.Split(',');
```

would split the string, currentLine, to individual data items using the delimiter comma, and store the string array in the array reference variable named fields.

Note that if the delimiter is tab, it is represented by '/t,' as in currentLine.Split('/t').

The Split method requires the delimiter character to be of Char type, not string. Hence, Split(",") is invalid because "," is a string, but Split(',') is valid because ',' is a character. Note you may convert the string "," to a character, as in currentLine.Split(Char.Parse(","));.

Let's modify the code presented earlier in Figure 8-4 to compute the percent score for each student. To do this, split each line from Exams.txt, and average the two scores to get the percent. Figure 8-7 shows a modified version of the code presented earlier, within the Click event handler of a second button, btnComputePercent.

Figure 8-7: Code to get individual data items

```csharp
37      private void btnComputPercent_Click(object sender, EventArgs e)
38      {
39          float percentScore;
40          string[] fields;
41          string fmtStr, currentLine;
42          // Create  StreamReader object to read Exams.txt
43          StreamReader examsReader = new StreamReader("Exams.txt");
44
45          lstDisplay.Items.Clear();          //Clear ListBox
46          // Print heading in the listbox
47          fmtStr = "{0,-15}{1,8}{2,8}{3,8:N1}";
48          lstDisplay.Items.Add(string.Format(fmtStr, "Name", "Exam1", "Exam2", "Percent"));
49
50          while (examsReader.EndOfStream == false)
51          {
52              currentLine = examsReader.ReadLine();
53              // Split currentLine to three string data items and store in a string array.
54              fields = currentLine.Split(',');
55
56              percentScore = (float.Parse(fields[1]) + float.Parse(fields[2]))/2;
57
58              //display the three data items and student total in the List  box
59              lstDisplay.Items.Add(string.Format(fmtStr, fields[0], fields[1], fields[2],
60                                                        percentScore));
61          }
62          examsReader.Close();
63      }
```

Let's look at some key statements:

Line 52: `currentLine = examsReader.ReadLine();`

As in the previous code, this line reads the current line, and stores the entire line as a single string in the string variable, currentLine. At the end of the first iteration through the loop, the string variable currentLine would have the following value:

Line 54: `fields = currentLine.Split(',')`

This is a key statement. Because the Split method returns a string array, the variable fields is declared as the reference to a string array, in line 40. At the end of the first iteration, the array fields would look like this:

fields	{string[3]}
[0]	"Mike Burns"
[1]	" 75"
[2]	" 92"

Line 56: `percentScore = (float.Parse(fields[1]) + float.Parse(fields[2]))/2;`

The two scores from the elements fields[1] and fields[2] are converted to float type and averaged to compute the percent. The percent is stored in the float variable percentScore.

Line 59: `lstDisplay.Items.Add(string.Format(fmtStr, fields[0], fields[1], fields[2], percentScore));`

The element, fields[0], gives the name; fields[1] and fields[2] give score1 and score2, respectively. The three items are displayed in the ListBox using the format string, fmtStr, defined in line 47 as "{0,-15}{1,8}{2,8}{3,8:N1}." The name is left justified using the "-" sign in the format code {0,-15}, and the percent is displayed with one decimal digit using the code N1 in {3,8:N1}.

Step 1-6: Open the form ReadTextFiles. Open the Click event handler of btnComputePercent. Add the missing code from lines 51–61 in Figure 8-7.
Put a break at the end of the loop on line 61.

Make sure the Font type for the ListBox is a fixed width type, like Courier New. Otherwise, the columns may not align evenly with properly formatted codes.

Step 1-7: Run the project. Click the Compute Percent button. When the program breaks, view the values of currentLine, fields array and percentScore. View the data for the first few iterations.

Watch Out for Hidden Blank Lines at the End of Text Files!

If you accidentally hit the Enter key to insert a blank line at the end of the text file, you will get an error at line 56 (percentScore = float.Parse(fields[1]) + float.Parse(fields[2]))/2;). This is because when a blank line is split, the resulting array does not have the elements like fields[1] and fields[2].

See It for Yourself!

Step 1-8: Insert a blank line at the end of Exams.txt, as shown below, and save the file. Now run the project and click ComputePercent button. You will get the error message that "index was outside the bounds of the array." Reopen Exams.txt, delete the blank line and save it.

```
Exams.TXT - Notepad
File  Edit  Format  View  Help
Mike Burns,   75,   92
Cathy Arnold, 86,   95
Wesley Mathews, 82, 90
Jerry Luke,   70,   89
Gina Rodriguez, 77, 88
|
```

It's time to practice! Do Exercise 8-1.

Review Question

8.3 The following data on customer orders (Customer#, Order#, Order Qty, Unit Price) is stored in a text file named Orders.txt.

0155, 76533, 3, 10.95
0140, 75446, 12, 8.50
0150, 75550, 1, 89.95
0145, 75648, 25, 2.50

Consider the following code that reads the data from the above file.

```
StreamReader OrdersReader = new StreamReader("Orders.txt");
string[] fields;
while (OrdersReader.EndOfStream == false)
{
    currentLine = OrdersReader.ReadLine();
    fields = currentLine.Split(',');
}
lstDisplay.Items.Add(fields[1]);
```

What would be the output displayed in the ListBox lstDisplay?

8.3 Reading Data from Files into Arrays

Reading data from files on secondary storage devices like hard disks and flash drives is relatively slow compared to accessing data from arrays and variables, which are stored in the internal memory. So, if data that is stored in a file is to be accessed multiple times in a program, it generally would be more efficient to read the data from the file only once, store it in an array (or arrays), and access the array(s) for subsequent use of the data. For very large data sets, the consumption of internal memory by arrays should be weighed against the improvement in access speed.

A typical application that uses arrays is looking up information multiple times, like the phone number or address for different customers. Another application is computing the average for a field and then identifying all records that have above average values.

Let's consider an expanded version of the ReadTextFiles form to display a list of all students whose percent score is above the class average as shown in Figure 8-8.

Figure 8-8: The ReadTextFiles form displaying above average students

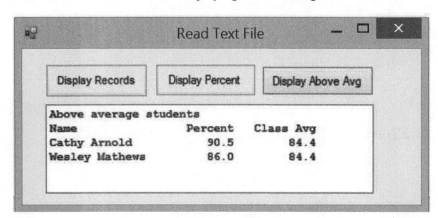

To find above average students, first you need to read all student records and compute the class average. Then, you need to access the same records again to identify who got scores above the class average. Rather than reading the file twice, read the file once, compute the class average, and store student names in a one-dimensional array and the percent scores in another array. Then, use these arrays to identify those who got above average scores.

What array size should you use to create these arrays? The array size limits the number of data items that can be stored in an array. If a program needs to read data from a file with an increasing number of records over time, then the array size should be as large as the maximum number of records the file is likely to have. Similarly, if the program needs to read from files of different sizes, then the array size should be as large as the size of the largest file. For ease of viewing the arrays, this example limits the sizes of the two arrays to 10, as specified in the following statements that create the arrays.

```
string[] names          = new string[10];
float[]  percentScores  = new float[10];
```

Storing data in the arrays involves the following process:

Read the **first** record from Exams.txt and store the name in the first element of the names array, and store the percent score in the first element of the percentScores array, as shown in Figure 8-9.

Read the **second** record from Exams.txt and store the name in the second element of the names array, and store the percent score in the second element of the percentScores array.

This process is repeated for each record in the file.

Figure 8-9: The names and percentScores arrays

names	{string[10]}	percentScores	{float[10]}
[0]	"Mike Burns"	[0]	83.5
[1]	"Cathy Arnold"	[1]	90.5
[2]	"Wesley Mathews"	[2]	86
[3]	"Jerry Luke"	[3]	79.5
[4]	"Gina Rodriguez"	[4]	82.5
[5]	null	[5]	0
[6]	null	[6]	0
[7]	null	[7]	0
[8]	null	[8]	0
[9]	null	[9]	0

Figure 8-10 shows the first part of the code that reads and stores the data in the two arrays and computes the class average.

Figure 8-10: Code to compute class average using arrays

```
65    private void btnAboveAvg_Click(object sender, EventArgs e)
66    {
67        //Declare two 1-D arrays
68        string[] names = new string[10];
69        float[] percentScores = new float[10];
70        // percentScores is the array of all percent scores
71        int recordCount, index = 0;
72
73        string currentLine;
74        string[] fields;
75        float percentScore, classTotal = 0, classAvg;
76        //percentScore is the percent score for a student;
77        StreamReader examsReader = new StreamReader("Exams.txt");
78
79        while (examsReader.EndOfStream == false)
80        {
81            currentLine = examsReader.ReadLine(); // read current line
82            fields = currentLine.Split(','); // split current line
83            percentScore = (float.Parse(fields[1]) + float.Parse(fields[2]))/2;
84
85            names[index] = fields[0]; //Store name into current element of names array
86            percentScores[index] = percentScore;   // Store percent into percentScores
87            index = index + 1;
88            classTotal = classTotal + percentScore; // Add current percent to classTotal
89        }
90        recordCount = index;
91        classAvg = classTotal / (float) recordCount;
```

Let us examine the key statements of the code.

Lines 81-83

```
81        currentLine = examsReader.ReadLine(); // read current line
82        fields = currentLine.Split(','); // split current line
83        percentScore = (float.Parse(fields[1]) + float.Parse(fields[2]))/2;
```

This is identical to the code you used earlier in Figure 8-7 to compute and display the percent scores. This code reads each record, splits the record to three string type data items that are stored in fields array, and computes the percent score. The fields array, as discussed earlier, looks as follows after the first iteration through the loop:

fields	{string[3]}
[0]	"Mike Burns"
[1]	"75"
[2]	" 92"

Line 85: `names[index] = fields[0]; //Store name into current element of names array`

To access each element of the names array, this statement uses the variable index, which is initialized to zero (line 71) and incremented by one within the loop (line 87).

The first time through the loop, names[index] refers to the first element, names[0], which is assigned the value of fields[0] that always represents the name from the current record. The statement in line 85 would be equivalent to

```
names[0] = fields[0];
```

So, after the first iteration, the names array would look as follows:

names	{string[10]}
[0]	"Mike Burns"
[1]	null
[2]	null
[3]	null
[4]	null
[5]	null
[6]	null
[7]	null
[8]	null
[9]	null

The second time through the loop, index is 1. So line 85 is equivalent to

```
names[1] = fields[0];
```

So, the name from the second record is stored in the second element of names, as shown below:

names	{string[10]}
[0]	"Mike Burns"
[1]	"Cathy Arnold"
[2]	null

The third, fourth and fifth times, index becomes 2, 3 and 4, respectively; so line 85 would be equivalent to

```
names[2] = fields[0];
names[3] = fields[0];
names[4] = fields[0];
```

After all records are read during the first five iterations of the loop, the array will have names stored in the first five elements as follows:

names	{string[10]}
[0]	"Mike Burns"
[1]	"Cathy Arnold"
[2]	"Wesley Mathews"
[3]	"Jerry Luke"
[4]	"Gina Rodriguez"
[5]	null
[6]	null
[7]	null
[8]	null
[9]	null

Line 86: `percentScores[index] = percentScore;//Store percent into percentScores`

This statement is similar to the previous one except that it stores the current student's percent score in the current element of percentScores array. Because the value of index increments by one each time through the loop, line 86 would be equivalent to

percentScores[0] = percentScore;
percentScores[1] = percentScore;
percentScores[2] = percentScore;
etc.

After the loop is completed, percentScores will have all five percents stored in the array as follows:

percentScores	{float[10]}
[0]	83.5
[1]	90.5
[2]	86
[3]	79.5
[4]	82.5
[5]	0
[6]	0
[7]	0
[8]	0
[9]	0

Line 88: `classTotal = classTotal + percentScore;`

This line calculates a running total of the percent scores for all student records that were read. The variable classTotal that accumulates the percent scores must be initialized (see line 75).

Line 90-91

```
90              recordCount = index;
91              classAvg = classTotal / (float) recordCount;
```

Because index is reinitialized to zero in the next part (not shown in Figure 8-10), its current value (that gives the number of records read) is stored in another variable, recordCount.

Step 1-9: Open the Click event handler of btnAboveAvg. Add the missing code from lines 85–91 in Figure 8-10.

Step 1-10: Put a break at the end of the loop on line 89, and run the project.
Click ComputeAvg button, and observe the values of **names, percentScores** and **fields** arrays each time through the loop.

Identifying Above Average Students

To identify students whose percent score is above the class average, the class average is compared to each element of percentScore array, using a loop. If an element is above class average, the corresponding element from the names array is displayed. For example if percentScore[2] is above average, names[2] is displayed.

This method works only if the two arrays are loaded consistently so that there is a one-to-one correspondence between the elements of names and those of percentScores (i.e., percentScores[0] is the percent for names[0], percent[1] represents the percent for names[1], etc).
Lines 93–106 in Figure 8-11 show the second part of the code, which displays the names and scores of students whose percent is above the class average. Lines 20–92 compute the class average, which is the same as Figure 8-10, discussed earlier.

Figure 8-11: Code to compute average and display above average students

```csharp
65      private void btnAboveAvg_Click(object sender, EventArgs e)
66      {
67          //Declare two 1-D arrays
68          string[] names = new string[10];
69          float[] percentScores = new float[10];
70          // percentScores is the array of all percent scores
71          int recordCount, index = 0;
72
73          string currentLine;
74          string[] fields;
75          float percentScore, classTotal = 0, classAvg;
76          //percentScore is the percent score for a student;
77          StreamReader examsReader = new StreamReader("Exams.txt");
78
79          while (examsReader.EndOfStream == false)
80          {
81              currentLine = examsReader.ReadLine(); // read current line
82              fields = currentLine.Split(','); // split current line
83              percentScore = (float.Parse(fields[1]) + float.Parse(fields[2]))/2;
84
85              names[index] = fields[0]; //Store name into current element of names array
86              percentScores[index] = percentScore;   // Store percent into percentScores
87              index = index + 1;
88              classTotal = classTotal + percentScore; // Add current percent to classTotal
89          }
90          recordCount = index;
91          classAvg = classTotal / (float) recordCount;
92
93          string fmtStr = "{0,-15}{1,12:N1}{2,12:N1}";
94          lstDisplay.Items.Clear();
95          lstDisplay.Items.Add("Above average students");
96          lstDisplay.Items.Add(string.Format(fmtStr, "Name", "Percent", "Class Avg"));
97
98          //Find Above average students
99          for (index = 0; index < recordCount; index++)
100         {
101             if (percentScores[index] > classAvg)
102                 lstDisplay.Items.Add(string.Format(fmtStr, names[index],
103                                     percentScores[index], classAvg));
104         }
105         examReader.Close();
106     }
```

Line 101: `if (percentScores[index] > classAvg)`

Because the loop variable assumes values 0 to 4, line 101 becomes equivalent to
```
if (percentScores[0] > classAvg)
if (percentScores[1] > classAvg)
if (percentScores[2] > classAvg)
etc.
```

Thus, the statement `if (percentScores[index] > classAvg)` compares each element of percentScores to classAvg.

Line 102: `lstDisplay.Items.Add(string.Format(fmtStr, names[index], percentScores[…]…);`

If any element of percentScores is greater than classAvg, then line 102 displays the name from the corresponding element of names array using the same index value.

Step 1-11: Add the missing code from lines 99–104 in Figure 8-11 to the Click event handler of btnAboveAvg to display above average students. Put breaks at lines 102 and 104.

Step 1-12: Run the project. When the program breaks, observe the values of classAvg and names and percentScore arrays.

It's time to practice! Do Exercise 8-2

ReadBlock Method of StreamReader

Instead of reading an entire line using the ReadLine method, you may read a specified number of characters, starting with a specified position in the current line, into a character array. For example, the following statement reads 10 characters starting with index 0 into a character array named studentArray:

examsReader.ReadBlock(studentArray, 0, 10);

ReadToEnd Method of StreamReader

The ReadToEnd method of StreamReader reads the entire file, starting with the current position, and returns a string. The following statement would read the entire file Exams.txt and display it in the ListBox, lstDisplay:

lstDisplay.Items.Add(examsReader.ReadToEnd);

Review Questions

8.4 The following is a sample of the order data (Customer#, Order#, Order Qty, Unit Price) stored in a text file named Orders.txt.

0155, 76533, 3, 10.95
0140, 75446, 12, 8.50
0150, 75550, 1, 89.95
0145, 75648, 25, 2.50

The following code is intended to read data from the above file and compute and display the extended cost (= Order Qty x Unit Price) for each order. The statement that computes the extended cost is incomplete.

```
StreamReader OrdersReader = new StreamReader("Orders.txt");
string[] fields;
float extCost;
while (OrdersReader.EndOfStream == false)
{
        currentLine = OrdersReader.ReadLine();
        fields = currentLine.Split(',');
        // insert code to compute extCost (= Order Qty x Unit Price)
        extCost =
        lstDisplay.Items.Add(extCost);
}
```

Insert the necessary code to compute the extCost.

8.5 The following is a sample of the order data (Customer#, Order#, Order Qty, Unit Price) stored in a text file named Orders.txt.

0155, 76533, 3, 10.95
0140, 75446, 12, 8.50
0150, 75550, 1, 89.95
0145, 75648, 25, 2.50

The following code is intended to read data from the above file, compute the extended cost (= Order Qty x Unit Price) for each order, and store the Order# and extended cost in two different arrays (orderNumbers and extCosts). The statements to store the Order# and extended cost into arrays are missing.

```
StreamReader OrdersReader = new StreamReader("Orders.txt");
string[] fields;
float extCost;
int index = 0;
string[] orderNumbers = new string[10];
float[]extCosts = new float[10];
while (OrdersReader.EndOfStream == false)
{
        currentLine = OrdersReader.ReadLine();
        fields = currentLine.Split(',');
        extCost = float.Parse(fields[2])*float.Parse(fields[3]);
        // insert code to store Order# into the array, orderNumbers
        ...
        // insert code to store extended cost into the array, extCosts
        ...
        index = index + 1;
}
```

Insert the missing code to store the Order# into the array, orderNumbers, and to store extended cost into the array, extCosts.

8.4 Additional Methods of Arrays

Array.Resize Method

When data are read from large files whose size may grow over time, storing data in arrays poses challenges in specifying the size of the array. An option would be to specify an initial size that is large enough to hold the current data but resize the array automatically whenever the number of records to be stored in the array exceeds the array size.

The Array.Resize() method has the following syntax:
    Array.Resize (ref *ArrayName*, *NewArraySize*)
where ref is a key word, *ArrayName* is the name of the array (must be a one-dimensional array) and *NewArraySize* is the new size of the array.

The Resize method creates a new array with the specified size, copies the data from the old array to the new array (if the new size is smaller, only a subset of the data is copied) and replaces the old array with the new.

For example, you would expand the size of the array, names, from its current size of 10 to a new size of 15, as follows:
　　　Array.Resize(ref names, 15);
The first 10 elements of the new array will have the data from the 10 elements of the old array. The remaining 5 elements are empty.

The statement Array.Resize (ref names, 6) copies the first six elements from the old array to the resized array. The rest of the data is discarded.

Tutorial 2: Looking Up Phone Numbers Using Arrays

This tutorial uses the **Resize** method in a new form called **LargeFiles** that looks up the telephone numbers of vendors from a large set of vendor names and phone numbers stored in two one-dimensional arrays. This form also uses the **IndexOf** method. The data are read into arrays from a file named **Vendor.txt** that contains the following seven fields:
　　　Vendor#, name, street, city, state, zip, phone number.

Here is a sample of data that are currently stored in the Vendor.txt file:

```
1,FCC,445 12th Street SW, Washington,DC,20554,888-225-5322
2,U.S.Copyright Office,101 Independence Ave, Washington,DC,20559,202-707-3000
3,CDC,1600 Clifton Road,Atlanta,GA,30333,800-232-4636
4,White House, 1600 Pennsylvania Avenue NW, Washington,DC,20500,202-456-1111
5,Labor Statistics,Postal Sq Bldg,2 Mass Ave,Washington,DC,20212,800-877-8339
6,AMTRAK,60 Massachusetts Avenue NE, Washington,DC,20002,800-523-6590
7,U.S. Air Force,1690 Air Force Pentagon,Washington,DC,20330-1670,800-423-8723
8,National Guard,111 South George Mason Dr,Arlington,VA,22204,800-864-6264
9,Office of Govt Ethics,1201 New York Ave,Washington,DC,20005,202-482-9300
10,IRS,1111 Constitution Ave NW, Washington, DC,20224,800-829-1040
```

A major feature of the code is that the size of the array is increased dynamically whenever the number of records read from the file equals the size of the array. Thus, the code is intended to read from files that contain a very large number of records and is expected to grow in size. The actual file used here (Vendor.txt), however, is very small to make it easy to track execution of the code and to test the loop.

The first part of the code that reads the seven data items from each line, and adds the names and phone numbers into two arrays, is shown in Figure 8-12.

Figure 8-12: The LargeFiles form with Array.Resize method

```csharp
20      string[] names, phoneNumbers; //Declare arrays.
21      int recordCount = 0;
22
23      private void LargeFiles_Load(object sender, EventArgs e)
24      {
25          string currentLine;
26          int arrayInitialSize = 10000, arrayIncrement = 5000;
27          string[]fields = new string[7]; // Declare an array named fields
28          StreamReader vendorReader = new StreamReader("Vendor.txt");
29
30          names = new string[arrayInitialSize];
31          phoneNumbers = new string[arrayInitialSize];
32
33          // Read each line, split it, and store name & phone number into two arrays.
34          while (vendorReader.EndOfStream == false)
35          {
36              // If the names array is full, increase the size of arrays.
37              if (recordCount == names.Length)
38              {
39                  Array.Resize(ref names, names.Length + arrayIncrement);
40                  Array.Resize(ref phoneNumbers, phoneNumbers.Length + arrayIncrement);
41              }
42              currentLine = vendorReader.ReadLine();
43              fields = currentLine.Split(','); // split the line
44
45              names[recordCount] = fields[1]; //Store name into the array, names
46              phoneNumbers[recordCount] = fields[6]; //Store phone number in phoneNumbers
47              recordCount = recordCount + 1;
48          }
49          vendorReader.Close();
50      }
```

The process of reading each line, splitting it into different data items and storing the seven data items in a temporary array (lines 42 and 43) is similar to the process we used earlier in the ReadTextFiles form. After a record is stored in the temporary array, fields, you store the vendor name (second element; index 1) and phone number (seventh element; index 6) into the arrays, names and phone numbers (lines 45 and 46).

Let's look at some key parts of the code:

Lines 20 and 21: The arrays, names and phoneNumbers, are declared at the class level so that they can be created and loaded in the Load event of the form and used in another method to find the phone number for a specified name. Similarly, the variable recordCount also is declared at the class level so that it can be used in two different event procedures.

Lines 30 and 31: The arrays are created with an initial size that is specified in line 26.

Lines 37–41: If the number of records loaded into arrays is equal to the size of the arrays, then the arrays are resized by adding a specified increment to the array size. This is a major difference from previous examples.

Step 2-1: Make sure that the file Vendor.txt is in the debug folder of your project folder, Ch8_FilesAndArrays. To make it easy to test the code, this file contains only 10 records.

Step 2-2: Open the form, LargeFiles, shown in Figure 8-13.

Figure 8-13: The LargeFiles form

Step 2-3: Open the load event handler of the form. Fill in the missing code in lines 39 and 40 from Figure 8-12.
To make it easy to test the loop, change ArrayInitialSize to 5 (from 10,000) and ArrayIncrement to 5 (from 5,000) in line 26.

Step 2-4: Make sure that the Program.cs file specifies LargeFiles as the startup form.
Put a break at line 48, and run the form.

Go through the loop five times, and observe the values of recordCount and Names.Length by hovering the cursor over them. The length of the names array remains at 5, which is the initial value, and recordCount increments by one each time.

Click Continue, and observe that Names.Length is increased to 10.

Array.IndexOf Method

The Array.IndexOf method, introduced in Chapter 7, returns the index of a specified data item in a one-dimensional array.

This method has the syntax,
    Array.IndexOf(*ArrayName, SearchValue*).
ArrayName is the name of a one-dimensional array, and *SearchValue* is the value you are trying to find in an array. The method returns the index of the first element that matches the specified value. If no value is found, the method returns -1.

For example,
    `Array.IndexOf(names, "Mike Burns")`

returns the index of the first element of the names array that matches the specified value, Mike Burns.

Figure 8-14 shows the code that uses this method to find a vendor's phone number by searching for a specified name.

Figure 8-14: Code to find the phone number using IndexOf method

```
53      private void btnFindPhone_Click(object sender, EventArgs e)
54      {
55          string searchName = txtVendorName.Text;
56          string fmtStr = "{0,-15}{1,8}";
57          lstDisplay.Items.Clear();
58
59          // Method 1: Use IndexOf method to find the record
60          int index = Array.IndexOf(names, searchName);
61                  //if there is no match, index would be -1.
62          if (index >= 0)
63              lstDisplay.Items.Add(string.Format(fmtStr, searchName,
64                                      phoneNumbers[index]));
65          else
66              MessageBox.Show("Sorry, name not found");
```

Line 60 uses the IndexOf method to get the index of the element in the Names array that matches the value of SearchName.

Lines 62–64 display the corresponding element from phoneNumbers array if the index is not -1. If not, a message is displayed.

Figure 8-15 shows an alternate method that uses a loop to compare the value of SearchName with each element in the Names array.

Figure 8-15: Using a loop to search a field

```
67
68          // Method 2: Use for loop to find the record
69          for (index = 0; index < recordCount; index++)
70          {
71              if (searchName == names[index])
72              {
73                  lstDisplay.Items.Add(String.Format(fmtStr, names[index],
74                                      phoneNumbers[index]));
75                  return;
76              }
77          }
78          MessageBox.Show("Sorry, there is no Vendor by that name");
79      }
```

Step 2-5: Add the code from Figure 8-15.
  Test the code by typing in the value AMTRAK, which is one of the vendor names in the Vendor.txt file.

Tutorial 3: Using a ComboBox to Select the Search Name

This tutorial creates a different version of the application where the vendor names are added to the ComboBox, and when the user selects a name, the phone number is automatically displayed in the ListBox on the form named SmallFiles, as shown in Figure 8-16. The use of ComboBox to select a vendor is appropriate only if the number of vendors is relatively small. It is time consuming to load a large data set into a ComboBox.

Figure 8-16: The SmallFiles Form with ComboBox to Select Name

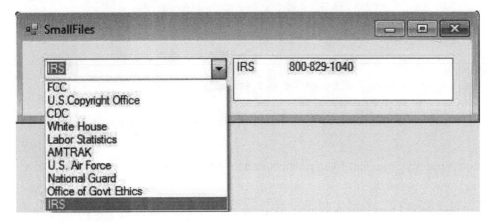

In this form called **SmallFiles**, you learn how to add items to a ComboBox from a file, and see an application of the SelectedIndexChanged event.

Step 3-1: Open the form, SmallFiles.

Two major aspects of the revised code are

1. The load event handler of the form, which reads data from Vendor.txt into two arrays, also adds vendor names to the ComboBox.
2. The SelectedIndexChanged event of the ComboBox finds the phone number for the selected name, whenever the user selects a vendor.

The complete code is shown in Figure 8-17. Because the use of ComboBox to select a vendor is appropriate only if the number of vendors is relatively small, the code does not include resizing of arrays. You use the same vendor file, **Vendor.txt**.

Figure 8-17: Code to add names to a ComboBox and find phone numbers

```csharp
19          string[] names, phoneNumbers;   //Declare arrays
20          int recordCount = 0;
21
22          private void SmallFiles_Load(object sender, EventArgs e)
23          {
24              string currentLine;
25              string[] fields = new string[7];    // Declare array named fields
26              StreamReader vendorReader = new StreamReader("Vendor.txt");
27
28              names = new string[10];
29              phoneNumbers = new string[10];
30              // Read each line, split & store name and phone number into two arrays.
31              while (vendorReader.EndOfStream == false)
32              {
33                  currentLine = vendorReader.ReadLine();
34                  fields = currentLine.Split(',');
35
36                  names[recordCount] = fields[1]; //Store name into the array, names
37                  phoneNumbers[recordCount] = fields[6];//Store phone# into phoneNumbers
38
39                  cboVendorNames.Items.Add(fields[1]);    // Add names to the ComboBox
40                  recordCount = recordCount + 1;
41              }
42              vendorReader.Close();
43          }
44
45          private void cboVendorNames_SelectedIndexChanged(object sender, EventArgs e)
46          {
47              // Get selected vendor name from the ComboBox
48              string searchName = cboVendorNames.SelectedItem.ToString();
49              string fmtStr = "{0,-15}{1,8}";
50              lstDisplay.Items.Clear();
51
52              // Use IndexOf method to find the record
53              int index = Array.IndexOf(names, searchName);
54                      //if there is no match, index would be -1.
55              if (index >= 0)
56                  lstDisplay.Items.Add(string.Format(fmtStr, searchName,
57                                              phoneNumbers[index]));
58              else
59                  MessageBox.Show("Sorry, name not found");
60          }
```

The statements related to the use of the ComboBox are as follows.

Line 39: cboVendorNames.Items.Add(fields[1])

This statement adds the vendor name to the ComboBox when the form is loaded.

Lines 45-60

These show the SelectedIndexChanged event handler of the ComboBox.

Line 48: `string searchName = cboVendorNames.SelectedItem.ToString()`

The SelectedItem is of object type. Therefore, it is converted to string before assigning it to the string variable searchName.

Step 3-2: Open the Load event handler of the form. Fill in the missing code in lines 36-39 from Figure 8-17.
Open the SelectedIndexChangedEvent of the ComboBox. Fill in the missing code in lines 53-59 from Figure 8-17.
Run the form and test it.

Review Questions

8.6   The following code creates an array named shows with the following initial values:
```
string[] shows = {"Mamma Mia","Rock of Ages","Les Miserables",
        "Motown The Musical", "The Lion King", "The Phantom of the Opera"};
```
Write the necessary code to add the names of the shows from the array to a ComboBox named cboShows. You may use a foreach loop to access each element of the array.

8.7   The following code creates two arrays named months and sales. The sales array shows the units sold for the corresponding month.
```
string[] months = { "Jan", "Feb", "Mar", "Apr", "May", "Jun", "Jul", "Aug", "Sep", "Oct", "Nov", "Dec" };
int[] unitsSold = { 572, 645, 693, 564, 580, 756, 960, 756, 663, 678, 745, 960};
```
Write the necessary code to find the units sold for a month name entered into a TextBox named txtMonthName, and display the units sold in a ListBox named lstDisplay.

8.5 Writing to Text Files

You can write data to a text file using the **StreamWriter** class available in the System.IO namespace of Framework Class Library.

Tutorial 4: Writing Scores to Text File

Let's develop a form that writes data from Text Boxes into a text file. Figure 8-18 shows the form named WriteToTextFile.

Figure 8-18: The WriteToTextFile form

The StreamWriter Object

The StreamWriter class can be used to write data into a text file. The StreamWriter object is created using the following syntax:
```
StreamWriter VariableName = new StreamWriter(@"Path and FileName", append flag);
```

If **append flag** is true, data are appended to existing data, if any. If append flag is false, existing data, if any, are overwritten. In both cases, if the specified file does not exist, a new file is created.

If no path is specified, it is assumed to be Bin\Debug folder of the project directory. The "@" prefix is not needed when the path is not specified. For example, to write data to Exams.txt with the append option, the StreamWriter object is created as follows:
```
StreamWriter examsWriter = new StreamWriter("Exams.txt", true);
```

The WriteLine Method of StreamWriter

The WriteLine method of StreamWriter writes a string as a separate line into a file. Use this method to write all three data items for each student into a separate line in Exams.txt file.

Declaring Variables at the Class Level

Because data are to be written multiple times (each time when the user clicks the Save button) to Exams.txt, you create the StreamWriter once when the form is loaded, use it each time the Save button is clicked, and close the StreamWriter when the form is closed. That means the variable (ExmasWriter) that references StreamWriter is used in three different methods:
- the **Load event** of the form,
- the Click event of the Save button, and
- the **FormClosing event** of the form.

To make the variable examsWriter available in all three methods, you need to declare examsWriter at the class level (outside the three methods), and create the StreamWriter object in the **Load event** handler, as shown in the code in Figure 8-19.

Figure 8-19: Code to declare and create StreamWriter object

```
14    public partial class WriteToTextFile : Form
15    {
16        public WriteToTextFile()
17        {
18            InitializeComponent();
19        }
20        // Declare variable ExamsWriter at class level:
21        StreamWriter examsWriter;
22        private void WriteToTextFiles_Load(object sender, EventArgs e)
23        {
24            ExamsSaveDialog.ShowDialog(); // Display Save As Dialog box
25            // Get the file name selected by the user, and
26            // save it in the variable named fileName
27            string fileName = ExamsSaveDialog.FileName;
```

The Load Event of Forms

The load event of forms is triggered whenever you open a form. You create the Load event handler by double clicking the form in design view.

Step 4-1: Open the form, WriteToTextFiles, shown in Figure 8-18.

Step 4-2: Add the using directive to include System.IO namespace in the list:

```
1  using System;
2  using System.Collections.Generic;
3  using System.ComponentModel;
4  using System.Data;
5  using System.Drawing;
6  using System.Linq;
7  using System.Text;
8  using System.Threading.Tasks;
9  using System.Windows.Forms;
10 using System.IO;              // Add this statement
11
12 namespace Ch8_FilesAndArrays
```

Step 4-3: Open the code window. Fill in the code in line 21 from Figure 8-19 to declare the variable examsWriter, at the class level.

Step 4-4: Fill in the code in lines 25-26 from Figure 9-19 to create the object examsWriter. Put a break on line 25.

Step 4-5: Run the form.
The program breaks at line 25, which creates the examsWriter object. The form is not visible at this time because the code is part of the Load event handler.
Click Run to complete loading the form.
Remove the break.

Because examsWriter is created in the Load event, it is available when the form is open, and it is ready to be used for writing within the Click event handler of the Save button, shown in Figure 8-20.

Figure 8-20: Code to write data to Exams.txt

```
private void btnSave_Click(object sender, EventArgs e)
{
    string name, score1, score2;
    // Store all data in string variables.
    name = txtName.Text;
    score1 = txtScore1.Text;
    score2 = txtScore2.Text;
    // Combine name & socres and store in Exams.txt
    examsWriter.WriteLine(name + "," + score1 + "," + score2);
    MessageBox.Show("Data Saved");
}
```

Note that the above code writes the data using the StreamWriter every time the user clicks the Save button without having to create the StreamWriter and close it each time.

To write the data, the name, score1 and score2 are combined into a single string,
(name + "," + score1 + "," + score2),

with commas between each data item. The commas are inserted to make it easy to split the string if the data are read back in a program.

Step 4-6: Open the Click event handler of Save button. Fill in the missing code from Figure 8-20 to write data to Exams.txt file.

Step 4-7: Run the form. Enter data for name, score1 and score2, and click Save. Click OK when you get the message "Data Saved."

Step 4-8: View the Bin\Debug folder of your project directory. Open the Exams.txt file that was created earlier in the form ReadTextFile. You will not find the record that you entered and was supposed to have been written to it. What happened?

The data are written to the external device (hard disk, flash drive, etc.) only when the StreamWriter is closed. Until then, data written by StreamWriter are held in the buffer temporary memory. You have not added the code to close the Stream Writer.

FormClosing Event of Forms

The StringWriter object should be available until the form is closed. So, you should close StreamWriter only when the form is closed. This can be done by closing StreamWriter within the **FormClosing** event handler of the form, which is triggered whenever the form is closed. Figure 8-21 shows the FormClosing event handler.

Figure 8-21: FormClosing event handler

```
private void WriteToTextFiles_FormClosing(object sender,FormClosingEventArgs e)
{
    if (examsWriter != null) // Check whether ExamsWriter exits
        examsWriter.Close();
}
```

Because FormClosing is not the default event handler of forms, double clicking the form won't create this event handler. You need to select the events button in the Properties window to display form events, as shown in Figure 8-22. Then, double click the event (FormClosing) to create the event handler.

Figure 8-22: Events of the form

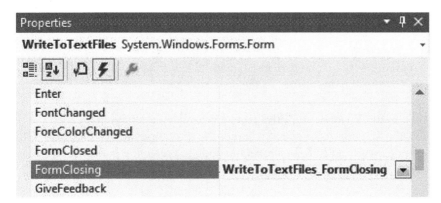

Step 4-9: Open the code window and create the FormClosing event handler.

Step 4-10: Fill in the code from Figure 8-21 to close examsWriter.

Step 4-11: Add the following code to the Click event handler of Exit button to close the form.
```
this.Close();   // "this" represents the current form
```

Step 4-12: Open Exams.txt. Make sure that the insertion point is on a new line after Gina Rodriguez's record. Close Exams.txt.
Run the project. Enter two sets of data (make up your own), and save each set.
Close the form.
Open Exams.txt. The file should have the two sets of data you entered, in addition to the five records that it already had.

Append versus Overwrite

The two sets of data that you entered were added to data that already existed in the file because the append flag was set to true in the following statement that created the StreamWriter:

```
25            ExamsWriter = new StreamWriter("Exams.txt", true);
```

If you set the append flag to false, all existing data will be overwritten by new data.

Step 4-13: In line 25, change the append flag to false as follows:

```
25            ExamsWriter = new StreamWriter("Exams.txt", false);
```

Add a third set of data and save it. Close the form.
Open Exams.txt and view it. You should see only the third set of data, because the previous records were overwritten.
Now, change the append flag back to true.

It's time to practice! Do Exercise 8-3

Directory Class

The Directory class in System.IO namespace provides several methods to work with directories (folders), including methods to create folders and to check whether a specified path exists. Figure 8-23 shows two commonly used methods of Directory class.

Figure 8-23: Methods of Directory class

Method	Function
Exists(path)	Returns true if the path exists and false if the path does not exist
CreateDirectory(path)	Creates the directories in the specified path

You may create Exams.txt file in the WriteToTextFiles form by specifying a path other than the one for the default directory. Use the above methods to check whether the directories in the specified path exist, and to create them if they do not exist. Figure 8-24 shows the code to create the path E:\Csharp\output if it does not exist, and to use that path to write the data.

Figure 8-24: Create a path using Directory class

```
string path = @"E:\Csharp\output\";
if (!(Directory.Exists(path)))
        Directory.CreateDirectory(path);
examsWriter = new StreamWriter(path + "Exams.txt", true);
```

If the drive E: is invalid, you will get an error. If the directory, Csharp, does not exist, it is created, and the directory, output, is created within Csharp. If Csharp folder exists but output folder does not exist, it will be created.

The last statement,
```
examsWriter = new StreamWriter(path + "Exams.txt", true);
```
uses the newly created path to write the data.

Step 4-14: Temporarily, replace the statement in line 25 by the code in Figure 8-24 to create a directory and use it to write the data. Run the project.
Enter a new record and save it. Verify that a new directory Csharp\output is created, and the file Exams.txt is created within it.
Remove the code that you inserted, and change line 25 back to its original form with no path specified.

Review Questions

8.8   The following form, discussed earlier, lets the user save multiple sets of data into a text file by clicking the Save button after entering each set of data.

The code shown below writes the name and the two scores from TextBoxes into the file Exams.txt.

```
StreamWriter examsWriter;
private void WriteToTextFiles_Load(object sender, EventArgs e)
{
        examsWriter = new StreamWriter("Exams.txt", true);
}
private void btnSave_Click(object sender, EventArgs e)
{
        string name, score1, score2;
        name = txtName.Text;
        score1 = txtScore1.Text;
        score2 = txtScore2.Text;
        examsWriter.WriteLine(name + "," + score1 + "," + score2);
}
private void WriteToTextFiles_FormClosing(…FormClosingEventArgs e)
{
        examsWriter.Close();
}
```

Tutorial 4: Writing Scores to Text File

The streamWriter named examsWriter is declared at the class level, created in the Load event of the form, used in the Click event of the button multiple times and closed within the FormClosing event. What would be the effect of moving the declaration statement (**StreamWriter examsWriter;**) to the beginning of the Load event handler of the form, as follows?

```
// StreamWriter examsWriter;
private void WriteToTextFiles_Load(object sender, EventArgs e)
    {
    StreamWriter examsWriter;
    examsWriter = new StreamWriter("Exams.txt", true);
}
```

8.9 Refer to the form described in Review Question 8.8. Consider the following two changes to the code:
  1. Move the declaration statement (StreamWriter examsWriter;),to btnSave_Click event handler, as shown below.
  2. Move the statement that creates examsWriter from the Load event handler to btnSave_Click event handler, as shown below.

```
// StreamWriter examsWriter;
private void WriteToTextFiles_Load(object sender, EventArgs e)
{
    // examsWriter = new StreamWriter("Exams.txt", true);
}

private void btnSave_Click(object sender, EventArgs e)
{
    StreamWriter examsWriter;
    examsWriter = new StreamWriter("Exams.txt", true);

    string name, score1, score2;
    name = txtName.Text;
    score1 = txtScore1.Text;
    score2 = txtScore2.Text;
    examsWriter.WriteLine(name + "," + score1 + "," + score2);
}

private void WriteToTextFiles_FormClosing(object sender, FormClosingEventArgs e)
{
    examsWriter.Close();
}
```

What would be the combined effect of the above two changes?

8.10 Refer to the form described in Review Question 8.8. Consider the following three changes as shown in the code displayed on the next page:
  1. Move the declaration statement (StreamWriter examsWriter;) to btnSave_Click event handler.
  2. Move the statement that creates examsWriter from the Load event handler to btnSave_Click event handler.
  3. Move the statement, examsWriter.Close(), to the end of btnSave_Click event handler.

```
    // StreamWriter examsWriter;
    private void WriteToTextFiles_Load(object sender, EventArgs e)
    {
        // examsWriter = new StreamWriter("Exams.txt", true);
    }


    private void btnSave_Click(object sender, EventArgs e)
    {
        StreamWriter examsWriter;
        examsWriter = new StreamWriter("Exams.txt", true);

        string name, score1, score2;
        name = txtName.Text;
        score1 = txtScore1.Text;
        score2 = txtScore2.Text;
        examsWriter.WriteLine(name + "," + score1 + "," + score2);

        examsWriter.Close();
    }
    private void WriteToTextFiles_FormClosing(object sender, FormClosingEventArgs e)
    {
        // examsWriter.Close();
    }
```

What would be the combined effect of the above three changes?

8.6 SaveFileDialog Control

The **SaveFileDialog** control allows users to select the path and file to save the data, rather than hardcoding the path/file as you did in previous examples.

To add a SaveFileDialog control to your project, double click *SaveFileDialog* under *Dialogs* group in the *Toolbox* window. The control is added to the tray at the bottom of the Designer window.

ShowDialog() method

This method displays a *Save As* dialog box to let users select the path and file.
This is how you would code to let the user specify the file:
    ExamsSaveDialog.ShowDialog();

FileName property

The FileName property stores the path and name of the file selected by the user. If the user selects Cancel from the Save As dialog box, the value of FileName property would be an empty string.

The following code gets the path and name of the file selected by the user from the FileName property and stores it in the variable filename:
```
    string fileName = ExamsSaveDialog.FileName;
```

The following code uses the selected file to create the StreamWriter object. The if statement checks whether the length of the file name is greater than zero—that is, if the user didn't cancel out:

```
if (fileName.Length > 0)
    examsWriter = new StreamWriter(fileName, true);
else
    MessageBox.Show("No file selected");
```

Step 4-15: Add *SaveFileDialog* control from the Toolbox to the WriteToTextFiles form. Select the control, and change its name to ExamFileDialog.

Step 4-16: To let the user select a file using the SaveFileDialog, change the Load event handler of the form, as shown in Figure 8-25.

Figure 8-25: Using SaveFileDialog to let the user select a file

```
private void WriteToTextFiles_Load(object sender, EventArgs e)
{
    ExamsSaveDialog.ShowDialog(); // Display Save As Dialog box
    // Get the file name selected by the user, and
    // save it in the variable named fileName
    string fileName = ExamsSaveDialog.FileName;

    //Create StreamWriter object
    if (fileName.Length > 0)     // if the user selected a file
        ExamsWriter = new StreamWriter(fileName, true);
    //true means append; false means overwrite
    else
    {
        MessageBox.Show("No file selected");
        btnSave.Enabled = false;
    }
}
```

Step 4-17: Run the project. Select Exams.txt file from the Debug folder, and click Save.
When asked whether you want to replace the file, select **Yes**. (The term replace is misleading. You will not lose the existing contents of the file because the append flag is set to true.)
When the program breaks, observe the value of fileName variable that includes the path and file name. Click Continue.
Enter a new record, and Save it.
Open Exams.txt to verify that the file contains all previous records and the new record that you added.

If the user doesn't select a file, and clicks *Cancel* instead, the value of the fileName variable will be an empty string. So, line 31 will be skipped and the ExamWriter object won't be created. This will result in an error in the FormClosing event handler, shown below, when the form is closed.

```
private void WriteToTextFiles_FormClosing(object sender, FormClosingEventArgs e)
{
    ExamsWriter.Close();
}
```

Step 4-18: Run the project. Intead of selecting a file, click Cancel in the *Save As* window.
Clikc OK when the "No file selected" message is displayed.
The program breaks at line 38. Verify that the fileName variable shows an empty string.
Click the *Continue* button on the Toolbar.
Click the *Exit* button on the form.
The program breaks in the FormClosing event handler with a NullReferenceException error, as shown below, because ExamWriter object was not created in the Load event handler of the form.

```
private void WriteToTextFiles_FormClosing(object sender, FormClosingEventArgs e)
{
    //if (examsWriter != null)  // check whether examsWriter exists
        examsWriter.Close();
}
```
⚠ **NullReferenceException was unhandled**

To correct this problem, modify the FormClosing event handler by adding an if statement that checks whether examsWriter exits, as follows:

```
private void WriteToTextFiles_FormClosing(object sender,FormClosingEventArgs e)
{
    if (examsWriter != null)    // check wether examsWriter exists
        examsWriter.Close();
}
```

Run the project again, and verify that closing the form doesn't give an error when the user cancels out of the Save As window.

InitialDirectory Property

The InitialDirectory property allows you to specify the directory initially displayed in the *Save As* dialog box. If not specified, the Document directory or the last used directory, if any, is displayed. Here is how you would specify the initial directory:

```
ExamsSaveDialog.InitialDirectory = @"E:\CSharp";
```

The Title Property

The title property lets you specify a title for the Save As dialog box. Here is an example:

```
ExamsSaveDialog.Title = "Exams Save As dialog box";
```

Step 4-19: Insert the necessary statements at the beginning of the Load event handler to specify the initial directory and title. Run the project and observe the effect.

The following is the complete code to write data into a text file selected using the SaveFileDialog control:

8.6 SaveFileDialog Control

```csharp
using System.IO;    // Add this statement

namespace Ch8_FilesAndArrays
{
    public partial class WriteToTextFile : Form
    {
        public WriteToTextFile()
        {
            InitializeComponent();
        }
        // Declare variable ExamsWriter at class level:
        StreamWriter ExamsWriter;
        private void WriteToTextFiles_Load(object sender, EventArgs e)
        {
            ExamsSaveDialog.ShowDialog(); // Display Save As Dialog box
            // Get the file name selected by the user, and
            // save it in the variable named fileName
            string fileName = ExamsSaveDialog.FileName;

            //Create StreamWriter object
            if (fileName.Length > 0)    // if the user selected a file
                ExamsWriter = new StreamWriter("Exams.txt", false);
            //true means append; false means overwrite
            else
            {
                MessageBox.Show("No file selected");
                btnSave.Enabled = false;
            }
        }

        private void btnSave_Click(object sender, EventArgs e)
        {
            string name, score1, score2;
            // Store all data in string variables.
            name = txtName.Text;
            score1 = txtScore1.Text;
            score2 = txtScore2.Text;
            // Combine name & socres and store in Exams.txt
            ExamsWriter.WriteLine(name + "," + score1 + "," + score2);
            MessageBox.Show("Data Saved");
        }

        private void WriteToTextFiles_FormClosing(object sender,
                                        FormClosingEventArgs e)
        {
            if (ExamsWriter != null) // Check whether ExamsWriter exits
                ExamsWriter.Close();
        }
```

```
58
59          private void btnExit_Click(object sender, EventArgs e)
60          {
61              this.Close();   // "this" represents the current form
62          }
63
64          private void btnClear_Click(object sender, EventArgs e)
65          {
66              txtName.Text = "";
67              txtScore1.Text = "";
68              txtScore2.Text = "";
69          }
70      }
71  }
```

It's time to practice! Do Exercise 8-4.

8.7 Passing Arrays to Methods

You may pass an entire array, or any other reference types, to a method just like you pass a value type like an integer or float. Tutorial 5 presents an application that passes an array to a method.

Tutorial 5: Passing Arrays to Compute Average

This tutorial creates a modified version of the ReadTextFiles form that reads the name and two exam scores from the text file Exams.txt. In the modified version, a separate method is used to read the names and score, and store the names and percent scores in arrays. These arrays are passed to two different event handlers.

Similar to ReadTextFile, the revised form named **PassingArrays** lets the user display the percent scores for each student and display the names of students with above average percent scores, as shown in Figure 8-26.

Figure 8-26: The PassingArrays form

Both tasks, display percent and display above average students, require reading each student record and computing the percent score for the student. In the ReadTextFile form, you repeated the code to read and compute percents within the Click event handlers of both buttons. In the revised form, PassingArrays, you create a method called ComputePercent() and call it from both event handlers, as shown in Figure 8-27. Avoiding duplication of code makes maintenance of code easier.

Figure 8-27: Structure chart showing ComputePercent method shared by two event handlers

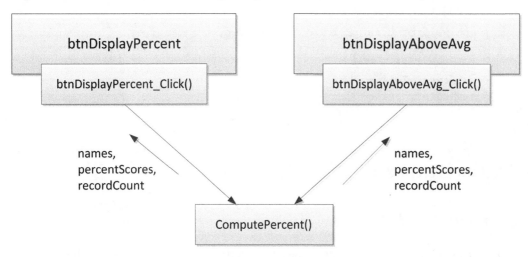

Figure 8-27 shows what happens conceptually. The names and percentScores are stored in arrays and passed from ComputePercent() method (which reads student records and computes percents) to both calling programs. But, how it is achieved in the code is almost the opposite: the reference variables for the two arrays are passed by value **to ComputePercent from the calling programs**. The ComputePercent method uses those references to access the arrays and fill the empty arrays with actual data.

For a called method to access an array, it must declare a reference variable for the array as a parameter in its header. For example, ComputePercent that updates the two arrays must declare two reference variables (with any valid names) for the arrays as shown below:

```
private int ComputePercent(string[] names, float[] percentScores)
```

Note that the reference variables are passed by value (indicated by lack of the key word "ref") because there is no need for the method to change the references to the arrays. ComputePercent uses the references, names and percentScores, to access the arrays and update them. When control is passed back to the calling program, it can access the same arrays. ComputePercent will also count the number of records and return it as an integer to the calling program, indicated by the return type *int*.

To invoke ComputePercent(), the calling program must create the arrays and pass the references to the called program, as shown below:

```
string[] names = new string[10];
float[] percentScores = new float[10];
int recordCount = ComputePercent(names, percentScores);
```

The complete code for ComputePercent method is shown in Figure 8-28. This method reads each student's name and two exam scores, computes the percent score, stores the names in the array referenced by the parameter **names** and stores the percent scores in the array referenced by **percentScores**.

Figure 8-28: The ComputePercent method

```csharp
20      private int ComputePercent(string[] names, float[] percentScores)
21      {   // Read data from file and load arrays. Return record count
22          int index = 0;
23          string currentLine;
24          string[] fields;
25          float percentScore;
26          StreamReader ExamReader = new StreamReader("Exams.txt");
27
28          while (ExamReader.EndOfStream == false)
29          {
30              currentLine = ExamReader.ReadLine();
31              fields = currentLine.Split(',');
32              percentScore = 100*(float.Parse(fields[1]) + float.Parse(fields[2]))/200;
33
34              names[index] = fields[0]; //Store name into names array
35              percentScores[index] = percentScore; // Store percent into percentScores
36              index = index + 1;
37          }
38          return index;   // index gives the number of records read
39      }
```

Again, as stated earlier, the reference variables for the two arrays are passed by value because there is no need for this method to change the reference to the arrays. Passing the reference variables allows the method to access the corresponding arrays and change the values of the arrays.

The Click event handler of btnDisplayPercent that calls ComputePercent is shown in Figure 8-29.

Figure 8-29: Event handler of DisplayPercent button calling ComputePercent

```csharp
40
41      private void btnDisplayPercent_Click(object sender, EventArgs e)
42      {
43          string[] names = new string[10];   //Declare names array
44          float[] percentScores = new float[10]; //Declare percentScores array
45
46          // Invoke ComputePercent to load the arrays and get record count
47          int recordCount = ComputePercent(names, percentScores);
48
49          lstDisplay.Items.Clear();
50          string fmtStr = "{0,-15}{1,8:N1}";
51          lstDisplay.Items.Add(string.Format(fmtStr, "Name", "Percent"));
52          // Get names and percent scores from arrays and display them
53          for (int index=0; index < recordCount; index ++)
54              lstDisplay.Items.Add(string.Format(fmtStr, names[index], percentScores[index]));
55      }
```

Line 47 calls ComputePercent method, which updates the two arrays.

Line 54 accesses the arrays to display them in the ListBox.

The Click event handler of btnDisplayAboveAvg also calls ComputePercent, as shown in Figure 8-30. This event handler uses the data from the arrays to compute the average and display names of above average students.

Figure 8-30: Event handler of DisplayAboveAvg button calling ComputePercent

```csharp
56
57      private void btnDisplayAboveAvg_Click(object sender, EventArgs e)
58      {
59          string[] names = new string[10];
60          float[] percentScores = new float[10];
61          int index = 0;
62          // Invoke ComputePercent to load the arrays and get record count
63          int recordCount = ComputePercent(names, percentScores);
64
65          // Compute class average:
66          float classTotal = 0, classAvg;
67          foreach (float percentScore in percentScores)
68              classTotal += percentScore;
69          classAvg = classTotal / recordCount;
70
71          lstDisplay.Items.Clear();
72          string fmtStr = "{0,-15}{1,8:N1}{2,12:N1}";
73          lstDisplay.Items.Add(string.Format(fmtStr, "Name", "Percent", "class Avg"));
74          // Display above average:
75          for (index = 0; index < recordCount; index++)
76          {
77              if (percentScores[index] > classAvg)
78                  lstDisplay.Items.Add(string.Format(fmtStr, names[index],
79                                                 percentScores[index], classAvg));
80          }
81      }
```

After calling ComputePercent (line 63), this method accesses the arrays in line 67 to accumulate the total and in line 78 to display names and percents.

Step 5-1: Open the form PassingArrays, shown in Figure 8-26.

> Fill in the missing parameters for ComputePercent in line 20 of Figure 8-28.
> Add the code in lines 34 and 35 to store the name and percent into arrays.
> Add the statements to call ComputePercent in the Click event procedures of btnDisplayPercent (line 47, Figure 8-29).
> Add the statements to call DisplayAboveAvg (line 63, Figure 8-30).

Step 5-2: Run the form and test the code.

8.8 Two-Dimensional (2-D) Arrays

A one-dimensional (1-D) array stores one set of data. A two-dimensional (2-D) array can store multiple sets of data, all of the same type. For example, the units sold in each quarter of a year for six different products could be stored in a 2-D array with 6 rows and 4 columns, where each cell represents an element of the array. Each row would represent a different product, and each column a different quarter, as shown in Figure 8-31.

Figure 8-31: A 2-D array

```
                     Columns (Quarters)
              index  0 (qtr 1)  1(qtr 2)  2(qtr 3)  3(qtr 4)
                0   ┌─────────┬─────────┬─────────┬─────────┐
                1   ├─────────┼─────────┼─────────┼─────────┤
   Rows         2   ├─────────┼─────────┼─────────┼─────────┤
   (Products)   3   ├─────────┼─────────┼─────────┼─────────┤
                4   ├─────────┼─────────┼─────────┼─────────┤
                5   └─────────┴─────────┴─────────┴─────────┘
```

Creating a 2-D Array

The syntax for creating a 2-D array is

>   *type* [,] *arrayName* = new *type* [*rowCount, columnCount*]

where *rowCount* is the number of rows and *columnCount* is the number of columns. Note that the declaration of variables that represent a 2-D array requires a comma inside the pair of square brackets.

The following statement creates an int type array, called sales, that has 6 rows and 4 columns:

```
int[,] sales = new int[6, 4];
```

Creating a 2-D Array with Initial Values

You may use the array initializer to create a 2-D array with initial values other than zero by specifying each set of values within curly braces, as shown below.

```
int [,] sales = { {16, 32, 23, 40}, {144, 182, 169, 205}, {54, 49, 72, 63} }
```

The previous statement creates the following 2-D array, named sales, with 3 rows and 4 columns.

		Columns			
		0	1	2	3
	0	16	32	23	40
Rows	1	144	182	169	205
	2	54	49	72	63

Referring to an Element in a 2-D Array

Each element in a 2-D array is referred to by the index of the row and the index of the column. Here are some examples:

sales[0, 0]—the element at row 0 and column 0, with the value 16
sales[0, 1]—the element at row 0 and column 1, with the value 32
sales[0, 2]—the element at row 0 and column 2, with the value 23
...

sales[1, 0]—the element at row 1 and column 0, with the value 144
sales[1, 1]—the element at row 1 and column 1, with the value 182
sales[1, 2]—the element at row 1 and column 2, with the value 169

Example: Using 2-D Arrays for Sales Data

This example uses the following data that consist of the product number and quarterly sales for a set of products:

1033, 16, 32, 23, 40
1009, 144, 182, 169, 205
1014, 54, 49, 72, 63
1005, 38, 62, 55, 77
1001, 7, 9, 8, 15
1025, 98, 110, 105, 130

The application that process these data use a form that allows the user to specify any product number and display the total sales for that product, as shown in Figure 8-32.

Figure 8-32: The TwoDimArrays Form

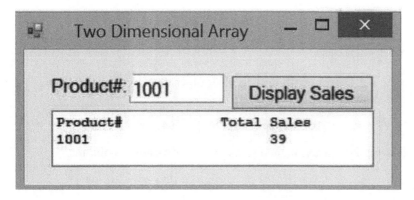

In this application, you need to store two types of data. Though the product number consists of only numeric digits, it is essentially text data because you do not do any numeric calculations with it. The sales data, however, are integers. An array has to be of a single type.

One option would be to create one string type 1-D array to store product numbers, and another int type 2-D array to store the sales.

A second approach would be to store product numbers and sales in a single 2-D array. Because all data in the text file, including sales, is stored as text, we use the second approach—that is, create a string type array that stores both product numbers and sales data. You will need to convert the sales data to int type when you read them from the array so that you can add them up.

The code has two major components:
1. Load the data from the text file to the 2-D array. This is done in the **load event** of the form.
2. Search the array to find the sales for a specified production number. This is done in the **Click event** handler of a button.

Because the array is used in two different methods, it is declared outside both methods at the class level. Figure 8-33 shows part 1 of the code.

Figure 8-33: Code to create a 2-D array and load data into it

```
20      //Declare a 2-D array with 10 rows and 5 columns:
21      string[,] sales = new string[10, 5];
22      int recordCount;    // Actual number of records read
23
24      private void TwoDimArray_Load(object sender, EventArgs e)
25      {
26          string currentLine;
27          string[] fields = new string[2];
28          int row = 0, column = 0;
29          //Create StreamReader
30          StreamReader salesReader = new StreamReader("Sales.txt");
31
32          //Read each line from the file, split and store in a row of the 2-D array.
33          while (salesReader.EndOfStream == false)
34          {
35              currentLine = salesReader.ReadLine();//Read product# and 4 sales data
36              fields = currentLine.Split(',');  //Store Product# & sales in a temp array
37
38              //Copy Product# and four sales to the five columns of the current row:
39              sales[row, 0] = fields[0];   // product number stored in column 0
40              sales[row, 1] = fields[1];   // Sales for qtr 1 stored in column 1
41              sales[row, 2] = fields[2];   // Sales for qtr 2 stored in column 2
42              sales[row, 3] = fields[3];   // Sales for qtr 3
43              sales[row, 4] = fields[4];   // Sales for qtr 4
44
45              // Or, use a loop to change the column of the array element:
46              for (column = 0; column <= 4; column ++)
47                  sales[row, column] = fields[column];
48
49              row = row + 1;
50          }
51          recordCount = row; // recordCount is the number of rows that has data in it.
52          salesReader.Close();
53      }
```

The code shown in Figure 8-33, which reads the data from the text file and stores them in the array, has three major steps:

Read current line into a string variable named currentLine (line 35).
Split it and store in a temp array named fields (line 36).
Copy from fields to a row of the 2-D array (lines 39–43 and lines 46–47 do the same thing).

The data are copied from the current row in the text file to the 2-D array, as shown in Figure 8-34.

Figure 8-34: Copying the current line to a temp array and then to a 2-D array

Current row in the text file: 1033, 16, 32, 23, 40
string variable, currentLine: "1033, 16, 32, 23, 40"

fields (1-D array):

	0	1	2	3	4
	"1033"	"16"	"32"	"23"	"40"

sales (2-D array):

Rows (Products)

	0	1	2	3	4
0	"1033"	"16"	"32"	"23"	"40"
1					
2					
3					
4					
5					

The code in Lines 39–43 copies the five elements of the 1-D array, **fields**, to the current row of the 2-D array, **sales**, using a separate statement for each element.

Consider line 39: **sales[row, 0] = fields[0];**. The first time through the while loop, the value of row is 0. So, fields[0] (the product#) is copied to sales[0,0]—that is, row 0 and column 0.

Similarly, line 40, **sales[row, 1] = fields[1]**, copies fields[1] (sales for qtr. 1) to sales[0,1].

The five statements together copy all five fields from the current line of Sales.txt file into the five columns of the current row to fill the first row, as shown above in Figure 8-34.

The second time through the loop, the value of row is 1. So, the statements in lines 39–43 copy data from the second line of Sales.txt to sales[1,0], sales[1,1],…, sales[1,4], filling the second row of sales array, as shown in the following table.

Sales (2-D array):

Rows (Products)

	0	1	2	3	4
0	"1033"	"16"	"32"	"23"	"40"
1	"1009"	"144"	"182"	"169"	"205"
2					
3					
4					
5					

The process continues until all lines of Sales.txt are read and copied into sales array.

Using a Loop to Copy from Fields to Sales

In the following code, the five statements in lines 39–43 are replaced by a single statement within a loop in lines 46–47:

```
46              for (column = 0; column <= 4; column ++)
47                  sales[row, column] = fields[column];
```

The first time through the loop, column is 0. So, line 47 is equivalent to line 39:
```
sales[row, 0] = fields[0];
```

The second time, column is 1. So, line 47 is equivalent to line 40:
```
sales[row, 1] = fields[1];
```

In effect, this loop generates the five statements in lines 39–43.

Tutorial 6: 2-D Array to Find the Sales for a Product

This tutorial reads the sales data from the file, Sales.txt, into a two-dimensional array, and uses the array to find the total sales for a product.

Part 1

Step 6-1: Open the project, Ch8_ArraysAndFiles. Open the form, TwoDimArray, shown in Figure 8-35.

Figure 8-35: The TwoDimArray form

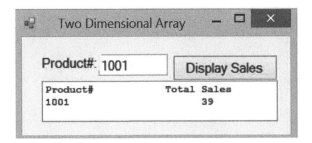

Step 6-2: Open the load event handler of the form.
Add the code in lines 46 and 47 from Figure 8-33 to copy the data from Sales.txt to the array sales.
Comment out lines 39–43 because lines 46 and 47 do the same thing.
Put a break at line 50.

Step 6-3: Run the form. When the program breaks, display Locals window (Debug, Windows, Locals), and view the values of fields array. Because sales array is declared at the class level, it is not shown in Locals window.

To view sales array, display **QuickWatch** window (Debug, QuickWatch), type sales for *Expressions*, click *Add Watch*, and click *Close*. Expand sales array to view the values, displayed as follows:

Name	Value
sales	{string[10, 5]}
[0, 0]	"1033"
[0, 1]	"16"
[0, 2]	"32"
[0, 3]	"23"
[0, 4]	"40"
[1, 0]	null
[1, 1]	null
[1, 2]	null

Click continue to watch the values in row 1 of sales array.

Click continue until all lines from Sales.txt are copied to the array.

Part 2

Search 2-D array to find the sales for a specified product number. This is done in the Click event handler of btnDisplaySales. The code is shown in Figure 8-36.

Figure 8-36: Code to search the 2-D array

```
54
55        private void btnDisplaySales_Click(object sender, EventArgs e)
56        {
57            int row = 0, column, rowSum=0;
58            string fmtStr = "{0,-20}{1,8}";
59            lstDisplay.Items.Clear();
60            lstDisplay.Items.Add(string.Format(fmtStr, "Product#", "Total Sales"));
61
62            for (row = 0; row < recordCount; row++)
63                // recordCount is the number of rows with data in it
64            {
65                // Check whether column 0 of current row matches Prodduct#
66                if (sales[row, 0] == txtProdNumber.Text)
67                {
68                    // Add columns 1 - 4
69                    rowSum = int.Parse(sales[row, 1]) + int.Parse(sales[row, 2]) +
70                            int.Parse(sales[row, 3]) + int.Parse(sales[row, 4]);
71                    // Alternate method using a loop:
72                    for (column = 1; column <= 4; column++)
73                        rowSum = rowSum + int.Parse(sales[row, column]);
74
75                    lstDisplay.Items.Add(string.Format(fmtStr, txtProdNumber.Text, rowSum));
76                    return;    // Exit the method, if a matching product is found
77                }
78        }
```

Let's look at the key statements:

Line 62: `for (row = 0; row < recordCount; row++)`

Here, recordCount is the actual number of rows with data in it. The "<" sign is used in the expression because the largest row number that has data is one less than the number of rows.

Line 66: `if (sales[row, 0] == txtProdNumber.Text)`

For each value of row, this statement compares Column 0 of the array (product#) with the product# that you entered into the TextBox.

Line 69: `rowSum = int.Parse(sales[row, 1]) + int.Parse(sales[row, 2]) +`
`           int.Parse(sales[row, 3]) + int.Parse(sales[row, 4]);`

This statement adds columns 1–4 (sales for quarters 1–4) of the current row.

Lines 72-73: `for (column = 1; column <= 4; column++)`
    `rowSum = rowSum + int.Parse(sales[row, column]);`

By varying the value of column from 1 to 4, the statements add columns 1–4 of the current row to rowSum.

Line 76: return

The return statement, in general, transfers control to the calling method. Here, the effect is to stop processing the current method so that the rest of the rows are not checked for a match.

Step 6-4: Open the Click event handler of btnDisplaySales. Fill in the code from lines 72 and 73 from Figure 8-36 to compute the total for each row. Comment out line 69. Put breaks at lines 73 and 78.

Step 6-5: Run the form. Enter 1014 for Product Number, and click the Display Sales button.

When the program breaks at line 78 for the first time, observe the values of row and rowSum in the Locals window.

View the data stored in sales array by displaying the QuickWatch window (Debug, Windows, Watch, Watch1). Click Continue, and observe the data for the second iteration in the Locals window. To switch to Locals window, click the Locals tab at the bottom left of the window.

Click Continue to go through the loop the third time when row 2 that contains product# 1014 is accessed. Because the expression (sales[2,0] == txtProdNumber.Text) is true, line 72 is processed, and the program breaks at line 73.

Observe the values of column and rowSum for each iteration of the loop by clicking Continue. The final value of rowSum (238) is displayed in the ListBox.

It's time to practice! Do Exercise 8-5.

Jagged Arrays

Jagged arrays are similar to two-dimensional arrays, except that the rows can have different lengths. It is a one-dimensional array where each element is an array that does not have a fixed size. Figure 8-37 shows an example of a jagged array with three rows.

Figure 8-37: Jagged array

		0	1	2	3
	0	16	32	23	40
Rows	1	144	182		
	2	54	49	72	

This is how you would create a jagged int type array with three rows:

    int[][] jaggedSales = new int[3][];

Note that the declaration of the variable includes two sets of square brackets, indicating that it is an array of arrays. Further, the number of columns is not specified.

You specify the number of columns for each row and the initial values (optional) as follows:

jaggedSales[0] = new int[4] {16, 32, 23, 40};
jaggedSales [1] = new int[2] {144, 182};
jaggedSales [2] = new int[3] {54, 49, 72};

You access an element of the array by specifying the row and column index. For example, jaggedSales [2][1] has the value 49.

Review Question

8.11 The following code creates a 2-dimensional array, and then displays one element of the array in a ListBox named lstDisplay.

```
int[,] sales = {{ 16, 32, 23, 40 }, { 144, 182, 1679, 205 },{ 54, 49, 72, 63 } };
lstDisplay.Items.Add(sales[1, 3]);
```

What would be the output displayed in the ListBox, lstDisplay?

Exercises

Exercise 8-1

The following is a sample of the order data (Customer#, Order#, OrderQty, UnitPrice) to be processed. Store the data in a text file named Order.txt. You may use Notepad to enter the data.

0155, 76533, 3, 10.95
0140, 75446, 12, 8.50
0150, 75550, 1, 89.95
0145, 75648, 25, 2.50
0130, 76360, 5, 17.90

Create a form, ProcessOrder, that reads each order record, computes the extended cost, and displays Cust#, Order#, OrderQty, UnitPrice and ExtendedCost in a ListBox, with the columns aligned.

Exercise 8-2

Add a new button named btnAboveAvgForExam1. Add the code to the Click event handler of the button to display students with above average scores for the first exam. List the student's name, exam 1 score, and class average.

Hint: When the file is read, store the score for exam 1 in an array, and compute the average for exam 1. Next, use a loop to list all students whose exam 1 scores were above the class average.

Exercise 8-3

Modify the form named ReadTextFile within the project Ch8_FilesAndArrays so that when the user presses the ComputePercent button, the Click event handler will write the name, score1, score2 and percentScore to a text file named PercentScores.txt. Add the necessary code to the end of the Click event handler of ComputePercent button.

Exercise 8-4

Modify Exercise 8-3 so that the user can select the file to write the name, score1, score2 and percentScore.

Exercise 8-5

Open the project Ch8_ArraysAndFiles from the Tutorial_Starts folder. Open the form named TwoDimArray. Add a button, btnQtr1Total, that computes the total sales for a specified quarter and displays it in the ListBox. Let the user enter the required quarter (a number 1–4) into a TextBox.

Rentals Assignment 2

The main goals of this assignment are to give you experience in
1. Designing and creating a user interface that includes different types of controls like RadioButtons and ComboBoxes
2. Reading from and writing to text files
3. Working with arrays

You are to develop an enhanced version of the system specified in Rental Assignment 1 in Chapter 3. "Ace Auto Rentals" is a small business that rents automobiles. Ace would like you to develop an application that allows them to process the rental charges as specified below:

Create a form, **EnterRentals**, that allows the user to do the following for each car rental:

1. Enter a five-digit customer#, customer name, rental date and number of days of rental.
2. Select the type of auto. The current options are

    SUV, minivan, luxury, full size, midsize and compact.

    This list is likely to change.

    To make it easy for the user to select an auto type and view the daily rental rate, first you must store each auto type and corresponding rate in a text file named **AutoTypes.txt** (SUV = $ 80, minivan = $70, luxury = $50, full size = $40, midsize = $35 and compact = $25).

    Every time you open the form, read the auto types and rates, and add the auto types to a ComboBox or ListBox, sorted by auto type, so that the user can select one from the list.

    In addition, store the auto types in an array named autoTypes, and store the rates in another array named dailyRates.

    When the user selects an autotype, find the rate for the selected auto type from the arrays and display the rate in a Label.

3. Select one of the discount categories. The current options are

    Normal, State Employees, Business and Favorite Customer.

    This list is **not** likely to change.

    Choose the appropriate type of GUI to let the user select a category.

4. Compute the rental charge, discount and total cost.
    Rental charge: Number of days x Daily rental rate
    Discount: Normal = 0%, State emps = 15%, Business = 10%, Favorite = 20%
    Total cost: Rental charge - Discount

5. Display the number of days, rental charge, discount and total cost, nicely formatted with proper labels.

6. Write the customer#, customer name, type of auto, number of days, rental charge, discount and total cost into a text file, **Rental.txt**. The user should be able to clear the data and enter data on rentals for additional customers without having to exit and re-run the form.

Validate the data to make sure that no input data are missing, date and number of days are valid and the user made a selection for auto type and discount.

Use the form to enter and save information for several rentals. Make sure that at least a few customers have more than one rental.

Store the files **Autotypes.txt** and **Rental.txt** in the default path for files (typically, the Bin/Debug folder). **Don't specify any path for files** (so that the program can be tested on a different computer by copying the project directory).

Create a second form, **DisplayRentals**, that allows the user to display all previous rentals (customer#, type of auto and rental date) for any specified customer#. Allow the user to type a customer# into a TextBox, and click a button to display the data. Format and align the columns nicely.

To keep it simple, you may read the data from the Rental.txt file each time to find the rentals. (A better way would be to read the data in the Form Load event handler, store the data in arrays one time and search the arrays to find the rentals. You can get 5% extra points if you use this method).

Create a third form, **Rentals**, that serves as a switchboard with buttons that allow the user to go to EnterRentals and DisplayRentals. You may open EnterRentals, for example from the form Rentals, as follows:

> First, declare a variable named enterRentals to represent EnterRentals form:
> ```
> EnterRentals enterRentals;
> ```
> Next, create an object of the type EnterRentals, and make it visible using the Show method (we will discuss this in more detail later):
> ```
> enterRentals = new EnterRentals();
> enterRentals.Show();
> ```

Make Rentals the startup form.

The factors that determine your grade include the following:

Use meaningful and easily understood names for forms, controls and variables used in the code. Use standard naming conventions. Do not use names like form1 and button1.

Make the code easy to understand. Use comments to specify what each part of the code does. Avoid unnecessary duplication of code.

Use the C# features you learned so far, effectively.

Help Desk Assignment 2

The main goals of this assignment are to give you experience in:
1. Designing and creating a user interface that includes different types of controls like RadioButtons and ComboBoxes
2. Reading from and writing to text files
3. Working with arrays

You will need to develop an enhanced version of the system specified in Help Desk Assignment 1 in Chapter 3. To refresh your memory, the system maintains information on the support provided to employees by the IT help desk in an organization. The help desk provides over-the-phone and on-site assistance with technical problems to employees from different departments at multiple locations. The department of the employee is charged for the services.

The charge for service consists of a fixed charge and a variable charge. The fixed charge is $10, and the standard variable charge is the product of duration of service in minutes and charge/minute. The charge/minute varies with the area in which the support is provided, as specified later.

An additional feature of this assignment is the priority charge. Typically, requests for services are put in a standard queue, and services are provided in the order in which requests are received. However, services that can wait may be put in a low priority queue to receive a 20% reduction in service charge. Urgent services that are put in a high priority queue are charged 40% over the standard service charge. Services that are put in the standard queue don't have any additional charge.

Create a project named **HDassignment2**, and develop a form named **EnterService** that allows the user to do the following for each service provided:

1. Enter the ID of the employee who requested the service, first name of the employee, last name, start time of service and duration of service in minutes.

2. Select the area of support. The current options are:
    computer system, ERP, BI, cloud, data warehouse
   This list is likely to change in the future.

   To make it easy for the user to select an area, first you must store each service area and corresponding charge per minute in a text file named **ServiceArea.txt**. The charge/min for different areas are:
     computer system: $1.00/min; ERP: $1.25/min; BI: $1.50/min;
     cloud: $1.25/min; data warehouse: $1.50/min

   Every time you open the form, read the service areas and charge/min, and add the areas to a ComboBox or ListBox, sorted by area, so that the user can select one from the list.

   In addition, store the service areas in an array named serviceAreas, and store the charge/min in another array named chargePerMin.

   When the user selects a service area, find the charge/min for the selected area from the arrays and display the charge/min in a Label.

3. Select one of the priority codes. The current options are:
    low, standard, high.
   This list is **not** likely to change. Choose the appropriate type of GUI to let the user select a priority code.

4. Compute the standard charge, priority charge and total charge.
     standard charge: fixed charge ($10) + (duration of service in minutes × charge/minute)

priority charge:	low: –20% of standard charge; high: +40% of standard charge; standard priority: no priority charge
Total charge:	standard charge + priority charge

5. Display the duration of service in minutes, standard charge, priority charge and total charge, nicely formatted with proper labels.

6. Write the employee ID, name (combined first and last names), area of service, date of service, duration of service, standard charge, priority charge and total charge into a text file named **Service.txt**.

   Hint: When you store the start time of the service in a DateTime type variable, the date is automatically stored along with the time. You just need to specify the proper format to get the date from the start time.

The user should be able to clear the data and enter data on additional services without having to exit and re-run the form.

Validate the data to make sure that no input data are missing, start time is a valid time, duration of service is a valid number and the user made a selection for support area and priority code.

Use the form to enter and save information for several services. Make sure that at least a few employees have more than one service.

Store the files **ServiceArea.txt** and **Service.txt** in the default path for files (typically the Bin/Debug folder). **Don't specify any path for files** (so that the program can be tested on a different computer by copying the project directory).

Create a second form, **DisplayService**, which allows the user to display all previous Services (employee name, service area and service date) for any specified employee ID. Allow the user to type an employee ID into a TextBox, and click a button to display the data. Format and align the columns nicely.

To keep it simple, you may read the data from the Service.txt file each time the user clicks the button to display the services for an employee. (A better way would be to read the data in the Form Load event handler, store the data in arrays once, and search the arrays to find the services for different employees. You can get 5% **extra points** if you use this method).

Create a third form, **Helpdesk**, that serves as a switchboard with buttons that allow the user to go to EnterService or DisplayService. You may open EnterService, for example, from the Helpdesk form as follows:

First, declare a variable named enterService to represent EnterService form:
```
EnterService enterService;
```
Next, create an object of the type EnterService, and make it visible using the Show method (we will discuss this in more detail later):
```
enterService = new EnterService();
enterService.Show();
```
Make Helpdesk the startup form.

The factors that determine your grade include the following:

Use meaningful and easily understood names for forms, controls and variables used in the code. Use standard naming conventions. Do not use names like form1 and button1.

Make sure the code is easy to understand. Use comments to specify what each part of the code does. Avoid unnecessary duplication of code. Use the C# features you learned effectively.

Chapter 9

Collections

A collection, like an array, can hold a group of elements (items). An element could be of any data type, like a simple number, a string, an array or an object. How do collections differ from arrays? This chapter describes the unique features of collections and presents their practical applications.

Learning Objectives

After studying this chapter, you should be able to:

- Describe the differences between collections and arrays; generic collections and nongeneric collections.
- Describe the major features of List, Dictionary, SortedList, SortedDictionary, LinkedList, Stack, Queue.
- Create Lists of different types, add items to a Lists and find an item.
- Create Dictionaries of different types, add items to Dictionaires and find an item.
- Work with the Keys collection and Values collection of a Dictionary.

Topics

9.1	Introduction to Collections		9.3	Dictionary<TKey, TValue>Collection
9.2	List<T>Collection			

9.1 Introduction to Collections

You already used collections when you worked with GUI controls like ComboBox and ListBox. To create a list of items in these controls, you added items to the **Items** collection of these controls using the Items.Add() method, without having to specify the index of the element as in a C# array. Collections provide several such methods that make it easy to work with them without having to know the underlying structure. Further, the size of a collection is increased automatically when you insert new items.

The classes provided by the .Net Framework to create collections can be grouped broadly into **generic (typed)** and **nongeneric (untyped)** collections. The generic collections are strongly typed—that is, they allow you to specify a single type that can be stored in the collection. Nongeneric collections allow elements of different types to be stored in a single collection. That means a nongeneric collection intended to be of a certain type like float may contain, for example, a string that was accidently added to it. Therefore, use of nongenerics is generally not recommended, unless there is a special reason for using them. Both types of collections may include strings, arrays, objects and value types like int and float.

Some generic types have corresponding nongenerics, and others do not. Table 9-1 shows some commonly used generic collections and their corresponding nongeneric collections, if any.

Additional features of the different collections listed in Table 9-1 include the following:

List does not require a unique key, but **Dictionary, SortedList** and **SortedDictionary** require a unique key.

SortedList and **SortedDictionary** maintain the data in sorted order, making insertion and deletion of items generally less efficient than unsorted collections.

Table 9-1: Commonly used collections

Generic (typed) collections	Corresponding nongeneric (untyped) collections	Implementation and other features
List<T> **T** represents the type	ArrayList	Implemented as an array; allows accessing items by index. Performance could be slightly lower compared to arrays, but easier to use.
Dictionary<TKey, TValue> **TKey** represents the key, and **TValue** represents the value	Hashtable	Implemented as a hash table. Values accessed by key.
SortedList<TKey, TValue>	SortedList	Implemented as two arrays, one for the key and another for values; allows access by index, and by key.
SortedDictionary<TKey, TValue>		Implemented as a tree. Generally, better than SortedList, when • adding and removing items into large sets • number of items unknown • populating from unsorted data
LinkedList<T>		A group of nodes, each node consisting of a data item and a link to the next node. Efficient to insert items, but inefficient to search.
Queue<T>	Queue	A FIFO (first in, first out) list
Stack<T>	Stack	A LIFO (last in, first out) list

We will look at two popular unsorted generic collections that have overall good performance:

1. List<T>, which does not require a unique key, and
2. Dictionary<TKey, TValue>, which requires a unique key.

Review Questions

9.1 List two features of C# collections that distinguish them from arrays.
9.2 What is a generic collection?

9.2 List<T> Collection

List is a generic (typed) collection of items, similar to an array. Elements, in general, could be of any data type like a simple number, a string, an array or an object; but within any one List, all items must be of the same type. For example, a List of arrays, where each item is an array that contains a product number and four quarterly sales, might look as follows:

Index

0	1033	16	32	23	40
1	1009	144	182	169	205
2	1014	54	49	72	63
3	1005	38	62	55	77
4	1001	7	9	8	15
5	1025	98	110	105	130

Methods and Properties of List

Common properties and methods of List are shown in Table 9-2.

Table 9-2: Common properties and methods of List

Property	Description
Capacity	The number of elements that the List can hold
Count	The number of elements in the List
Method	
Add(item)	Adds an element to the end of the List and returns the index of the element added
Clear()	Removes all elements from the List
Contains(item)	Returns true if the item is found in the List; if not, returns false
BinarySearch(item)	Searches a List for the item. Returns the index of the item, if found. If not, a negative integer is returned.
Insert(index, item)	Inserts the item at the specified index
Remove(item)	Removes the first occurrence of the specified item
RemoveAt(item)	Removes the item at the specified index
Sort()	Sorts the elements in the List

Creating a List

The statement to create a List has the following syntax:

    List<*Type*> *listName* = new List<*Type*>();

where *listName* is the name of the variable that references the List object, and *Type* is the type of items in the List. Note that a pair of parentheses is required in the *new* expression after the type.

Examples

```
List<int> unitsSold = new List<int>(); // Creates a List of int type named unitsSold

List<string> movies = new List<string>();// Creates a List of string type named movies

List<int[]> classGrades = new List<int[]>(); // Creates a List of int arrays named
                                                                    classGrades

List<Array> examScores = new List<Array>(); // Creates a List of arrays named examScores

List<Customer> customers = new List<Customer>(); // Creates a List of Customer type
```

Note that names like unitsSold and movies in the above examples are the names of variables that reference the List, although, for the sake of brevity, we describe them as the names of Lists.

In general, the term "List" is used in place of "variable that references the List object."

Adding Items to a List

The Add() method adds a specified item to the end of the List. When the number of items in a List exceeds its initial capacity, the capacity is doubled automatically.

Examples Using the Lists Created Above

```
unitsSold.Add(205);  // adds 205 to the end of the unitsSold List

movies.Add("Pocahontas");

int[] studentGrades = {85,91,80}; //creates the array to be inserted into a List
classGrades.Add(studentGrades); //Adds the array studentGrades to classGrades
```

Accessing an Item

One way to access each element is by specifying the index of the element.

Examples

unitsSold[0] returns the int type number stored in the first element of unitsSold

classGrades[0] returns the array stored in the first element of classGrades

You also may use the foreach loop to access every element of the List.

Tutorial 1: Looking Up Sales Data Using a List

Let's look at an example, similar to the one discussed in Chapter 8, to look up sales data for a specified product number. The product number and sales for quarters 1–4 are read from the text file, Sales.txt, shown in Figure 9-1. The first data item in each row is the product number, which is followed by sales for quarters 1–4.

Figure 9-1: Sales.txt File

1033, 16, 32, 23, 40
1009, 144, 182, 169, 205
1014, 54, 49, 72, 63
1005, 38, 62, 55, 77
1001, 7, 9, 8, 15
1025, 98, 110, 105, 130

In Chapter 8, we used a 2-D array to store the data. In this application, you will use a generic List to store the data and to help find the total annual sales for a specified product. The product number is selected from a ComboBox, and the total sales displayed in a ListBox, as shown in Figure 9-2.

Figure 9-2: The GenericList form

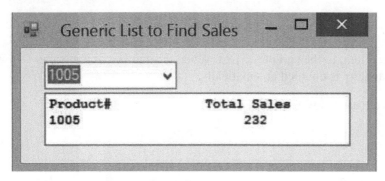

The coding for this form consists of three parts:
1. Create a List named sales.
2. Read each row from Sales.txt, split it and store it in a temporary array, and add the array to the List. Add the product number to the ComboBox. Both tasks are done in the Load event handler of the form.
3. When the user selects a product number from the ComboBox, find the corresponding item from the List. This is done in the SelectedIndexChanged event of the ComboBox.

Creating the List

Step 1-1: Open the project Ch9_Collections from Tutorial_Starts folder. Open the form GenericList, shown in Figure 9-2.

Step 1-2: Create a List of string arrays, referenced by the variable, sales, as shown in line 21 of Figure 9-3. This is declared at the class level so that the List can be used in the Load event and in the SelectedIndexChanged event. The List is of string[] (string array) type, though the product number and units sold are integers, because the data read from the Text file are string type.

Figure 9-3: Code to create the List, referenced by the variable sales

```
17          public GenericList()
18          {
19              InitializeComponent();
20          }
21      private List<string[]> sales = new List<string[]>();
```

Figure 9-4 shows part 2 of the code that reads each row from Sales.txt, splits it and stores it in a temporary array, and adds the array to the List. Except for line 36, the code is the same as what you used in other examples to read each line from a text file and store it in an array called fields. The code also adds the product number to the ComboBox.

Figure 9-4: Code to add items to the List and ComboBox

```
22
23          private void GenericList_Load(object sender, EventArgs e)
24          {
25              string currentLine;
26              string[] fields = new string[2];
27              //Create StreamReader
28              StreamReader SalesReader = new StreamReader("Sales.txt");
29
30              //Read each line from the file, split and store in an array
31              while (SalesReader.EndOfStream == false)
32              {
33                  currentLine = SalesReader.ReadLine(); // Read a product# and four sales data
34                  fields = currentLine.Split(','); // Store Product# and Sales in a temp array
35
36                  sales.Add(fields);                  // Add the entire array to the List, sales.
37                  cboProdNumber.Items.Add(fields[0]); // Add product# to the ComboBox
38              }
39              SalesReader.Close();
40          }
```

Line 36 adds the array fields to the List. The array consists of the product number and sales for quarters 1-4. Figure 9-5 shows the List with the first row of the file added to it.

Figure 9-5: sales List after the first item is added

Index

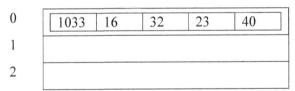

0
1
2

Step 1-3: Insert the code from line 36 of Figure 9-4 into the Load event handler of the form.

Step 1-4: Make sure that the file Sales.txt is in the Debug folder of the project.
Open Program.cs file and specify GenericList as the startup form.
Put a break at line 38 at the end of the loop, and run the form.
When the program breaks at line 38, click Continue to add a second item.
Display Watch window (Debug, Windows, Watch, Watch1) and type **sales** for Name.
Expand sales List to view the two arrays that were added to sales, as shown in Figure 9-6.

Figure 9-6: sales List after adding two items

Click *Continue* two more times. After the fourth time, you should see the Capacity doubled to 8 from the initial capacity of 4. Figure 9-7 shows the List with six rows of the file added to it.

Figure 9-7: sales List after adding all six rows

Index					
0	1033	16	32	23	40
1	1009	144	182	169	205
2	1014	54	49	72	63
3	1005	38	62	55	77
4	1001	7	9	8	15
5	1025	98	110	105	130

Finding an Item in the List Using the Index

When both the ComboBox and the List are not sorted (or when both are sorted), both have the product numbers stored in the same order because they were added in the same order from the Sales.txt file. That means the index of the selected product in the ComboBox is the same as the index of the corresponding item in the List. So, the SelectedIndex from the ComboBox can be used as the index for the List to find the item.

For example, if you select product number 1014 from the ComboBox, the SelectedIndex would be 2, and sales[2] will give you the corresponding array of four quarterly sales.

Figure 9-8 shows the code to find the quarterly sales using the SelectedIndex from the ComboBox, and find the sum of sales for the four quarters.

Figure 9-8: Code to find an item using the index of the selected item

```csharp
        private void cboProdNumber_SelectedIndexChanged(object sender, EventArgs e)
        {
            string fmtStr = "{0,-20}{1,8}";
            lstDisplay.Items.Clear();
            lstDisplay.Items.Add(string.Format(fmtStr, "Product#", "Total Sales"));

            // If both the List and ComboBox are not sorted, or if both are sorted, then
            // you can use the SelectedIndex from the ComboBox to find the item from the List:
            int selectedIndex = cboProdNumber.SelectedIndex;
            string[] productSale = sales[selectedIndex];

            // compute total sales for quarters 1 - 4
              int annualSales = 0;
            for (int quarter = 1; quarter <= 4; quarter++)
                annualSales = annualSales + int.Parse(productSale[quarter]);

            lstDisplay.Items.Add(string.Format(fmtStr, cboProdNumber.SelectedItem, annualSales));
```

Line 50: `int selectedIndex = cboProdNumber.SelectedIndex;`

The above statement gets the index of the selected product from the ComboBox and stores it in selectedIndex.

Line 51: `string[] productSale = sales[selectedIndex];`

sales[selectedIndex] gets the item (array) from the position specified by the selectedIndex. This string array is stored in the reference variable productSale of string array type. The string array, productSales, would have five elements with indices 0 - 4 representing the product number and sales for quarters 1 through 4.

Lines 55-56 use a loop to add up the four quarterly sales to get the total sales for the year.

Step 1-5: Add code in lines 50 and 51 of the SelectedIndexChaned event from Figure 9-8.
Add the code in lines 55 and 56 of the SelectedIndexChanged event to compute the annual sales.

Step 1-6: Make sure the Sorted property of the ComboBox is set to its default value of False.
Put a break on line 54. Run the form. Select product number 1005.
When the program breaks, display the Locals window (Debug, windows, Locals) and verify that the selectedIndex is 3 and the List displays the corresponding array of sales.
Continue running. The total sales should be 232.

Step 1-7: Set the Sorted property of ComboBox to True. Run the form and select the same product number, 1005. You will see the wrong total sales because the index of 1005 has changed to 1 in the sorted ComboBox, but in the List, the index of 1005 is still 3.

Finding an Item in the List Using foreach Loop

When either the ComboBox or the List is sorted, and the other is not sorted, the order of the two sets of product numbers is different. In this case, you need to use a loop to match the product number in each item (array) in the List to the selected product number, as shown in Figure 9-9.

Figure 9-9: Code to find an item using foreach loop

```
59    // If either the List or the ComboBox is sorted and the other not sorted, then
60    // loop through each item in sales collection to find the selected Product:
61    // This works also when both are sorted, and both are not sorted
62    string selectedProduct = cboProdNumber.SelectedItem.ToString();
63
64    // 1. Using foreach loop to search the List:
65    foreach (string[] prodSales in sales) // prodSale represents an item (array) in the List.
66    {
67        if (prodSales[0] == selectedProduct) // element 0 is the Product#
68        {
69            // compute total sales for quarters 1 - 4
70            annualSales = 0;
71            for (int quarter = 1; quarter <= 4; quarter++)
72                annualSales = annualSales + int.Parse(prodSales[quarter]);
73
74            lstDisplay.Items.Add(string.Format(fmtStr, cboProdNumber.SelectedItem, annualSales));
75            return;      // Exit the method, if a matching product is found
76        }
77    }
```

Line 65: `foreach (string[] prodSales in sales)`

Because each item in the List is an array of string types, each item is assigned to the variable prodSales, which also is an array of string types.

Line 67: `if (prodSales[0] == selectedProduct)`

prodSales[0] is the first element of the array, which is the product#. It is compared to the product number selected in the ComboBox.

Line 71 - 72: `for (int quarter = 1; quarter <= 4; quarter++)`
            `annualSales = annualSales + int.Parse(prodSales[quarter]);`
Elements 2 through 5 (index 1 - 4) of the prodSales array contain the quarterly sales, which are added together to compute the annual sales.

Step 1-8: Add code from Figure 9-9 to the SelectedIndexChanged event handler. Make sure the Sort property of the ComboBox is set to True. Comment out the code that uses the previous method in lines 50–58.

Step 1-9: Remove the break from line 54. Put a break on line 76 and run the form.
Observe the values of the prodSales array, including prodSales[0], which is the product number. Repeat clicking the Run button and observe the values. Note that this method gives the correct annual sales when the ComboBox is sorted, although the data in the product numbers in the sales array are not sorted.

It's time to practice! Do Exercise 9-1.

Next, you will create the same application using a Dictionary collection.

Review Questions

9.3   The file, Price.txt, contains the Product# and price of several products, as shown below:
Price.txt
"1033", 89.95
"1009", 21.50

"1014", 159.00
"1005", 2.50
"1001", 55.70
"1025", 33.45

The following code is intended to read the data from Price.txt and store them in a List of Arrays named prices:

```
List<string[]> prices = new List<string[]>();
private void GenericList_Load(object sender, EventArgs e)
{
    string currentLine;
    string[] fields;
    StreamReader PriceReader = new StreamReader("Price.txt");
    while (PriceReader.EndOfStream == false)
    {
        currentLine = PriceReader.ReadLine();
        fields = currentLine.Split(',');
        XXXXXX   // missing code to add product number and price to
                 the List, prices
    }
    PricesReader.Close();
}
```

In the above code, "XXXXXX" represents missing code to add the product number and price to the List, prices. Write the missing code.

9.4 A List of Arrays named prices is created as follows:
List List<string[]> prices = new List<string[]>();

Assume that each row from the text file, Price.txt (see Review Question 9.3), is read and stored in the List as a string array.
Write the code to find the price for product# "1001," and store the price in a string variable named price.

9.3 Dictionary<TKey, TValue> Collection

A Dictionary is a collection of keys and values. TKey and TValues specify the type of the key and of the value, respectively. For example, a dictionary that stores a set of product numbers and corresponding prices might look as follows:

Key (Prod#)	Value (Price)
1033	89.95
1009	21.50
1014	159.00
1005	2.50
1001	55.70
1025	33.45

Thus, a major difference between a List and a Dictionary is that a Dictionary requires a unique key that makes it easier to find an item using the key. Similar to a List, a Dictionary can store data of any type.

A Dictionary is implemented as a hash table that uses an algorithm to determine the location of an item based on the key. Overall, the Dictionary is efficient in finding items and in inserting/deleting items, though the performance may vary with the type of the key. However, it is less efficient than Lists for looping through with foreach. Figure 9-10 shows some commonly used properties and methods of the Dictionary.

Figure 9-10: Commonly used properties and methods of Dictionary

Property	Description
Count	The number of key/value pairs in the Dictionary
Item	The value associated with the specified key
Keys	A collection containing the keys in the Dictionary
Values	A collection containing the values in the Dictionary
Method	
Add(key, value)	Adds a key/value pair to the Dictionary
Clear	Removes all keys and values
Remove(key)	Removes the value for the specified key
ContainsKey(key)	Determines if the Dictionary contains the specified key
ContainsValue(value)	Determines if the Dictionary contains the specified value
TryGetValue(key, *out* variable)	Returns the value, if the key is found; if not, returns false

Creating a Dictionary

The syntax to create a Dictionary object, and assign it to a reference variable, is
    Dictionary<*TKey, TValue*> *dictionayName* = new Dictionary <*TKey, TValue*>();
where *TKey* and *TValue* specify the type of the key and of the value, respectively, and *dictionaryName* is the name of the variable that references the Dictionary.

A value can be of any data type. But using reference types, like array, for keys may pose a problem because this may not generate unique hash keys for storing the values using the standard method of generating keys. Similarly, using objects for keys becomes more complex.

Examples

```
Dictionary<string, float> prices = new Dictionary<string, float>();
```
—Creates a Dictionary named prices with key of string type and value of float type

```
Dictionary<int, Array> sales = new Dictionary<int, Array>();
```
—Creates a Dictionary named sales with key of int type and value of Array type

```
Dictionary<short, Customer> customers = new Dictionary<int, Customer>();
```
—Creates a Dictionary named customers with key of short type and value of Customer type

Names like prices and sales are the names of variables that reference the Dictionary object, although, for the sake of brevity, we describe them as the names of Dictionary.

In general, the term "Dictionary" is used in place of "variable that references the Dictionary object."

Adding Items to a Dictionary

To add an item to a Dictionary, you add a key and a value, as in the following examples:

```
prices.Add("Litebook 15", 399.95);
```
—Adds an item with "Litebook 15" as the key and 399.95 as the value

```
float[] prodSales = {55.5, 29.0, 109.5, 85.0};
sales.Add(4, prodSales);
```
—Adds an item with 4 as the key and prodSales array as the value

Accessing an Item

An item in a Dictionary is accessed by specifying the key, as in the following examples:

```
prices["Litebook 15"]
```
—Returns the value 399.95

```
sales[4]
```
—Returns the array {55.5, 29.0, 109.5, 85.0};

Tutorial 2: Dictionary with Product# as Key and Price as Value

The application developed in this tutorial stores a string type product number as the key and a float type price as the value of the Dictionary. The product numbers and corresponding prices are read from the file Price.txt shown in Figure 9-11, and stored in the Dictionary, prices<string, float>.

Figure 9-11: Price.txt file

"1033", 89.95
"1009", 21.50
"1014", 159.00
"1005", 2.50
"1001", 55.70
"1025", 33.45

A ComboBox is used to let the user select a product number and display the corresponding price from the Dictionary, as shown in Figure 9-12.

Figure 9-12: The DictionaryValueType form to find product price

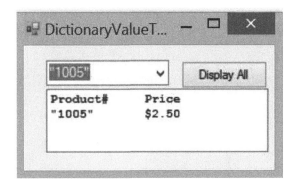

The coding has three parts:
1. Create a Dictionary named prices.
2. Read each row from Price.txt, split it and store it in a temporary array, and add the product number as the key and price as the value to the Dictionary. In addition, add the product number to the ComboBox. This is done in the Load event handler of the form.
3. When the user selects a product number from the ComboBox, find the corresponding item from the Dictionary. This is done in the SelectedIndexChanged event of the ComboBox.

Creating a Dictionary

Step 2-1: Open the form DictionaryValueType, shown in Figure 9-12.

Step 2-2: Add the code to create a Dictionary named prices with key of string type, representing the product number and value of float type, representing the price, as shown in line 21 in Figure 9-13. The Dictionary is declared at the class level so that the Dictionary can be used from the Load event and from the SelectedIndexChanged event.

Figure 9-13: Code to create the Dictionary

```
17      public DictionaryValueType()
18      {
19          InitializeComponent();
20      }
21      private Dictionary<string, float> prices = new Dictionary<string, float>();
```

Figure 9-14 shows part 2 of the code that reads each row from Prices.txt, splits it and stores it in a temporary array, and adds the product number as the key and price as the value. Except for line 36, the code is similar to what you used to store the data in a List. The code also adds the product number to the ComboBox.

Figure 9-14: Code to add items to the Dictionary and ComboBox

```
23      private void DictionaryValueType_Load(object sender, EventArgs e)
24      {
25          string currentLine;
26          string[] fields = new string[2];
27          //Create StreamReader
28          StreamReader PriceReader = new StreamReader("Price.txt");
29
30          //Read each line from the file, split and store in an array.
31          while (PriceReader.EndOfStream == false)
32          {
33              currentLine = PriceReader.ReadLine(); // Read product# & price
34              fields = currentLine.Split(',');
35              // Add the key (Prod#) and value (price) to the Dictionary:
36              prices.Add(fields[0], float.Parse(fields[1]));
37              cboProdNumber.Items.Add(fields[0]);  // Add product# to ComboBox
38          }
39          PriceReader.Close();
40      }
```

Line 36: `prices.Add(fields[0], float.Parse(fields[1]));`

Adds the first element (product#) of fields array as the key and the second element (price) as the value.

Step 2-3: Add the code in line 36 from Figure 9-14 to create the Dictionary.

Step 2-4: Make sure the file Price.txt is in the Debug folder of the project.
Put a break at line 38 at the end of the loop, and run the form.
When the program breaks at line 38, click *Continue* to add another item.
Display Watch window (Debug, Windows, Watch, Watch1) and type prices for Name.
Expand prices and the two key-value pairs (index 0 and 1) to view the first two keys and values, as shown in Figure 9-15.

Figure 9-15: prices Dictionary in Watch window after adding two items

Name	Value
⊟ prices	Count = 2
⊟ [0]	{[1033, 89.95]}
Key	"1033"
Value	89.95
⊞ Non-Public members	
⊟ [1]	{[1009, 21.5]}
Key	"1009"
Value	21.5

Finding an Item from the Dictionary

Unlike a List, the value for an item in a Dictionary can be found by specifying the key. In this example, when the user selects a product number from the ComboBox, it is used as the key to find the corresponding price.

Figure 9-16 shows the code to find the price for the selected product number by using the product number as the key.

Figure 9-16: Code to find the price for a product

```
41
42      private void cboProdNumber_SelectedIndexChanged(object sender, EventArgs e)
43      {
44          string fmtStr = "{0,-10}{1,8:C}";
45          lstDisplay.Items.Clear();
46          lstDisplay.Items.Add(string.Format(fmtStr, "Product#", "Price"));
47
48          string selectedProdcut = cboProdNumber.SelectedItem.ToString();
49          // Use the key (Product#) to get the value (price).
50          float price = prices[selectedProdcut];
51          lstDisplay.Items.Add(string.Format(fmtStr, selectedProdcut, price));
52      }
```

Line 50: `float price = prices[selectedProdcut];`

The selected product number (selectedProduct) is used as the key in **prices[selectedProdcut]**, which returns the corresponding value (price).

Unlike in a List, this method of finding the value using the key works even when the ComboBox is sorted, because you are not using the SelectedIndex of the ComboBox to find an item.

Step 2-5: Add the code in line 50 from Figure 9-16 to the SelectedIndexChaned event of the ComboBox to find the price for the product number selected by the user. Uncomment line 51.

Step 2-6: Remove the break from line 38. Run the form, and test it by selecting product numbers from the ComboBox.

KeyValuePair Collection

Each item in the Dictionary is of KeyValuePair type and is a member of the KeyValuePair collection. A KeyValuePair consists of a key and a value. Figure 9-17 shows the code to display every key and value using a foreach loop to loop through the collection of KeyValuePairs.

Figure 9-17: Using a KeyValuePair to hold each item from the Dictionary

```
53
54      private void btnDisplayAll_Click(object sender, EventArgs e)
55      {
56          string fmtStr = "{0,-10}{1,8:C}";
57          lstDisplay.Items.Add(string.Format(fmtStr, "Product#", "Price"));
58
59          foreach (KeyValuePair<string, float> kvp in prices)
60          {
61              lstDisplay.Items.Add(string.Format(fmtStr, kvp.Key, kvp.Value));
62          }
63      }
```

Line 59: `foreach (KeyValuePair<string, float> kvp in prices)`

Each item in the Dictionary is of KeyValuePair type. Line 59 declares variable kvp to be of KeyValuePair type, with the key as string type and the value as float type. The foreach loop gets each KeyValuePair from prices and assigns it to the variable kvp, one at a time.

Line 61: `lstDisplay.Items.Add(string.Format(fmtStr, kvp.Key, kvp.Value));`

The Key and Value properties of the KeyValuePair type variable, kvp, are used to display the key and value in the ListBox, as follows:

```
Product#      Price
"1033"        $89.95
"1009"        $21.50
"1014"       $159.00
"1005"         $2.50
"1001"        $55.70
"1025"        $33.45
```

Step 2-7: Add the code in lines 59–62 from Figure 9-17 to the Click event handler of btnDisplayAll. Run the form and test the code.

Keys Collection and Values Collection

The Keys collection is a collection of keys from all items of the Dictionary. Similarly, there is a separate Values collection, which is a collection of values from all items. Figure 9-18 shows the first few items in the two collections of the Dictionary.

Figure 9-18: Keys collection and Values collection

Keys Collection	Values Collection
"1033"	89.95
"1009"	21.50
"1014"	159.00
"1005"	2.5

You may use the foreach loop to access the key or the value from each item directly without using a KeyValuePair to access the entire item. Figure 9-19 shows the code that uses a foreach loop to access each price from the Values collection, and compute the average price.

Figure 9-19: Using the Values collection to access each price and compute the average price

```
64
65        private void btnAvgPrice_Click(object sender, EventArgs e)
66        {
67            float totalPrice = 0;
68            foreach (float price in prices.Values)
69                totalPrice += price;
70
71            float avgPrice = totalPrice / prices.Count;
72            lstDisplay.Items.Add("Average Price: " + avgPrice.ToString("C2"));
73        }
```

Line 68: foreach (float price in prices.Values)

```
totalPrice += price;
```

Gets individual prices from the Values collection, and computes the sum.

Step 2-8: Add the code from lines 68–71 of Figure 9-19 to the Click event handler of btnAvgPrice. Run the form and test the code.

You may use the same method to access each key from the Keys collection.

The following is the complete code that uses a Dictionary with product number as the key and price as the value.

```csharp
21      private Dictionary<string, float> prices = new Dictionary<string, float>();
22
23      private void DictionaryValueType_Load(object sender, EventArgs e)
24      {
25          string currentLine;
26          string[] fields = new string[2];
27          //Create StreamReader
28          StreamReader PriceReader = new StreamReader("Price.txt");
29
30          //Read each line from the file, split and store in an array.
31          while (PriceReader.EndOfStream == false)
32          {
33              currentLine = PriceReader.ReadLine(); // Read product# & price
34              fields = currentLine.Split(',');
35              // Add the key (Prod#) and value (price) to the Dictionary:
36              prices.Add(fields[0], float.Parse(fields[1]));
37              cboProdNumber.Items.Add(fields[0]);   // Add product# to ComboBox
38          }
39          PriceReader.Close();
40      }
41
42      private void cboProdNumber_SelectedIndexChanged(object sender, EventArgs e)
43      {
44          string fmtStr = "{0,-10}{1,8:C}";
45          lstDisplay.Items.Clear();
46          lstDisplay.Items.Add(string.Format(fmtStr, "Product#", "Price"));
47
48          string selectedProdcut = cboProdNumber.SelectedItem.ToString();
49          // Use the key (Product#) to get the value (price).
50          float price = prices[selectedProdcut];
51          lstDisplay.Items.Add(string.Format(fmtStr, selectedProdcut, price));
52      }
53
54      private void btnDisplayAll_Click(object sender, EventArgs e)
55      {
56          string fmtStr = "{0,-10}{1,8:C}";
57          lstDisplay.Items.Add(string.Format(fmtStr, "Product#", "Price"));
58
59          foreach (KeyValuePair<string, float> kvp in prices)
60          {
61              lstDisplay.Items.Add(string.Format(fmtStr, kvp.Key, kvp.Value));
62          }
63      }
64
65      private void btnAvgPrice_Click(object sender, EventArgs e)
66      {
67          float totalPrice = 0;
68          foreach (float price in prices.Values)
69              totalPrice += price;
70
71          float avgPrice = totalPrice / prices.Count;
72          lstDisplay.Items.Add("Average Price: " + avgPrice.ToString("C2"));
73      }
```

It's time to practice! Do Exercise 9.2.

Dictionary with an Array as Its Value

Next, we look at an example of a Dictionary with an array as its value.

Tutorial 3: Dictionary with Product# as Key and Sales Array as Value

In this application, you store a product number of string type as the key and an array of four quarterly sales as the value in a Dictionary. The product numbers and corresponding quarterly sales are read from the file Sales.txt, shown in Figure 9-20, and stored in the Dictionary, sales<string, Array>.

Figure 9-20: Sales.txt File

1033, 16, 32, 23, 40
1009, 144, 182, 169, 205
1014, 54, 49, 72, 63
1005, 38, 62, 55, 77
1001, 7, 9, 8, 15
1025, 98, 110, 105, 130

Figure 9-21 shows the Dictionary with the product number and sales.

Figure 9-21: sales Dictionary<string, array> showing sample data for key and value

Key	Value			
1033	16	32	23	40
1009	144	182	169	205
1014	54	49	72	63
1005	38	62	55	77
1001	7	9	8	15
1025	98	110	105	130

The product number is selected from the ComboBox, and the total sales displayed in a ListBox, as shown in Figure 9-22.

Figure 9-22: The DictionaryArrayType form

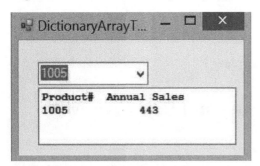

Similar to the last two applications that used Dictionary<string, float> and List<string[]>, the coding for this form has three parts:

1. Create a Dictionary named sales<string, array>.
2. Read each row from Sales.txt, split it and store it in a temporary array, and add the product number as the key and the array of four quarterly sales as the value to the Dictionary. Add the product number to the ComboBox. This is done in the Load event handler of the form.
3. When the user selects a product number from the ComboBox, find the corresponding value (array of quarterly sales) from the Dictionary, and compute the total sales. This is done in the SelectedIndexChanged event of the ComboBox.

Creating a Dictionary

Step 3-1: Open the form DictionaryArrayType, shown in Figure 9-22.

Step 3-2: Add the code in line 21 to create a Dictionary named sales with key of string type representing the product number and value of Array type representing the array of four sales, as shown in Figure 9-23.

The first half of the code that reads data from the file and stores each line in an array should be familiar code. We will focus on lines 36 and 40, which contain new statements.

Line 36: `Array.Copy(fields, 1, salesArray, 0, 4);`

This statement uses the Array.Copy method, discussed in Chapter 7, to copy the four quarterly sales from elements 1–4 of the array fields into a new array named salesArray.
To refresh your memory, the general syntax of Array.Copy is

> Array.Copy(source array, starting index in the source array, target array,
> starting index in target array, number of elements to copy)

Line 40: sales.Add(fields[0], salesArray);

This statement adds product number from fields[0] as the key and salesArray as the value.

Figure 9-23: Code to add items to the Dictionary and ComboBox

```
21      private Dictionary<string, string[]> sales = new Dictionary<string, string[]>();
22
23      private void DictionaryArrayType_Load(object sender, EventArgs e)
24      {
25          string currentLine;
26          string[] fields;
27          StreamReader SalesReader = new StreamReader("Sales.txt");
28          //Read each line from the file, split and store in an array.
29          while (SalesReader.EndOfStream == false)
30          {
31              currentLine = SalesReader.ReadLine(); // Read product# & four sales.
32              fields = currentLine.Split(',');
33
34              // Copy four sales from fields into a another array, salesArray.
35              string[] salesArray = new string[4];
36              Array.Copy(fields, 1, salesArray, 0, 4);
37              //Syntax: Array.Copy(fromArray, fromIndex, toArray, toIndex, length
38
39              // Add the key (Prod#) and value (salesArray) to the Dictionary:
40              sales.Add(fields[0], salesArray);
41              cboProdNumber.Items.Add(fields[0]);   // Add product# to the ComboBox
42          }
43          SalesReader.Close();
44      }
```

Step 3-3: Add the code in lines 36 and 40 from Figure 9-23 to the Load event of the form.

Step 3-4: Put a break at line 42. Run the form. When the code breaks at line 42, click *Continue* to add the second item. Display the Watch window (Debug, Windows, Watch) and type sales for Name. Expand sales, Key and Value to view their values, as shown in Figure 9-24.

Figure 9-24: sales Dictionary with the Key and value for first two items

Name	Value
⊟ sales	Count = 2
⊟ [0]	{[1033, System.String[]]}
Key	"1033"
⊟ Value	{string[4]}
[0]	" 16"
[1]	" 32"
[2]	" 23"
[3]	" 40"
⊞ Non-Public members	
⊟ [1]	{[1009, System.String[]]}
Key	"1009"
⊟ Value	{string[4]}
[0]	" 144"
[1]	" 182"
[2]	" 169"
[3]	" 205 "

Figure 9-25 shows the code to find the quarterly sales for a selected product number by using the product number as the key.

Figure 9-25: Code to find sales for a product

```
45
46          private void cboProdNumber_SelectedIndexChanged(object sender, EventArgs e)
47          {
48              string fmtStr = "{0,-10}{1,8}";
49              lstDisplay.Items.Clear();
50              lstDisplay.Items.Add(string.Format(fmtStr, "Product#", "Annual Sales"));
51
52              string selectedProduct = cboProdNumber.SelectedItem.ToString();
53              // Use the key (Product#) to get the value (quarterly sales array).
54              string[] salesArray = sales[selectedProduct];
55
56              int annualSales = 0;
57              foreach (string qtrSale in salesArray)
58                  annualSales = annualSales + int.Parse(qtrSale);
59              // Or, you may use a for loop:
60              //for (int quarter = 0; quarter <= 3; quarter++)
61              //    annualSales = annualSales + int.Parse(salesArray[quarter]);
62              lstDisplay.Items.Add(string.Format(fmtStr, selectedProduct, annualSales));
63          }
```

Line 54: sales[selectedProduct] uses selectedProduct as the key to get the value, which is an array. The array returned is stored in salesArray.

Lines 57–58 use a foreach loop to add all four elements of salesArray.

Step 3-5: Add the code in lines 54–58 from Figure 9-25 to the SelectedIndexChanged event of the ComboBox to find the sales for the selected product number. Run the form and test the code.

It's time to practice! Do Exercise 9.3.

Review Question

9.5   How does a Dictionary differ from a List?

Exercises

Figure 9-26 shows data from the Vendor.txt file that are stored in the Debug folder in Ch9_Collections project directory.

Figure 9-26: Vendor.txt file (vendor#, name, street, city, state, zip, phone)

1,FCC,445 12th Street SW, Washington,DC,20554,888-225-5322
2,U.S.Copyright Office,101 Independence Ave, Washington,DC,20559,202-707-3000
3,CDC,1600 Clifton Road,Atlanta,GA,30333,800-232-4636
4,White House, 1600 Pennsylvania Avenue NW, Washington,DC,20500,202-456-1111
5,Labor Statistics,Postal Sq Bldg,2 Mass Ave,Washington,DC,20212,800-877-8339
6,AMTRAK,60 Massachusetts Avenue NE, Washington,DC,20002,800-523-6590
7,U.S. Air Force,1690 Air Force Pentagon,Washington,DC,20330-1670,800-423-8723
8,National Guard,111 South George Mason Dr,Arlington,VA,22204,800-864-6264
9,Office of Govt Ethics,1201 New York Ave,Washington,DC,20005,202-482-9300
10,IRS,1111 Constitution Ave NW, Washington, DC,20224,800-829-1040

Exercise 9-1: List

Create a form named **VendorsList** that allows the user to find and display the zip code and phone number for any specified vendor, as follows:

Read vendor records from **Vendor.txt** (Figure 9-26), and load them into a List of arrays, named vendors, and add the vendor names into a sorted ComboBox.

When the user selects a vendor name from the ComboBox, find and display the name and corresponding zip code and phone number in a ListBox.

Exercise 9-2: Dictionary

Create a form named **VendorsDictionary1** that allows the user to find and display the phone number for any specified vendor, as follows:

Read each vendor record from the Vendor.txt file (Figure 9-26). Add the name as the **key**, and the phone number as the **value** into a Dictionary named vendorPhones. Add the name to a sorted ComboBox.

When the user selects a vendor name from the ComboBox, find and display the name and corresponding phone number in a ListBox.

Exercise 9-3: Dictionary

Create a form named **VendorsDictionary2** that allows the user to find and display the zip code and phone number for any specified vendor, as follows:

Read each vendor record from the Vendor.txt file (Figure 9-26), split it and store it in an array. Copy the zip code and phone into another array (you may use Array.Copy). Add the name as the **key,** and the array of zip code and phone number as the **value** into a Dictionary named vendors. Add the name to a sorted ComboBox.

When the user selects a Vendor name from the ComboBox, find and display the name and corresponding zip code and phone number in a ListBox.

Chapter 10

Graphical User Interface: Additional Controls

Chapter 5 introduced several controls that make it easy for users to enter data and help minimize data entry errors: RadioButton, CheckBox, ScrollBar, ComboBox and ListBox. In this chapter, you learn how to use two additional controls that help enhance the user interface:

1. The **ListView** control that allows you to display and work with multiple data items in a single row
2. The **Tab** control that lets you create a form with multiple pages, each page with its own set of controls

Three more user interface controls are introduced in Chapter 11: MenuStrip, ContextMenuStrip and ToolStrip controls.

Learning Objectives

After studying this chapter, you should be able to:

- Describe how a ListView differs from a ListBox.
- Setup a ListView and add items.
- Access items from a ListView.
- Create a form with Tab Pages.

Topics

10.1	ListView: Features	10.3	Accessing Data from ListView
10.2	Adding Items to ListView	10.4	Forms with Tab Pages

10.1 ListView: Features

A ListView, similar to the ComboBox and ListBox, lets you display multiple rows of data, and allows you to select the data from a row. A major difference is that ListView lets you add multiple data items to a row, as shown in Figure 10-1, which displays the id, name and unit price of each show in a theater using a ListView. By contrast, a row in a ComboBox and ListBox consists of only a single string. If you want to add multiple data items to a row, you must concatenate them together to form a single string, which makes it difficult to access individual data items from a row. A ListView lets you directly access an individual data item like the ticket price.

Similar to a ListBox, the ListView lets you select one or more items from a List. ListViews also can display items in different formats, including images and CheckBoxes.

To help understand the use of ListView, let's look at an application where the user selects a show from a ListView named **lvwShows** by clicking on the corresponding row, as shown in Figure 10-1. The total price (extended price) is computed by multiplying the unit price from the selected row by the number of tickets from the ScrollBar.

Figure 10-1: Example of a ListView

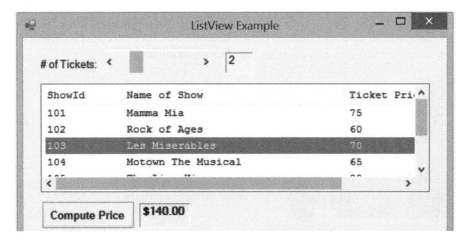

The data shown in Figure 10-1 are read from a Text file, Show.txt, shown in Figure 10-2, and added to the ListView, lvwShows.

Figure 10-2: Show.txt file

101,Mamma Mia,75
102,Rock of Ages,60
103,Les Miserables,70
104,Motown The Musical,65
105,The Lion King,90
106,The Phantom of the Opera,80
107,Pippin,75

Review Question

10.1   How does a ListViewItem differ from a ListBox?

10.2 Adding Items to ListView

Similar to the ComboBox and ListBox, the ListView has an **Items** collection that contains all items (rows) in the ListView.

ListViewItem

The items (rows) within the Items collection are of **ListViewItem** type (not string type), which may consist of **SubItems** that represent different data items (columns), like the id, name and ticket price for a show.

To add a row to a ListView, create a **ListViewItem** object and add it to Items collection.

To create a ListViewItem object named showsLVI, you may store the subitems in a string array and pass the array as an argument to the ListViewItem class, as follows:

```
String[] fieldsArray = {"101", "Mama Mia", "75"};
ListViewItem showsLVI = new ListViewItem(fieldsArray);
```

After you create the ListViewItem object, you add it to the Items collection of the ListView. The following statement adds the ListViewItem, showsLVI, to the ListView, lvwShows:

```
lvwShows.Items.Add(showsLVI);
```

You also may create a ListViewItem with just the first subitem, and then add the remaining subitems individually. But it is more convenient here to add all subitems together as a single array.

View Property

To display subitems of a row in tabular format with column headings, as shown in Figure 10-1, you need to set the View property to Details, as shown in Figure 10-3.

Figure 10-3: The View property of ListView

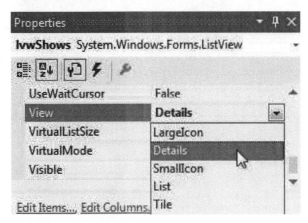

Column Collection

To display column headings, click on Columns (Collection) property of the ListView to open the ColumnHeader Collection Editor window. Click Add to add the first header and change its Text property to the desired heading. You also may change the Width property. You may repeat this for each heading, and click *OK* when done.

The code for reading each line from the Show.txt file using a loop, splitting it, storing it in a string array, and adding it to the Items collection is shown in Figure 10-4.

Figure 10-4: Code to add items to ListView

```csharp
21          private void ListView_Load(object sender, EventArgs e)
22          {
23              string currentLine;
24              string[] fields = new string[3];
25              StreamReader PriceReader = new StreamReader("Show.txt");
26              // Read each line, split it, and store in an array:
27              while (PriceReader.EndOfStream == false)
28              {
29                  currentLine = PriceReader.ReadLine();
30                  fields = currentLine.Split(',');
31
32                  // Use the array to create a ListViewItem (row).
33                  ListViewItem ShowsLVI = new ListViewItem(fields);
34
35                  // Add the entire ListViewItem to the ListView control
36                  lvwShows.Items.Add(ShowsLVI);
37              }
38              PriceReader.Close();
39          }
```

Line 33 creates a ListViewItem object using the array, fields, that contains the data from the current row. The ListViewItem object is referenced by the variable, showsLVI.

Line 36 adds the ListViewItem object to the ListView, lvwShows.

Table 10-1 presents a list of the properties and events that are important in working with ListViews.

Table 10-1: Properties and Events of ListView

Property	Description
CheckBoxes	If it is set to True, a CheckBox appears on each item (row), making it easier to select multiple items without using the control key. The default is False.
CheckedItems	If CheckBoxes property is set to True, CheckedItems collection would return the collection of checked items.
FullRowSelect	When set to True, you can select a row by clicking on any data item, and the entire row will be highlighted. The default is False, which requires you to click on the first data item of the row to be selected, and only that sub item is highlighted.
Items	The collection of ListViewItems.
SelectedItems	The collection of selected items.
SelectedIndices	The collection of indices of selected items.
Sorting	When set to Ascending (Descending), the items are sorted in ascending (descending) order using the data item in the first column. The default is None.
View	Determines the appearance of ListViewItems. When set to Details, data is displayed in a grid format. Other options include LargIcon, SmallIcon, List and Tile.
Events	
ColumnClick	Triggered when a column header is clicked.

Tutorial 1: Selecting a Show from a ListView

This tutorial creates the application that lets the user select a show, and computes the total price of tickets.

Step 1-1: Open the project named Ch10_GUIadditionalControls from Tutorial_Starts folder. Open the form, ListViewApp.

Add a ListView, as shown in Figure 10-1, and name it **lvwShows**.

Step 1-2: Set the View property of the ListView to Details, as shown in Figure 10-3. In addition, set the **FullRowSelect** property to True.

Step 1-3: Set column names using the Columns (Collection) property, as specified under "Column Collection."

Step 1-4: Open and view the Text file named Show.txt from the Debug folder of the project directory. Close the file.

Step 1-5: Add the code in lines 33 and 36 from Figure 10-4 to the Load event handler of the form. Run the project, and test the code.

Step 1-6: Make sure that the variable numOfTickets is declared at class level so that it can be accessed from multiple methods, as follows:

```
19
20          int numOfTickets;
```

Step 1-7: Open the ValueChanged event handler for the ScrollBar (Properties, events, ValueChanged). View the code that stores the value in the class level variable, numOfTickets, and displays the value in the Label.

Figure 10-5: The ValueChanged event handler of ScrollBar

```
40
41      private void hsbNumOfTickets_ValueChanged(object sender, EventArgs e)
42      {
43          numOfTickets = hsbNumOfTickets.Value;
44          lblNumOfTickets.Text = numOfTickets.ToString();
45      }
```

Review Question

10.2   What is the difference between a List View and a ListViewItem?

10.3 Accessing Data from a ListView

You may access a specific data item using the **Items** and **SubItems** collections by specifying the index of the item (row) and the index of the sub item (column). For example, to access the ticket price in the third column (index 2) for "Rock of Ages" in the second row (index 1), use the Items and SubItems collections as follows:

    lvwShows Items[1].SubItems[2].Text.

SelectedItems Collection

Similar to the ListBox, the ListView has the **SelectedItems** collection that includes all items (rows) selected by the user. Each selected item can be accessed using its index:

SelectedItems[0] gives the first item (row) selected by the user
SelectedItems[1] gives the second item selected by the user, and so forth.

For example, the ticket price from the single selected row (assuming only one show is selected) is given by

```
lvwShows.SelectedItems[0].SubItems[2].Text;
```

If multiple data items from a row are to be accessed, you may copy the entire row to a variable of ListViewItem type, and access individual data items from the variable, as follows:

```
ListViewItem showsLVI = lvwShows.SelectedItems[0];
float unitPrice = float.Parse(showsLVI.SubItems[2].Text);
```

In the above code, showsLVI is a variable of ListViewItem type.

Computing the Extended Price for the Selected Show

The code to compute the extended price (number of tickets x ticket price) is given in Figure 10-6.

Figure 10-6: Code to get ticket price from selected row and compute extended price

```
46
47      private void btnComputePrice_Click(object sender, EventArgs e)
48      {
49          float unitPrice = float.Parse(lvwShows.SelectedItems[0].SubItems[2].Text);
50          // If the CheckBoxes property of the ListView is set to True,
51          // use CheckedItems collection instead of SelectedItems as follows:
52          // float unitPrice = float.Parse(lvwShows.CheckedItems[0].SubItems[2].Text);
53
54          // Alternate method:Store the selected row in a variable & get price from it:
55          ListViewItem showsLVI = lvwShows.SelectedItems[0];
56          unitPrice = float.Parse(showsLVI.SubItems[2].Text);
57                  // gets price from SubItems[2].
58          float totalPrice = numOfTickets * unitPrice;
59          lblAmtDue.Text = totalPrice.ToString("C2");
60      }
```

Step 1-8: Add the code from Figure 10-6 to the Click event handler of btnComputePrice. Run the form and test the code.

Use appropriate properties from Table 10-1 to add CheckBoxes to the items. Let the user select an item by checking the CheckBox for the item. Make necessary changes in the code to compute the total price. Test the code

It's time to practice! Do Exercise 10-1.

ColumnClick Event: Selecting a Column

In addition to selecting a row, you also may select a column by clicking the column header, which triggers the ColumnClick event of the ListView. You may use this event to take an action when the user selects a column. Figure 10-7 shows the code within the ColumnClick event handler of the ListBox, to display the data item at the selected column of the selected row, in a Label named lblCurrentCell.

Figure 10-7: Code to get data from selected column

```
61
62     private void lvwShows_ColumnClick(object sender, ColumnClickEventArgs e)
63     {
64         int index = e.Column;
65         if (lvwShows.SelectedItems.Count > 0)
66             lblCurrentCell.Text = lvwShows.SelectedItems[0].SubItems[index].Text;
67     }
```

In line 64, the Column property of the parameter **e** gives the index of the selected column. The content of the cell in the selected column of the selected row is displayed in the Label, lblcurrentCell, as shown in Figure 10-8.

Figure 10-8: The ListViewApp form with content of current cell displayed

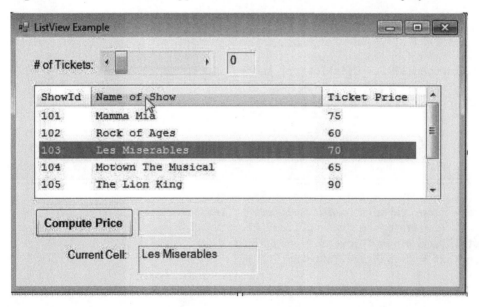

Step 1-9: Add the code from Figure 10-7 to the ColumnClick event handler of the ListView. Run the form. Select a row, and click on a column header to test the code.

Accessing Every Row of a ListView

A convenient way to access the data in every row of a column is to use a loop. For example, the code within the Click event handler of btnComputeAvg, shown in Figure 10-9, adds the price from each row using a foreach loop, computes the average price and displays the average in a Label named lblAvgPrice.

Figure 10-9: Using a foreach loop to access every row of ListView

```
68
69          private void btnComputeAvg_Click(object sender, EventArgs e)
70          {
71              float sumOfPrices = 0;
72              foreach (ListViewItem ShowsLVI in lvwShows.Items)
73                  sumOfPrices = sumOfPrices + float.Parse(ShowsLVI.SubItems[2].Text);
74              float avgPrice = sumOfPrices / lvwShows.Items.Count;
75              lblAvgPrice.Text = avgPrice.ToString("C");
76          }
```

Figure 10-10 shows the form with the Button and Label to compute and display the average.

Figure 10-10: The ListViewApp form with a button to compute average

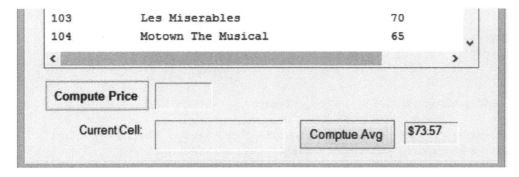

Step 1-10: Add the code shown in Figure 10-9 to the Click event handler of btnComputeAvg. Run the form, and test the code.

The following is the complete code for the ListViewApp form:

```csharp
        int numOfTickets;
        private void ListView_Load(object sender, EventArgs e)
        {
            string currentLine;
            string[] fields = new string[3];
            StreamReader PriceReader = new StreamReader("Show.txt");
            // Read each line, split it, and store in an array:
            while (PriceReader.EndOfStream == false)
            {
                currentLine = PriceReader.ReadLine();
                fields = currentLine.Split(',');

                // Use the array to create a ListViewItem (row).
                ListViewItem showsLVI = new ListViewItem(fields);

                // Add the entire ListViewItem to the ListView control
                lvwShows.Items.Add(showsLVI);
            }
            PriceReader.Close();
        }

        private void hsbNumOfTickets_ValueChanged(object sender, EventArgs e)
        {
            numOfTickets = hsbNumOfTickets.Value;
            lblNumOfTickets.Text = numOfTickets.ToString();
        }

        private void btnComputePrice_Click(object sender, EventArgs e)
        {
            float unitPrice = float.Parse(lvwShows.SelectedItems[0].SubItems[2].Text);
            // If the CheckBoxes property of the ListView is set to True,
            // use CheckedItems collection instead of SelectedItems as follows:
            // float unitPrice = float.Parse(lvwShows.CheckedItems[0].SubItems[2].Text);

            // Alternate method:Store the selected row in a variable & get price from it:
            ListViewItem showsLVI = lvwShows.SelectedItems[0];
            unitPrice = float.Parse(showsLVI.SubItems[2].Text);
                    // gets price from SubItems[2].
            float totalPrice = numOfTickets * unitPrice;
            lblAmtDue.Text = totalPrice.ToString("C2");
        }

        private void lvwShows_ColumnClick(object sender, ColumnClickEventArgs e)
        {
            int index = e.Column;
            if (lvwShows.SelectedItems.Count > 0)
                lblCurrentCell.Text = lvwShows.SelectedItems[0].SubItems[index].Text;
        }
```

```
68
69      private void btnComputeAvg_Click(object sender, EventArgs e)
70      {
71          float sumOfPrices = 0;
72          foreach (ListViewItem ShowsLVI in lvwShows.Items)
73              sumOfPrices = sumOfPrices + float.Parse(ShowsLVI.SubItems[2].Text);
74          float avgPrice = sumOfPrices / lvwShows.Items.Count;
75          lblAvgPrice.Text = avgPrice.ToString("C");
76      }
```

It's time to practice! Do Exercise 10-2 and Exercise 10-3.

Review Questions

10.3   The following ListView named lvwShows highlights the item selected by the user at runtime.

ShowId	Name of Show	Ticket Price
101	Mamma Mia	75
102	Rock of Ages	60
103	Les Miserables	70
104	Motown The Musical	65
105	The Lion King	90
107	Pippin	75
108	The Phantom of the Opera	80

What would be the output displayed in lblDisplay, after executing the following statement?

```
lblDisplay.Text = lvwShows.SelectedItems[0].SubItems[1].Text;
```

10.4   The following ListView named lvwShows displays information on different shows:

ShowId	Name of Show	Ticket Price
101	Mamma Mia	75
102	Rock of Ages	60
103	Les Miserables	70
104	Motown The Musical	65
105	The Lion King	90
107	Pippin	75
108	The Phantom of the Opera	80

What would be the output displayed in the Label, lblDisplay, after executing the following code?

```
foreach (ListViewItem showsLVI in lvwShows.Items)
    lblDisplay.Text = showsLVI.SubItems[2].Text;
```

10.4 Forms with Tab Pages

The tab control lets you create a multiple pages on a form, each page with its own set of controls. For example, consider a form similar to the DictionaryValueType form from Chapter 9, which lets you find the price for a specified product, and lets you list every product that is priced above a specified limit. Rather than doing both tasks on a single page of the form, you can have the controls for the two different tasks on two different tab pages of a single form, as shown in Figure 10-11.

Figure 10-11: The ProductPrices form with two tab pages

An alternative to using two tab pages for two different tasks is to use two different forms. But there is a significant difference between using two different forms and using two pages on a single form. When you use two or more tab pages on a single form, you can access controls on any one tab page from another page by using the name of the control. That is, tab pages act like different sections of a form. By contrast, to access controls on one form from another form, you need to create an instance of the page, as discussed in Chapter 11.

Each tab page on a form is a member of the TabPages collection of the TabControl. So, you can refer to a tab page by its index, starting with 0 for the first page, 1 for the second page, and so on. The **SelectedIndex** property of the TabControl allows you to identify the tab page that is currently selected.

Selecting a page from a TabControl fires the **SelectedIndexChanged** event, just like when you select an item from a ComboBox. So, you can code this event handler to take actions, if any, when the user selects a tab page.

Tutorial 2: Multipage Form with TabControl

Let's create a form that has two pages, as shown in Figure 10-11.

Step 2-1: Create a new project named Ch10_MultiPageForms. Change the name of the form to ProductPrices.

Step 2-2: Add the TabControl to the form from Containers group on the Toolbar. The TabControl displays two TabPages, as shown in Figure 10-12.

Select the TabControl by clicking on the header area, which includes tabs, tabPage1 and tabPage2. When you click on the name tabPage1 (or tabPage2), it might look like that page is selected. But clicking anywhere on the header area selects only the TabControl.
Set the name property of the TabControl to tabProductPrices.

Figure 10-12: The TabControl

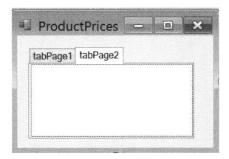

Step 2-3: To select tabPage1, click on tabPage1 in the header area, and then click inside the page (with dotted border).
Change the name of the page from tabPage1 to tbpLookUpPrices and set the Text property to Look Up Prices.
Similarly, change the name of tabPage2 to tbpDisplayProducts and set the Text property to Display Products.

Step 2-4: Select the page Look Up Prices. Add a ComboBox named cboProdNumber and a Label named lblPrice. The tab page should look as in Figure 10-11.

Step 2-5: Select the page Display Products. Add a button named DisplayProducts, a TextBox named txtPrice and a ListBox named lstDisplay. Add a Label to display the prompt "Enter Price Limit." The form should look as follows:

Because we already discussed the code in Chapter 9, we will not add the code for the two pages, but will focus on using tab control related features.

Next, we create the code to set focus to cboProdNumber, when the user selects Look Up Price page, and set focus to txtPrice, when the user selects Display Products page. This is done by coding the **SelectedIndexChanged** event of the TabControl, which is triggered when the user selects a tab page.

Step 2-6: Create the SelectedIndexChanged event handler of the TabControl (Select the TabControl by clicking the header area. Select events in Properties window and double click SelectedIndexChanged). Add the code from Figure 10-13 to the event handler.

Figure 10-13: SelectedIndexChanged event handler of tabProductPrices

```
19
20          private void tabProductPrices_SelectedIndexChanged(object sender, EventArgs e)
21          {
22              if (tabProductPrices.SelectedIndex == 0)
23                  cboProdNumber.Focus();
24              else if (tabProductPrices.SelectedIndex == 1)
25                  txtPriceLimit.Focus();
26          }
```

Step 2-7: Run the form. Click the page Display Products. The TextBox should have the focus. Click the tab page Look up Price. The ComboBox should have the focus.

Review Question

10.5   What is the difference between using multiple tab pages to do different tasks on a single form versus using multiple forms?

Exercises

Exercise 10-1
Modify ComputePrice_Click event procedure in the ListViewApp form so that the name of the selected show and the total price are displayed in the Label lblAmtDue, as in

    Les Miserables: $140.00

Exercise 10-2

Add a button on the ListViewApp form, and add the necessary code to the Click event handler to display all shows that start with the letter "M." Display the names of such shows in a ListBox.

Exercise 10-3

This exercise is a modified version of the IceCream form, shown below, from Exercise 6-1 in Chapter 6.

Chapter 10: Graphical User Interface: Additional Controls

This exercise uses a ListView (instead of the ComboBox) to display sundae names and prices, as described below:

1. Create a text file named Sundaes.txt, with sundae names and the following prices:

   Ice Cream, 3.5
   Custard, 3.75
   Yogurt, 4.00
   …

   Make up the rest of the sundae names and prices.

2. Replace the ComboBox with a ListView. Read the sundae names and unit prices from the Sundaes.txt file, and display them in the ListView in two columns (in place of the ComboBox) used in Exercise 5-1). This should be done in the Load event handler of IceCream form. Let the user select a Sundae from the ListView.

3. Let the user select the number of scoops using a scroll bar, and select nuts using a CheckBox.

4. A clear button should clear the ScrollBar and selection from the ListView so that no item is selected.

5. Have the program use the price of the selected sundae from the ListView to compute and display the amount due in a Label.

   Amount due = Unit price × Number of scoops + Cost of nuts ($0.50), if any

Extra Credit

Let the user select more than one sundae. Compute the cost assuming the number of scoops selected by the user applies to each selected sundae.

Chapter 11

Multi-form Applications and Menus

So far, we have only looked at forms that work independently without interacting with other forms. However, it may be necessary for a form to interact with other forms in a variety of ways, including opening/closing another form, accessing a control on another form, or sharing data or methods between forms.

In this chapter, you will learn how to code one form to show another form, access controls on another form and share methods between forms. Sharing data between forms is not covered here because proper sharing of data involves knowledge of additional object-oriented programming concepts.

Learning Objectives

After studying this chapter, you should be able to:

- Create a multi-form application.
- Share a *method* between different forms.
- Open a form programmatically from the Main method.
- Create Menus and Toolstrips.

Topics

11.1	Introduction		11.4	Closing a Parent Form
11.2	Accessing an Existing Form from Another		11.5	Menus
			11.6	Toolstrips
11.3	Sharing a Method between Multiple Forms			

11.1 Introduction

To help understand multi-form applications, consider an application that includes a form named **FinancialPlanner,** very similar to the FinancialPlanner form introduced in Chapter 4. Given yearly investments and growth rate, this form determines how long it takes for the investment value to reach the target amount. The investment value for every year until the value exceeds the target amount is displayed on a separate form named **DisplayInvestments**, as shown in Figure 11-1.

Figure 11-1: The DisplayInvestment form

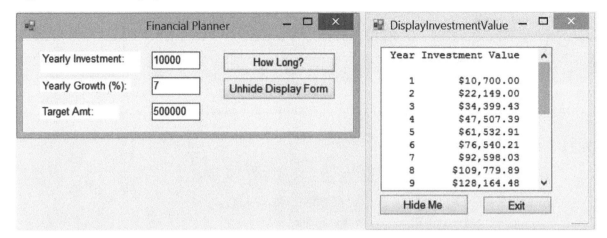

11.2 Accessing an Existing Form from Another

Accessing a form typically involves getting data from controls on the form, or displaying data in controls. Before you can access the controls, you need to open the form.

Opening an Existing Form from Another

Opening an existing form involves the following steps:

1. Declare a variable whose type is the form class you want to open. Note that the form you want to open, like DisplayInvestment, is a class.
   To declare a variable named displayInvestment of the type, DisplayInvestment form, use the following code:
   ```
   DisplayInvestment displayInvestment;
   ```
   Note that the variable starts with the lowercase letter "d," and the data type starts with the uppercase letter "D."

2. Create an instance of the form (an object) and store the reference to the object in the variable.
   For example,
   ```
   displayInvestment = new DisplayInvestment();
   ```
   The right side of the equal sign creates the object, whose reference is assigned to the variable displayInvestment.

   You may combine the statements from steps 1 and 2 as follows:
   ```
   DisplayInvestment displayInvestment = new DisplayInvestment();
   ```

3. Use the Show() method of the object to make it visible.
   ```
   displayInvestment.Show();
   ```

Accessing a Control on Another Form

To access a control on a form from other forms, you need to make the form public. To do this, open the Designer window for the form by double clicking the file name in Solution Explorer, as shown in Figure 11-2.

Figure 11-2: Opening the Designer window of a form

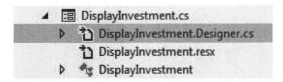

Next, change the control from "Private" to "Public," as shown in Figure 11-3.

Figure 11-3: Designer window of the form

```
23  ⊞            Windows Form Designer generated code
83
84               public System.Windows.Forms.ListBox lstDisplay;
85               private System.Windows.Forms.Button btnHide;
86               private System.Windows.Forms.Button btnExit;
```

A public control on another form can be accessed using the variable that references the object. For example, the following code would add a blank line in the ListBox, lstDisplay, on the form, DisplayInvestment using the variable, displayInvestment:

   displayInvestment.lstDisplay.Items.Add("");

Figure 11-4 shows the complete code in the form FinancialPlanner to compute investment values and display them in the ListBox on the form DisplayInvestments.

Figure 11-4: Code to access a control on another form

```
19      //Declare a variable of type DisplayInvestment form.
20      DisplayInvestment displayInvestment;
21
22      private void btnHowlong_Click(object sender, EventArgs e)
23      {
24          // Create an instance of DisplayIvestment form and
25          // store its reference in the variable displayInvestment:
26          displayInvestment = new DisplayInvestment();
27          // Load and show the form:
28          displayInvestment.Show();
29
30          // Get input data from Text Boxes:
31          double yearlyInvestment = double.Parse(txtInvestment.Text);
32          float YearlyGrowthPercent = float.Parse(txtGrowthRate.Text);
33          double targetAmt = double.Parse(txtTargetAmt.Text);
34
35          // Print heading
36          string formatCode = "{0,5}{1,17}";
37          // Add heading to the ListBox lstDisplay in DisplayInvestment form.
38          displayInvestment.lstDisplay.Items.Add(string.Format(formatCode,
39                                              "Year", "Investment Value"));
40          displayInvestment.lstDisplay.Items.Add("");
```

```
41
42          // Set year and investment value to zero
43          int year = 0;
44          double investmentValue = 0;
45          // Use while loop to compute investment value for each year
46          while (investmentValue < targetAmt)
47          {
48              year = year + 1;
49              investmentValue = (investmentValue + yearlyInvestment) *
50                                          (1 + YearlyGrowthPercent / 100);
51              // Add year & investment value to the lstDisplay in DisplayInvestment form
52              displayInvestment.lstDisplay.Items.Add(string.Format(formatCode,
53                                          year, investmentValue.ToString("C")));
54          }
55      }
```

Line 20: The variable, displayInvestment, is declared at the class level so that it can be used in multiple methods.

Lines 26–28: Creates an instance of the DisplayInvestment form and opens it.

Lines 31–53: Computes investment values and displays them in the ListBox lstDisplay. This is essentially the same as the code we discussed in Chapter 4, except that the ListBox is on the form DisplayInvestment.

So, the statements in lines 38, 40 and 52 that display results in the ListBox use the variable displayInvestment to access the ListBox, as in

```
        displayInvestment.lstDisplay.Items.Add(string.Format(formatCode,
                                        "Year", "Investment Value"));
```

Tutorial 1: Multi-form Financial Planning Application

Let's develop a multi-form fianancial planning application.

Step 1-1: Open the project, project Ch11_MultiFormApps1.

Step 1-2: Open the form, FinancialPlanner.

Step 1-3: Add the missing code in lines 20, 26 and 28 from Figure 11-4, to open the DisplayInvestment form.
Uncomment the statements in lines 38, 40 and 52, which write data into the DisplayInvestment form.

Step 1-4: Open Program.cs and change the startup form to FinancialPlanner. Run the form and test the code.

Step 1-5: Click the "How Long" button multiple times, each time with a different set of input data for yearly investment, growth rate and target amount. Note that every time you click the button, a new instance of the DisplayInvestment form is displayed.

Creating Multiple Instances of the Same Form

As observed in Step 1-5, every time you click the "How Long" button, the statements in line 26 and 28,
```
displayInvestment = new DisplayInvestment();
displayInvestment.Show();
```
will create and open a new instance of the DisplayInvestment form. This allows you to simultaneously display investment values for different sets of input data.

Hiding versus Closing a Form

You may hide a form without closing it. Hiding a form keeps the form and its properties and settings in memory. You don't lose the data entered/displayed in controls at runtime. It just is not visible.
You hide a form using the Hide method of the form. For example,
   this.Hide()
hides the current form.

The following code within the FinancialPlanner form hides an instance of the DisplayInvestment form:
   displayInvestment.Hide();

Closing a form removes the form from memory. So, you lose all current settings and data.

Unhiding a Form

The Show method makes a hidden form visible. The statement
   displayInvestment.Show();
shows the instance of the DisplayInvestment form represented by the variable displayInvestment.

Step 1-6: Add the following code to the Click event handler of btnHide button on the DisplayInvestment form.
   this.Hide();

Step 1-7: Add the following code to the Click event handler of btnUnHideDisplayForm button on the FinancialPlanner form.
   displayInvestment.Show();

Step 1-8: Run the FinancialPlanner form. Click "How Long?" button to open DisplayInvestment, and hide it by clicking the "Hide Me" button. Unhide it by clicking "Unhide Display Form." Note that the contents of the ListBox are not lost when you hide the form. Keep the form open.

Step 1-9: Change the target amount to 10000 and open another instance of the DisplayInvestment form. Now, hide both instances of DisplayInvestment. Next, Click "Unhide Display Form" button. Note that only the latest (second) instance of the DisplayInvestment form is visible. This is because the variable displayInvestment in the statement displayInvestment.Show() holds reference only to the latest instance.

Show versus ShowDialog

You may open a form using the Show method or the ShowDialog method.

When C# processes a statement that calls the Show method, it opens the form and continues processing the statements that follow the call. However, with the ShowDialog method, C# opens the form and stops processing the statements that follow the call until the dialog window is closed. So, ShowDialog is not appropriate if a form has to open a form and then access a control on that form, as in the current example.

You may use the ShowDialog method if you want to display information for the user or get input from the user before processing other statements in the code. For example, you may use a dialog window to ask the user for authentication information, like a password, before proceeding further.

In addition, if a form is opened using ShowDialog, the form is opened in **modal** mode, which means that the user cannot work with other open forms until the dialog window is closed, thus forcing the user to take an action. The Show method, on the other hand, opens a form in **nonmodal** mode, which makes other forms accessible to the user.

Step 1-10: Change the Show method to ShowDialog in Line 28, and put a break in line 31, as shown in Figure 11-5.

Figure 11-5: The ShowDialog method

```
28          displayInvestment.ShowDialog();
29
30          // Get input data from Text Boxes:
31          double yearlyInvestment = double.Parse(txtInvestment.Text);
```

Step 1-11: Run the form and click the "How Long?" button. The DisplayInvestment form opens, but no data are displayed because processing stops at line 28.
Close the DisplayInvestment form. Note that the program breaks at line 31, indicating that processing continues when you close the dialog window.
Change line 28 back to displayInvestment.Show();.

Review Questions

11.1 A form named DisplayOrder has a Label named lblTotalCost that displays the total cost. Another form within the same project needs to get the data from the Label, lblTotalCost, at runtime. Write the statement(s) to open DisplayOrders, get the total cost and save it in a variable named totalCost. Assume that lblTotalCost is Public.

11.2 How does hiding a form differ from closing a form?

11.3 What is the difference between the Show and ShowDialog methods of a form?

11.3 Sharing a Method between Multiple Forms

When a method needs to be used by more than one form within a project, it would be easier to maintain it if there is only one copy of the method within the project, rather than one in each form. Keeping such methods within any one form makes maintenance difficult, if the form is deleted due to future changes in the application. It would be better to keep shared methods in a separate class within the project.

The application shown in Figure 11-6 is used to illustrate how a method is shared between multiple forms. It uses three different forms:

1. A form named LookUpPrice that finds the price for a specified product number. To make repeated searches efficient, the data are stored in a Dictionary when the form is loaded, as we did in the form DictionaryValueType that was discussed in Chapter 9.
2. A second form named DisplayProducts that displays the product number and prices for a group of products with prices below a specified limit. Again, the data are read into a Dictionary when the form is loaded.
3. A form named ProductsMain that serves as a switchboard to let users open the other two forms.

Figure 11-6: A main form that opens two other forms

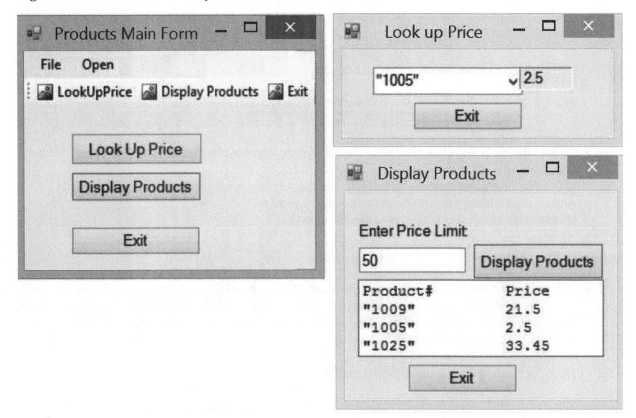

Because both LookUpPrice and DisplayProducts need to read and store the data in a Dictionary, we use a common method called LoadDictionary within a separate class and called from both forms, as shown in Figure 11-7.

Figure 11-7: Sharing a method between two forms

Tutorial 2: Sharing the LoadDictionary Method between Forms

This tutorial develops the application shown in Figure 11-7, which shares a method between two forms.

Step 2-1: Open the project, Ch11_MultiFormApps2, from the Tutorial_Starts folder.
The Price.txt file shown in Figure 11-8 is stored in the Debug folder of the current project.

Figure 11-8: Price.txt file

    "1033", 89.95
    "1009", 21.50
    "1014", 159.00
    "1005", 2.50
    "1001", 55.70
    "1025", 33.45
    "1040", 130.50
    "1035", 109.00

Step 2-2: Add a class named SharedMethods (select *Project, Add Class*).
Add a new method named LoadDictionary that reads the product numbers and prices from the text file Price.txt, and load them into a Dictionary<string, float>, as shown in Figure 11-9. You may copy the code from Load event handler of DictionaryValueType form in Ch9_Collections project.
Make sure you add the using directive **using System.IO;**, as shown in line 6, which specifies the namespace that contains the StreamReader class.

Figure 11-9: Code to add items to the Dictionary

```csharp
6   using System.IO;
7   namespace Ch11_MultiFormApps2
8   {
9       class SharedMethods
10      {
11          public static void LoadDictionary(Dictionary<string, float> prices)
12          {
13              string currentLine;
14              string[] fields = new string[2];
15              //Create StreamReader
16              StreamReader PriceReader = new StreamReader("Prices.txt");
17
18              //Read each line from the file, split and store in an array.
19              while (PriceReader.EndOfStream == false)
20              {
21                  currentLine = PriceReader.ReadLine(); // Read a product# and price
22                  fields = currentLine.Split(','); // Store in an array
23                  // Add the key (Prod#) and value (price) to the Dictionary:
24                  prices.Add(fields[0], float.Parse(fields[1]));
25              }
26              PriceReader.Close();
27          }
28      }
29  }
```

Static Methods

Note that the method LoadDictionary is declared as a static method. A static method is not associated with a specific instance (object) of the class. It is associated only with the class. So, it is invoked by the class name, followed by the member access operator (.) and the method name, as in

>   SharedMethods.LoadDictionary

To call a **nonstatic method**, you have to create an object of the class and use the object name in place of the class name.

Public Methods

LoadDictionary is also a public method. Public methods can be called from other forms. The LoadDictionary method needs to be called from the forms LookUpPrice and DisplayProducts.

Passing Dictionary as a Parameter to a Method

The header of the LoadDictionary method in line 11 declares a parameter (a reference variable) named prices that is of Dictionary type, so that prices can be shared with the calling program. The rest of the code to read data and load into the Dictionary is essentially same as the code in the form DictionaryValueType that we discussed in Chapter 9.

Next, create the two forms that invoke this method.

The coding for the **LookUpPrice** form has the following three steps:

1   Declare a variable named prices that references a Dictionary at the class level. The variable needs to be declared at the class level so that it can be used in the Load event to add product numbers to

the ComboBox from the Dictionary, and used in the SelectedIndexChanged event of the ComboBox to find the price for a selected product.

2. Call the LoadDictionary method (from SharedMethods class) to load data into the Dictionary. This is done within the Load event handler of the form. In addition, add the product number to the ComboBox.

3. When the user selects a product number from the ComboBox, find the corresponding item from the Dictionary. To do this, you use the SelectedIndexChanged event of the ComboBox.

Step 2-3: Add the code from Figure 11-10 to the LookUpPrices form.

Specify LookUpPrice as the startup form in the Program.cs file. Run the form and test the code.

Figure 11-10: Code to load data into the Dictionary and find price

```
19      private Dictionary<string, float> prices = new Dictionary<string, float>();
20
21      private void LookUpPrice_Load(object sender, EventArgs e)
22      {
23          // Call LoadDictionary method from sharedMethods class.
24          SharedMethods.LoadDictionary(prices);
25          // Add product numbers (key) to the ComboBox:
26          foreach (string prodNumber in prices.Keys)
27              cboProdNumber.Items.Add(prodNumber);
28      }
29
30      private void cboProdNumber_SelectedIndexChanged(object sender, EventArgs e)
31      {
32          string selectedProduct = cboProdNumber.SelectedItem.ToString();
33          // Use the key (Product#) to get the value (price).
34          float price = prices[selectedProduct];
35          lblPrice.Text = price.ToString();
36      }
```

Line 19 creates a Dictionary, and assigns its reference to a variable named prices.

Line 24 invokes the LoadDictionary method, passing the variable, prices, as an argument. When control returns to the Load event handler, the Dictionary referenced by prices will have the data stored in it.

Lines 26 and 27 add the product numbers to the ComboBox by looping through each item in the Keys collection of the Dictionary.

Line 34 finds the price from the Dictionary using the selected product as the key.

It's time to practice! Do Step 2-4.

Step 2-4: Do Exercise 11-1 to add the code to the DisplayProducts form to do the following:
Declare a variable to reference the Dictionary, at the class level.
Call the LoadDictionary method to load the data into the Dictionary.
Display products that have prices below the limit.
Run the form and test the code.

Step 2-5: Open ProductsMain form, which opens the other two forms.

Step 2-6: Add the code from Figure 11-11 to the Click event handlers of the buttons.

Figure 11-11: Code to open forms

```
19      private void btnLookUp_Click(object sender, EventArgs e)
20      {
21          LookUpPrice lookUpPrice = new LookUpPrice();
22          lookUpPrice.Show();
23      }
24      private void btnDisplayPrices_Click(object sender, EventArgs e)
25      {
26          DisplayProducts displayProducts = new DisplayProducts();
27          displayProducts.Show();
28      }
29      private void mnuExit_Click(object sender, EventArgs e)
30      {
31          this.Close();
32      }
```

Lines 21 and 22 create an instance of the form LookUpPrice, and use the Show method of the object to show the form. Similarly, lines 26 and 27 open the DisplayProducts form.

Step 2-7: Specify PrductsMain as the start up form in the the Program.cs file.
Run the main form and test the code.

The complete code for the three forms (ProductsMain, DisplayProducts and LookUpPrice) and the SharedMethods class that contains the LoadDictionary method is shown below:

Chapter 11: Multi-form Applications and Menus

Code for ProductsMain form

```csharp
13  public partial class ProductsMain : Form
14  {
15      public ProductsMain()
16      {
17          InitializeComponent();
18      }
19      private void btnLookUp_Click(object sender, EventArgs e)
20      {
21          LookUpPrice lookUpPrice = new LookUpPrice();
22          lookUpPrice.Show();
23      }
24      private void btnDisplayPrices_Click(object sender, EventArgs e)
25      {
26          DisplayProducts displayProducts = new DisplayProducts();
27          displayProducts.Show();
28      }
```

Code for the DisplayProducts form

```csharp
13  public partial class DisplayProducts : Form
14  {
15      public DisplayProducts()
16      {
17          InitializeComponent();
18      }
19      private Dictionary<string, float> prices = new Dictionary<string, float>();
20
21      private void DisplayPrices_Load(object sender, EventArgs e)
22      {
23          // Call LoadDictionary method from sharedMethods class.
24          SharedMethods.LoadDictionary(prices);
25      }
26
27      private void btnDisplayProducts_Click(object sender, EventArgs e)
28      {
29          lstDisplay.Items.Add("Product#          Price");
30          foreach (KeyValuePair<string, float> kvp in prices)
31          {
32              float priceLimit = float.Parse(txtPriceLimit.Text);
33              if (kvp.Value < priceLimit)
34                  lstDisplay.Items.Add(kvp.Key + "             " + kvp.Value);
35          }
36      }
```

Code for the LookUpPrice form

```csharp
13  public partial class LookUpPrice : Form
14  {
15      public LookUpPrice()
16      {
17          InitializeComponent();
18      }
19      private Dictionary<string, float> prices = new Dictionary<string, float>();
20
21      private void LookUpPrice_Load(object sender, EventArgs e)
22      {
23          // Call LoadDictionary method from sharedMethods class.
24          SharedMethods.LoadDictionary(prices);
25          // Add product numbers (key) to the ComboBox:
26          foreach (string prodNumber in prices.Keys)
27              cboProdNumber.Items.Add(prodNumber);
28      }
29
30      private void cboProdNumber_SelectedIndexChanged(object sender, EventArgs e)
31      {
32          string selectedProdcut = cboProdNumber.SelectedItem.ToString();
33          // Use the key (Product#) to get the value (price).
34          float price = prices[selectedProdcut];
35          lblPrice.Text = price.ToString();
36      }
```

Code for SharedMethods class with the LoadDictionary method

```csharp
9   class SharedMethods
10  {
11      public static void LoadDictionary(Dictionary<string, float> prices)
12      {
13          string currentLine;
14          string[] fields = new string[2];
15          //Create StreamReader
16          StreamReader PriceReader = new StreamReader("Prices.txt");
17
18          //Read each line from the file, split and store in an array.
19          while (PriceReader.EndOfStream == false)
20          {
21              currentLine = PriceReader.ReadLine(); // Read a product# and price
22              fields = currentLine.Split(','); // Store in an array
23              // Add the key (Prod#) and value (price) to the Dictionary:
24              prices.Add(fields[0], float.Parse(fields[1]));
25          }
26          PriceReader.Close();
27      }
28  }
```

Review Questions

11.4 How would you share a method between different forms in a project? That is, what change(s) do you need to make in the method, and where will you create the method?

11.5 What is the difference between a static method and a nonstatic method?

11.4 Closing a Parent Form

When you close a parent form (a form that creates an instance of another form), like ProductsMain, the instances of all other forms that were created in the main form also will be closed. This could be a problem in some situations. Let's look at an example.

Login Form

As in many applications, assume that the user needs to enter a valid password before accessing the form ProductsMain. We use a form named Login, as shown in Figure 11-12, to let the user enter a password.

Figure 11-12: Login form

The user enters a password and clicks the OK button. If the password is valid, we want the form ProductsMain to be displayed and the Login form to be closed; if not, a message is to be displayed. Cancel should close the Login form.

If you create and show an instance of the ProductsMain form from the Login form, you cannot close the Login form without also closing the ProductsMain form. Closing the Login form will automatically close the ProductsMain form.

So, how do you close the Login form but keep ProductsMain open? You open both the Login form and the ProductsMain form from the Main method in Program class, as shown in Figure 11-13.

11.4 Closing a Parent Form

Figure 11-13: Main method in Program class

```csharp
    static class Program
    {
        static void Main()
        {
            Application.EnableVisualStyles();
            Application.SetCompatibleTextRenderingDefault(false);

            Login login = new Login();
            login.ShowDialog(); // Open Login form as a dialog box.
            // Processing of Main stops till Login form is closed

            if (login.DialogResult == DialogResult.OK)
                Application.Run(new ProductsMain());
        }
    }
```

The statements in Lines 16 and 17 create an instance of the Login form and open it as a dialog box. So, after the Login form is opened, processing of Main stops until the Login is closed. Let's look at what happens on the Login form. Figure 11-14 shows the code in the Login form.

Figure 11-14: Code to verify password on the Login form

```csharp
        private void btnOK_Click(object sender, EventArgs e)
        {
            if (txtPassword.Text == "Csharp101")
            {
                this.DialogResult = DialogResult.OK;
                this.Close();
            }
            else
            {
                MessageBox.Show("Sorry, invalid passwd");
                txtPassword.Focus();
                txtPassword.SelectAll();
            }
        }
        private void btnCancel_Click(object sender, EventArgs e)
        {
            this.Close();
        }
```

If the user enters the correct password and clicks the OK button on the Login form, the **DialogResult** property of the Login form is assigned the value **DialogResult.OK**, as shown in Line 24 in Figure 11-14, and the form is closed. Processing of Main method would continue at line 20, which checks the value of the DialogResult property of the Login form.

If the password is invalid, the user is given the option to re-enter it. If the user closes the Login form without entering the correct password, the DialogResult property would have the default value **DialogResult.Cancel**.

When the Login form is closed, processing of Main continues at line 20 of the Main method:
```csharp
            if (login.DialogResult == DialogResult.OK)
```

This statement checks whether the value of DialogResult property of the Login form is DialogResult.OK. If it is, the statement in line 21,
```
Application.Run(new ProductsMain());
```
opens the ProductsMain form.

Note that the statement Application.Run(new ProductsMain()) may be replaced by the following pair of statements:
```
ProductsMain ProductsMain = new ProductsMain();
ProductsMain.ShowDialog();
```

In summary, if the user enters a valid password, the ProductsMain form is opened by the method Main() rather than by the Login form. So, closing the Login form will not have any effect on the ProductsMain form.

Step 2-8: Open the Login form, shown in Figure 11-15.

Figure 11-15: The Login form

Step 2-9: Add the code in lines 24 and 25 from Figure 11-14 to the Click event handlers of the OK button.

Step 2-10: Add the code in lines 16–21 from Figure 11-13 to the Main method in the class, Program.

Step 2-11: Run the project and test the code with both valid and invalid passwords, and by clicking the Cancel button.

In summary, to open the ProductsMain form based on authentication in the Login form, do not open the parent form, ProductsMain, from the Login form. Open both from the Main method in the Program class. That allows closing the Login form without closing ProductsMain.

11.5 Menus

Buttons on a form provide a convenient way to access a limited number of major functions that are performed frequently. Menus provide access to additional functions. We will look at two types of menus:
1. The top-level menu (main menu) that is displayed at the top of the form to let you select form-wide functions
2. Context menus (shortcut menus) that are displayed when you right click an object, to let you select functions that apply to that object

A top-level menu is added to a form by dragging and dropping the MenuStrip object from the Toolbox to the form. Let's create the menus shown in Figure 11-16 on the ProductsMain form to understand the process.

Figure 11-16: Menus on ProductsMain form

Step 2-12: Add the MenuStrip object from the *Menus and Toolbars* group on the Toolbox. You will see the object in the component tray. Change the name of the MenuStrip from menuStrip1 to MainMenu.

Step 2-13: Click on MainMenu in the component tray. The Menu Designer is displayed at the top of the form with the *Type Here* field, as shown in Figure 11-17.

Figure 11-17: The Menu Designer

Step 2-14: Type **File** in the *Type Here* field. This creates a new menu item of the type ToolStripMenuItem. Change the Text property to &File to make "F" the **access key** for this menu. That means you can press Alt+F to access this menu.

Step 2-15: Type **Exit** in the *Type Here* field in the dropdown menu below File. Because we add code to the Click event handler of this menu item, change its name to **mnuExit** to make it more meaningful. For the **ShortCutKeys** property, select Ctrl and x. That means pressing Ctrl+x will have the same effect as selecting this menu item.

Adding Code to the Click Event Handler of Menu Items

Each menu item acts like a button. You can add code to the Click event handlers of the menu items.

Step 2-16: Double click Exit menu item to open the Click event handler. For code, type
this.Close();

Next, you create the second menu, **Open**, and its dropdown menu with two options: **Look Up Price** and **Display Products**.

Step 2-17: Type **Open** in the *Type Here* field. In the Text property, change the Text to &Open to make "O" the **access key** for this menu. That means you can press Alt+O to access this menu.

Step 2-18: Type **Look Up Price** in the *Type Here* field in the dropdown menu below Open. Because we add code to the Click event handler of this menu item, change its name to **mnuLookUpPrice**. For the **ShortCutKeys** property, select Ctrl and L. That means pressing Ctrl+L will have the same effect as selecting this menu item.

Step 2-19: Type **Display Products** in the *Type Here* field below **Look Up Price.** Change the name of the menu item to **mnuDisplayProducts.** Set Ctrl+D as the ShortCut for this menu item.

When you have different groups of menu items, you may add a **separator** between groups by selecting it from the dropdown menu of the *Type Here* field, as shown in Figure 11-18:

Figure 11-18: Menu with a separator

Adding Code to the Click Event Handlers of the Menu Items Look Up Price and Display Products

Note that the menu item mnuLookUpPrice and the button btnLookUpPrice perform the same function of opening the form, LookUpPrice. So, rather than duplicating the code in the Click event handlers of both the button and the menu, we will create a method named LookUpPrice() and call it from both event handlers.

Calling Event Handlers

You may wonder, why not call the the Click event handler btnLookUp_Click directly from the Click event handler mnuLookUpPrice_Click, instead of creating a new method and calling it from both event handlers? This is not desireable because if future changes in the user interface require deleting the btnLookUp button, the menu won't work. In general, calling event handlers from other programs makes maintenance more difficult.

Step 2-20: Create a method named OpenLookUpPrice(), and call it from the Click event handler of the menu item, mnuLookUpPrice, as shown in Figure 11-19:

Figure 11-19: The OpenLookUpPrice method

```
33          private void OpenLookUpPrice()
34          {
35              LookUpPrice lookUpPrice = new LookUpPrice();
36              lookUpPrice.Show();
37          }
38          private void mnuLookUpPrice_Click(object sender, EventArgs e)
39          {
40              OpenLookUpPrice();
41          }
```

Step 2-21: Delete the code from the Click event handler of btnLookUpPrice and add the code to call the OpenLookUpPrice method, as shown in Figure 11-20:

Figure 11-20: The modified event handler

```
19          private void btnLookUp_Click(object sender, EventArgs e)
20          {
21              // Modified code:
22              OpenLookUpPrice();
23          }
```

Submenus

You may create a submenu for a menu item like Display Products by typing the submenu item in the *Type Here* field displayed to the right.

Checked Menu Items

A checked menu item has a CheckBox next to it. When the user clicks the menu item, the CheckBox alternates between the states checked and unchecked. To create a checked menu item, set the Checked property of a standard menu item to True.

A checked menu item may be used for on/off type of actions, like hiding and unhiding a control, and to indicate which item is selected from a group of menu items.

You may add code to the Click event handler of a checked menu item just like any other event handler.

It's time to practice! Do Exercise 11-2.

Context Menus

Typically, a context menu is associated with a control, and it is displayed by right clicking the control. For example, right clicking a TextBox at runtime displays its standard context menu items, like undo, cut, copy, and paste.

The process of creating a ContextMenu is very similar to the process of creating a top-level menu (main menu). A major difference is that you need to associate the context menu to a specific control.

The context menu is created using the ContextMenuStrip control. Let's create a context menu to let the user change the BackColor and ForeColor of the ListBox on the form DisplayProducts.

Step 2-22: Drag and drop the ContextMenuStrip object from the *Menus and Toolbars* group on the DisplayProducts form. You will see the object in the component tray. Change the name of the ContextMenuStrip from ContextMenuStrip1 to DisplayContextMenu.

Step 2-23: To associate the ContextMenu to the ListBox, set the ContextMenu property of the ListBox, lstDisplay, to DisplayContextMenu.

Step 2-24: Click on DisplayContextMenu in the component tray. The Menu Designer is displayed at the top of the form with the *Type Here* field.
Type BackColor in the *Type Here* field. Change the name of the field to cnuBackColor (the prefix "cnu" represents ContextMenu).
Set ShortCutKeys to Ctrl+B.

The ContextMenu is used to change the BackColor and ForeColor of the ListBox, lstDisplay, using the ColorDialog control, which should be added to form.

Step 2-25: Add the ColorDialog control to the form, and change its name to DisplayColorDialog. Double click BackColor menu item from the ContextMenu and enter the code shown in Figure 11-21 to the Click event handler of cnuBackColor.

Figure 11-21: Click event handler of ContextMenu

```
37
38          private void cnuBackColor_Click(object sender, EventArgs e)
39          {
40              DisplayColorDialog.ShowDialog();
41              lstDisplay.BackColor = DisplayColorDialog.Color;
42          }
```

Line 40 displays the ColorDialog window, and line 41 assigns the selected color to the BackColor property of the ListBox.

Step 2-26: Run the project and test the code.

It's time to Practice! Do Exercises 11-3.

Review Question

11.6   Describe a context menu.

11.6 ToolStrips

The ToolStrip control lets you create a dockable Toolbar that acts as a container for buttons and a variety of other controls, including Labels, TextBoxes and ComboBoxes. Unlike menu items in a top-level menu (main menu), the buttons on a Toolbar are readily visible.

Both an image and Text may be displayed on a button, as shown in Figure 11-22. By default, the ToolStrip is docked at the top of the form. You may use the Dock property of ToolStrip to choose other options, including *Left* and *Right*. Another feature of ToolStrip is that when there is not enough room to display all controls on a ToolStrip, controls are moved to a dropdown menu in the overflow area. In Figure 11-22, the Exit button in the overflow area is available by clicking the dropdown arrow at the end.

Figure 11-22: The ToolStrip control

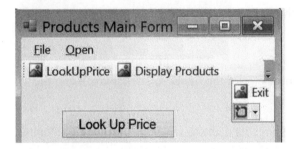

To add a Toolbar, drag and drop the ToolStrip control on the form.

Let's create a Toolbar that essentially duplicates the functionality of the three buttons on the ProductsMain form, as shown in Figure 11-22.

Step 2-27: Drag and drop the ToolStrip control on the form. Change the name of the ToolStrip object in the component tray to ProductToolStrip. By default, the ToolStrip is docked at the top of the form. You may use the Dock.

Step 2-28: Click on the ToolStrip to display the designer at the top of the form. Click on dropdown handler on the control to view the list of controls that can be added, as shown in Figure 11-23.

Figure 11-23: Controls on ToolStrip

Select the Button control. Change the name of the button to tsbLookUpPrice and the Text property to Look Up Price.

To display both image and Text, change **DisplayStyle** to ImageAndText.

To display an image on the button, click on the ellipses on the right of Image property to open the *Select Resources* window, and select an image that represents the task of opening.

Adding Code to the Click Event Handler of ToolStrip Buttons

The Click event handler is created by double clicking the button.

Step 2-29: Add the code from Figure 11-24 to call the OpenLookUpPrice method from the Click event handler of the button.

Figure 11-24: Click event handler of tsbLookUpPrice

```
43      private void tsbLookUpPrice_Click(object sender, EventArgs e)
44      {
45          OpenLookUpPrice();
46      }
```

The modified code for ProductsMain form with menus and ToolStrip is shown below:

```
19      private void btnLookUp_Click(object sender, EventArgs e)
20      {
21          // Modified code:
22          OpenLookUpPrice();
23      }
24      private void btnDisplayPrices_Click(object sender, EventArgs e)
25      {
26          DisplayProducts displayProducts = new DisplayProducts();
27          displayProducts.Show();
28      }
29
30      private void btnExit_Click(object sender, EventArgs e)
31      {
32          this.Close();
33      }
34      private void mnuExit_Click(object sender, EventArgs e)
35      {
36          this.Close();
37      }
38      private void OpenLookUpPrice()
39      {
40          LookUpPrice lookUpPrice = new LookUpPrice();
41          lookUpPrice.Show();
42      }
43      private void mnuLookUpPrice_Click(object sender, EventArgs e)
44      {
45          OpenLookUpPrice();
46      }
47      private void tsbLookUpPrice_Click(object sender, EventArgs e)
48      {
49          OpenLookUpPrice();
50      }
```

Review Question

11.7 Different methods that can be used to let the user perform a task in a C# form include (1) clicking a standard button, (2) selecting an item from a standard top-level menu (main menu), (3) selecting an item from a Context menu, and (4) clicking a button on a custom Toolbar.

Consider a form that allows the user to perform different tasks associated with processing payroll (compute pay, deductions, etc.) and save the data. These tasks are performed by clicking standard buttons on the form. In addition, it is desired to let the user change the appearance (color and font size) of a ListBox on the form.

Which one of the above methods is appropriate to let the user change the back color and/or the font size of the ListBox?

Exercises

Exercise 11-1

Create the form DisplayProducts, shown in Figure 11-25. Create a Dictionary in this form, and call LoadDictionary method to load the Dictionary with data. Use the Dictionary to display all products with price below the specified limit.

Figure 11-25: The Display Products form

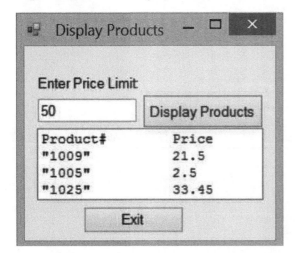

Exercise 11-2

Create a new method named OpenDisplayProducts to open the DisplayProducts form in Ch11_MultiformApps2. Call this method from the Click event handlers of btnDisplayProducts button and mnuDisplayProducts menu item.

Exercise 11-3

Add another menu item, ForeColor, to the Context menu on the DisplayProducts form to let the user change the ForeColor of the ListBox

Chapter 12

Databases

Database systems, in general, make data access faster and easier than file-based systems. They incorporate features that make it easier to share data, reduce redundancy and enhance security. In this chapter, you will learn how to access databases from C# programs.

Popular database management systems (software that manage databases) include Oracle, Microsoft SQL Server, DB2 and Microsoft Access. We use SQL Server database in this book; however, the techniques and steps presented in this chapter generally are applicable to all databases.

Learning Objectives

After studying this chapter, you should be able to:

- Describe how a data table differs from a database table.
- State the functions of the Table Adapter and Binding Source.
- Create a Data Source.
- Display data in a DataGridView and access the data.
- Display data in Details view.
- Find records using the key.
- Select records using the Binding Source and Table Adapter.
- Work with untyped data sets.
- Use Command object and DataReader to display data.
- Display data in a ListView and access the data.

Topics

12.1	Introduction		12.7	Selecting Records Using Binding Source
12.2	Creating Database Objects		12.8	Selecting Records Using Table Adapter
12.3	Displaying Data in a DataGridView		12.9	Untyped Datasets: Displaying Records
12.4	Accessing Data from a DataGridView		12.10	Untyped Datasets: Select Records
12.5	Displaying Data in Details View		12.11	Untyped Datasets: Add/Edit/Delete Record
12.6	Finding the Record for a Selected Key		12.12	Command Object and DataReader

12.1 Introduction

A relatively easy way to work with data from database tables is to display data in controls like TextBoxes, ComboBoxes and DataGridViews using a two-step process:

1. Transfer the data from **database tables,** which typically are stored on the disk, to **data tables**, which are temporary tables maintained in memory.
2. Bind (link) controls on a form to the data table to display the data.

A data table may contain data from one or more database tables. Data tables generally are placed inside a container called **dataset**. Thus, a dataset is a collection of one or more data tables.

The **table adapter** and **binding source** are two important .Net Framework classes that help you to work with databases.

The **table adapter** connects the data table to the database, and retrieves data from and saves data to the database.

The **binding source** links the user interface controls on the form to a data table. In addition, it provides methods and properties to perform additional functions like selecting records from the data table, and moving between records.

Figure 12-1 represents the relationship between database tables, data table and controls on a form.

Figure 12-1: Relationship between database tables, data table and controls on a form

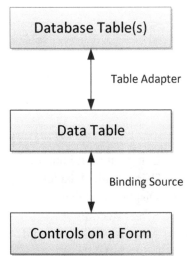

We will use the wizard to create the necessary objects and to help you work with the data.

12.2 Creating Database Objects

To create the database object, first, you specify the **data source**—that is, the source of data, its type and location. The data source could be a database on the local computer or on a server, or a web service or other object.

Next, you drag the data table to the form to create the controls that display the data and objects that bind the controls to the data table.

Specifying the Data Source

You specify the data source using the **Data Source Configuration Wizard**, which in turn generates the data tables and other objects, like the binding source.

Tutorial 1: Display Employee Records Using the Wizard

Let's create an application that lets you work with a human resource database that includes an Employee table.

Step 1-1: Create a new project named Ch12_Databases. Rename the form to EmployeeTabularView.

Step 1-2: Copy the SQL Server database file HR.mdf from Tutorial_Starts/DataFiles folder to a local directory outside the project directory, Ch12_Database. **Appendix B** provides instructions on how to create a new database in Visual Studio.

Step 1-3: To start the Data Source Configuration Wizard, click *Project* from the menu, and click *Add New Data Source*. You will see the Configuration Wizard window, as shown in Figure 12-2.

Figure 12-2: Selecting data source type

Step 1-4: Select *Database* and click *Next*. You will be asked to select a database model.

Step 1-5: For database model, select *Dataset*. You will be asked to choose the data connection, as shown in Figure 12-3.

Figure 12-3: Choosing data connection

Step 1-6: If the dropdown list shows the database file name, HR.mdf, you may go to step 1-8. If not, click the *New Connection* button. If the *Add Connection* window is displayed, click the *Change* button. The *Choose Data Source* window, shown in Figure 12-4, opens.

Figure 12-4: Adding connection

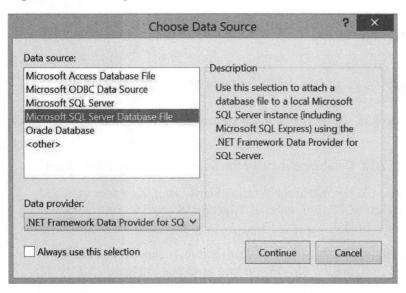

Step 1-7: Select *Microsoft SQL Server Database File* for *Data Source*. The dropdown list for *Data Provider* should show *.Net Framework Data Provider for Microsoft SQL Server*.

Click *Continue*. The following *Add Connection* window appears:

Click *Browse*, and select HR.mdf from the local directory. Click *Open*.

Click *OK*. If the database is not compatible with the current instance of SQL Server, you will be asked whether you want to upgrade the database. Select *Yes*.

Step 1-8: Click *Next*. You may be asked whether you want to copy the file to your project. **Make sure you select *No*.**

If you answer *Yes,* Visual Studio will copy the database file to the project directory, and by default, changes made to the database at runtime will be lost the next time the project is run.

Click *Next*. You will be asked whether you want to save the connection string.

Accept the default (Yes), and Click *Next*. You will be asked to select the database objects, as shown in Figure 12-5.

Chapter 12: Databases 339

Figure 12-5: Selecting database objects

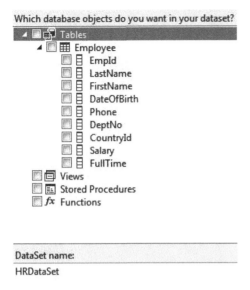

Step 1-9: Select Employee table by checking the CheckBox next to it. This will select all fields of the Employee table.

If you want to display only certain fields, uncheck the fields to be excluded. You also may select a view that combines data from multiple tables.

Leave the dataset name as HRDataSet. Though the current data are only from the Employee table, we will be adding data from other tables later. So, the dataset will have data tables from the entire HR database.

Click *Finish.*

If the Server Exaplorer window is not displayed, select *View, Server Explorer*. The Server Explorer window will show HR.mdf under Data Connections, as shown in Figure 12-6.

Figure 12-6: Server Explorer

▲ 🗗 Data Connections
   ▷ 🗐 HR.mdf

Step 1-10: Display the DataSource window: Click *View, Other Windows, Data Sources*. The DataSource window will be displayed on the left, as shown in Figure 12-7.

Figure 12-7: DataSource window

To view the fields and their properties, right click Employee table in Server Explorer window, and select *Open Table Definition*. The T-SQL window displays the SQL that creates the table. Close dbo.Employee[Design] window.

12.3 Displaying Data in a DataGridView

The DataGridView control provides a simple way to display and edit an entire table in a scrollable window. An alternative is to view one record at a time in controls like TextBoxes and Labels that display a single data item. To display a data table and work with it on a form, you follow two simple steps:

1. Choose between DataGridView (to view the entire table) and Details (to view one record at a time) from the dropdown list for the data table, as shown in Figure 12-8.
2. Drag and drop the data table on the form.

Figure 12-8: Display options

First, we will display employee data in a DataGridView.

Step 1-11: Choose DataGridView from the dropdown list for Employee data table.

Step 1-12: Drag and drop Employee table on the form.

> You will see a grid with the field names, a navigation bar at the top, and several objects, including the HRDataSet, employeeTableAdapter and employeeBindingSource, in the section called component tray, as shown in Figure 12-9. All these objects were created automatically by the wizard when you dropped the Employee table on the form.

Figure 12-9: DataGridView bound to the Employee table

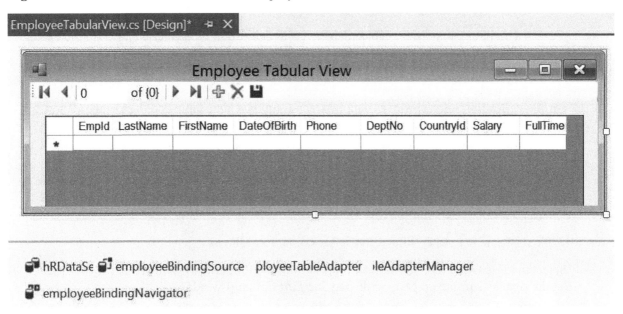

You may change the width, header, name, format or any other properties of columns by clicking on *edit column* (at the bottom of Properties window) to open the edit column window.

To understand the functions of the various objects displayed in the component tray, let's take another look at the relationship between them. The relationship, introduced earlier in Figure 1-10, is presented in Figure 12-10.

Figure 12-10: Relationship between components

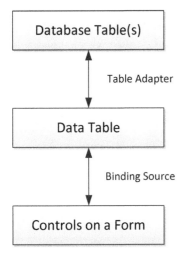

The objects on the form include

> **HRDataSet:** A dataset contains one or more data tables. Currently, HRDataSet contains only the Employee data table. When you run the project, the employeeTableAdapter will load the data from Employee database table into Employee data table.

employeeTableAdapter: Moves data between the database table (Employee), and the data table that also is named Employee, using **Fill** and **Update** (or **UpdateAll**) methods:

> The **Fill** method of the table adapter copies data from the database table into the data table.
>
> The **Update/UpdateAll** method saves the data from the data table into the database table.

employeeBindingSource: Moves data between Employee data table and the DataGridView on the form.

employeeBindingNavigator: Provides the navigation bar at the top of the form.

TableAdapterManager: If you are editing multiple related tables on a form, you can use the UpdateAll method of TableAdapterManager to update all tables in the correct order.

Step 1-13: First, specify Currency format for Salary column as follows:

Select the DataGridView, and select *Columns* from the Properites window. Click the ellipses (…) on the right side of the property to display the *Edit Columns* window.

Select Salary, *DefaultCellStyle*, and click the ellipses (…) on the right side of the property, as shown in the following figure.

Select *Format* from the *CellStyle Builder* window, and click the ellipses on the right to open the *Format String Dialog* window.

Select *Currency*, and specify 0 for *Decimal Places*.

Next, run the project. You will see the employee records displayed in the grid, as in Figure 12-11.

To test the form, make changes to a record, click Save and close the form. Now reopen the form to make sure the changes were saved. Test to make sure adding a new record and deleting a record work correctly.

Figure 12-11: Employee table in the DataGridView

EmpId	LastName	FirstName	DateOfBirth	Phone	DeptNo	CountryId	Salary	FullTime
11002	Iwuji	Chris	11/1/1992	920-426-4266	1	62	$46,200	☑
11005	Mathews	Ruth	3/24/1985	646-233-1122	4	22	$44,600	☑
11007	Rodriguez	Gina	5/30/1993	920-233-2233	3	10	$48,700	☐
11010	Hope	Chris	11/23/1990	630-426-7696	3	34	$55,400	☑
11015	Nair	Shashi	7/3/1982	920-424-3153	1	12	$65,200	☑
11016	Hope	John	9/30/1991	251-602-4321	2	34	$58,500	☐

Data-Bound Controls

A data-bound control is a user interface control that is linked to a data source. When you dragged and dropped the Employee data table to the form, the wizard added the DataGridView to the form and bound it to the Employee data table through the binding source. As you can see from Figure 12-12, the Data Source property of the DataGrid is set to employeeBindingSource.

Figure 12-12: DataSource property of DataGridView

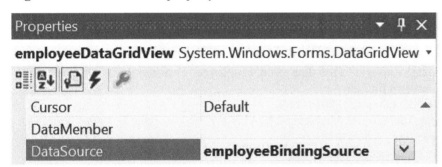

The employeeBindingSource, in turn, is linked to the Employee table in the HRDataSet, as indicated by the DataMember and DataSource properties of the binding source, shown in Figure 12-13.

Figure 12-13: DataMember and DataSource properties of binding source

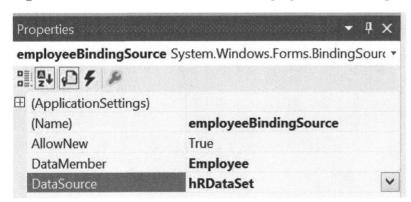

Auto-generated Code

Figure 12-14 shows the code generated by the wizard.

Figure 12-14: Code generated by the wizard

```csharp
20    private void EmployeeTabularView_Load(object sender, EventArgs e)
21    {
22        // TODO: This line of code loads data into the 'hRDataSet.Employee' table.
23        this.employeeTableAdapter.Fill(this.hRDataSet.Employee);
24    }
25
26    private void employeeBindingNavigatorSaveItem_Click(object sender, EventArgs e)
27    {
28        this.Validate();
29        this.employeeBindingSource.EndEdit();
30        this.tableAdapterManager.UpdateAll(this.hRDataSet);
31    }
```

The statement in line 23,

```
this.employeeTableAdapter.Fill(this.hRDataSet.Employee);
```

uses the Fill method of the table adapter to load data from the database table into the data table, Employee.

Saving changes from the DataGridView to the database table is a two-step process:

1. Save changes from the DataViewGrid to the data table, Employee. This is done using the EndEdit method of the binding source in line 29:
   ```
   this.employeeBindingSource.EndEdit(),
   ```
2. Save changes from the data table to the database table. This is done using the UpdateAll method of tableAdapterManager in line 30:
   ```
   this.tableAdapterManager.UpdateAll(this.hRDataSet);
   ```

View or Edit Connection String

Step 1-14: To view or edit connection string information, double click Settings under properties in the Solution Explorer, as shown in Figure 12-15.

Figure 12-15: Viewing ConnectionString information

You will see the connection string information, as shown in Figure 12-16.

Figure 12-16: ConnectionString information

Name	Type	Scope	Value
HRConnectionString	(Connectio...	Application	Data Source=(LocalDB)\v11.0;AttachDbFilename=|DataDirectory|

Step 1-15: Click in the right-most column under *Value*, and click on the ellipses to open the Connection Properties window that allows you to edit the information.
Click *Cancel*, and close the Settings window.

The Structure of the Dataset

Adding a data source also adds the file HRDataset.xsd (xsd for XML Schema Document) to the Solution Explorer window.

Step 1-16: To view the structure of the tables in the dataset, double click HRDataset.xsd in Solution Explorer. You will see HRDataSet.xsd window that shows the Employee data table, and the table adapter associated with it, as shown in Figure 12-17.

Figure 12-17: Dataset Schema window

Step 1-17: Select EmployeeTableAdapter to view its properties.
The properties window shows the connection string and the Select, Update, Delete and Insert commands used by the table adapter.
Click the "+" sign on the left of Select Command to view the SQL used by the table adapter to select records from the database. You also may view the SQL for other commands.

Review Questions

12.1 What is the difference between a dataset table and a database table?

12.2 What is the function of the table adapter?

12.3 What is the function of the binding source?

12.4   What does the Fill method of the table adapter do?

12.5   What does the Update method of the table adapter do?

12.6   What is a data bound control?

12.7   How is the data in a data bound control linked to a field in a data table?

12.4 Accessing Data Items from a DataGridView

In addition to viewing and editing database tables, you can access individual data items using the properties and events of the DataGridView. Figure 12-18 shows some of the major properties and events of the DataGridView.

Figure 12-18: Properties and events of DataGridView

Property	Description
ColumnCount	The number of columns.
Columns	Collection of all columns.
CurrentRow	Row containing the current cell. Each row consists of a collection of cells represented by the **Cells** collection.
CurrentCell	Current cell. The **Value** property gives the value of a cell.
Enable Adding (Editing, Deleting) - This is an item in the **task panel** that is displayed when you click the arrow (smart tag) at the upper right corner.	When checked, user can add (edit, delete) columns. By default, it is checked.
RowCount	The number of rows, including the row for new records.
Rows	Collection of all rows. Access a cell using the column index or the column name (e.g., Rows[3].Cells[4]; or Rows[3].Cells["Phone"]). Get the value (of object type) associated with a cell using Value property (e.g., Rows[3].Cells["Phone"].Value).
SelectedColumns	Collection of columns selected by the user.
SelectedRows	Collection of rows selected by the user.
Events	
CellClick	When a cell is clicked.
CurrentCellChanged	When the current cell changes.

Let's look at some examples of how you would use these properties and events to work with the table.

Rows and Cells Collections

You may access any cell using the **Rows** and **Cells** collections by specifying its column and row indices. For example, the data item in row 2 and column 8 is given by

```
employeeDataGridView.Rows[1].Cells[7].Value.ToString()
```

Note that the index of row 1 is 0, index of row 2 is 1, and so on.

Value Property

The **Value** property gives the value associated with a cell. The ToString method converts the Value, which is of object type, to string.

You also may use the column name in place of the column index, as in

```
employeeDataGridView.Rows[1].Cells["Salary"].Value.ToString()
```

Column Names

The above statement assumes that the column name for Salary field has been changed to Salary from its original name **salaryDataGridViewTextBoxColumn**, which is shown in Figure 12-19. Though the column headers (HeaderText property) in the DataGridView are the same as the field names, the column names are not, as shown in Figure 12-19. So, if you use column names in code, you may want to change the names of those columns so that the column names are the same as the field names and column headers. As described earlier, you display the *Edit Columns* window by clicking the ellipses to the right of Columns property of the DataGridView.

Select the DataGridView, and select *Columns* from the Properites window. Click the ellipses (...) on the right side of the property to display the *Edit Columns* window.

Figure 12-19: Column properties of DataGridView

Displaying the Value of a Selected Cell

You may automatically display the value of a cell when the user selects it by clicking it, using the **CurrentCellChanged** event handler and the **CurrentCell** property. Figure 12-20 shows the code to display the value of the current cell in a Label.

Figure 12-20: CurrentCellChanged event handler of DataGridView

```
33      private void employeeDataGridView_CurrentCellChanged(object sender, EventArgs e)
34      {
35          lblCurrentCell.Text = employeeDataGridView.CurrentCell.Value.ToString();
36      }
```

Step 1-18: Add a Label named lblCurrentCell and another Label that displays the prompt, "Current Cell," as shown in Figure 12-21.

Create the CurrentCellChanged event handler of the DataGridView and add the code from Figure 12-20. Click a cell to test the code. The Label should show the data item from the selected cell.

Figure 12-21: Employee form showing the average salary

EmpId	LastName	FirstName	DateOfBirth	Phone	DeptNo	CountryId	Salary	FullTime
11002	Iwuji	Chris	11/1/1992	920-426-4266	1	62	$46,200	☑
11005	Mathews	Ruth	3/24/1985	646-233-1122	4	22	$44,600	☑
11007	Rodriguez	Gina	5/30/1993	920-233-2233	3	10	$48,700	☐
11010	Hope	Chris	11/23/1990	630-426-7696	3	34	$55,400	☑
11015	Nair	Shashi	7/3/1982	920-424-3153	1	12	$65,200	☑
11016	Hope	John	9/30/1991	251-602-4321	2	34	$58,500	☐

Current Cell: Hope   Display Phone 630-426-7696   Compute Avg $53,185.71

You may display the value from any column in the selected row using the **CurrentRow** property of the DataGridView, and the **Columns** collection of the CurrentRow. For example, the following statement displays the phone number from the current row, in a Label named lblPhone:

lblPhone.Text = employeeDataGridView.CurrentRow.Cells[4].Value.ToString();

Step 1-19: Add a Label named lblPhone, and a button named btnDisplayPhone, as shown in Figure 12-21. Add the above statement to the Click event handler of the button. See Figure 12-22. Test the code.

Accessing Data from Every Row in a Column

Now let's look at accessing every row in a column, using a loop. You will compute the average salary and display it in a Label when the user clicks a button, as shown in Figure 12-21. The code to display the phone number and compute the average is shown in Figure 12-22.

Figure 12-22: Code to access every row

```
38      private void btnDisplayPhone_Click(object sender, EventArgs e)
39      {
40          lblPhone.Text = employeeDataGridView.CurrentRow.Cells[4].Value.ToString();
41      }
42      private void ComputeAvg_Click(object sender, EventArgs e)
43      {
44      float sumOfSalary = 0, avgSalary;
45      int numOfRows = employeeDataGridView.RowCount-1;
46                  // RowCount includes the row for new record.
47      // Use for loop to vary row index and get salary from Salary column.
48      for (int rowIndex = 0; rowIndex < numOfRows; rowIndex ++)
49          sumOfSalary = sumOfSalary + float.Parse(employeeDataGridView.
50                      Rows[rowIndex].Cells["Salary"].Value.ToString());
51          // To compute total only for selected rows, use SelectedRows[]
52          avgSalary = sumOfSalary/numOfRows;
53      lblAvg.Text = avgSalary.ToString("C");
54      }
```

The statement in line 50 assumes that the name of the column for salary has been changed to Salary. Further, note that the RowCount property in line 45 includes the row for new record.

Step 1-20: Change the name of the Salary column to salary: Click edit columns under properties to open the edit column window. Select Salary, and change the *Name* property (under *Design*) to salary.

Step 1-21: Add a Label named lblAvg and a button named btnComputeAvg. Add the code from Figure 12-22 to the Click event handler of btnComputeAvg.

Run the project and test the code.

12.5 Displaying Data in Details View

The DataGridView displays an entire table in a scrollable window. This is convenient when the number of records and fields in a table is relatively small.

The details view displays one record at a time, with each data item displayed in a separate control, like Label, TextBox or CheckBox. This makes editing/adding/deleting individual records easier, and is particularly suitable when there is a large number of records and fields.

To display employee data in details view, select Details from the dropdown list for Employee data table, as shown in Figure 12-23.

Figure 12-23: Display options

For each field of Employee table, you may select the type of control (TextBox, NumericUpDown, etc.) from the dropdown list for the field, shown in Figure 12-24. The default control is TextBox, except for DateOfBirth, which has DateTimePicker, and FullTime, which has CheckBox. We will use the default controls for this form.

Figure 12-24: Selecting the type of control to display a field

Next, drag and drop the Employee table to the form to add the controls and the components needed to bind the controls to the database table.

Step 1-22: Add a new form named EmployeeDetailsView.

Step 1-23: Select Details from the dropdown list for Employee data table.

Step 1-24: Drag and drop the Employee table on the form. Change the Text property of the CheckBox for FullTime field to "Full Time?"

Rather than dragging and dropping an entire table, you may drag and drop one field at a time if you do not want all fields on the form.

Controls are added to the form, as shown in Figure 12-25. In addition, the five components, including the DataSet and table adapter, are added to the component tray.

Figure 12-25: Controls and components added to the form by the wizard

Note that each form needs its own set of five components.

Auto-generated Code

The generated code to Fill the data table and to Update the database table are identical to that for the DataGridView.

Step 1-25: Specify EmployeeDetailsView as the start-up form in the Program.cs file. Run the form. Use the navigation bar to view records, add, delete, and save changes.

12.6 Finding the Record for a Selected Key

When a table contains a large number of records, searching and finding specific records is particularly important in Details view, which displays one record at a time.

We will create a separate form that focuses on selecting records rather than adding/editing/deleting records.

Searching a Key (Unique) Field

A simple way to let the user select the value for a key field (or for any unique field) to find the corresponding record is by displaying the values of the key in a ComboBox or ListBox.

Tutorial 2: Find Employee Records Using the Wizard

In this tutorial, you create a form that allows the user to select an employee's id from a ComboBox and display the employee record, as shown below:

To create the form, drag and drop the Employee table in Details mode as before, except that we use a ComboBox (instead of the standard TextBox) for EmpId.

Step 2-1: Add a new form named **EmployeeSearch**.

Step 2-2: Select Details from the dropdown list for Employee data table.

For EmpId field, select ComboBox for type of control, as shown in Figure 12-26.

Figure 12-26: Selecting ComboBox to display employee ids

Step 2-3: Drag and drop the Employee table on the form.

The form looks the same as the EmployeeDetails form you created previously, except that the EmpId field is a ComboBox.

Step 2-4: Run the new form. The ComboBox for EmpId shows the id for the current record, but the ComboBox does not have the list of EmpId's from all records, as we want.

DataBindings Property of Controls

The reason each control displays the data from a field of the current record is that each control is bound to a field using its DataBindings property. The DataBindings property of the ComboBox that displays EmpId is shown in Figure 12-27.

Figure 12-27: DataBindings property of ComboBox

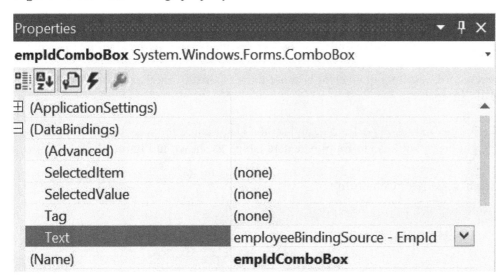

Figure 12-27 shows that the Text property of the ComboBox is bound to the EmpId field in the Employee data table (through employeeBindingSource). That means the data from EmpId field in the current record would be assigned to the Text property of the ComboBox, and any changes to the Text property are saved in the EmpId field of the table.

Similarly, the Text property of the TextBox for LastName is bound to the LastName field in the Employee data table. The Text property of each TextBox generated from the table is bound to the corresponding field.

Step 2-5: Select the ComboBox for EmpId and view the DataBindings property. Repeat for LastName TextBox.

View the DataBindings property of the DateTimePicker for DateOfBirth. Note that the Value of the control is bound to DateOfBirth field.

View the DataBindings property of the CheckBox. Note that the CheckState property is bound to the FullTime field.

Displaying a List of Values in a ComboBox

Binding the Text property of a ComboBox to a field shows only the data from the current record. **The purpose of the EmpId ComboBox is not to show the EmpId from the current record, but to display all EmpIds so the user can select one to find the corresponding record.** How do you do that?

First, you need to set two important properties of the ComboBox (or ListBox):
Set **DataSource** to the binding source.
Set **DisplayMember** to the field.
Setting these two properties would display a list of all EmpIds in the ComboBox.

Binding tables that do not have a binding source: Note that data tables that are not dragged and dropped to a form won't have a corresponding binding source. If there is no binding source for a table, you select
Other Data Source, Project Data Sources, HRDataSet, Employee.
Selecting the table for DataSource, in turn, will create the binding source for the table.

Next, you need to remove the DataBindings for the ComboBox by setting the DataBindings to *None*. If not, when you select an EmpId from the list, the new value will replace the existing value for current record, resulting in an error because there is already another record that has the selected value of EmpId.

Step 2-6: Select the ComboBox that displays EmpId, and set the DataSource property to employeeBindingSource (that links to Employee data table), as shown in Figure 12-28.

Figure 12-28: DataSource property of ComboBox

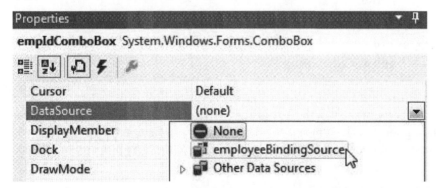

Next, set the Display Member to EmpId, as shown in Figure 12-29.

Figure 12-29: DisplayMember property of ComboBox

Step 2-7: Set the DataBindings property of EmpId ComboBox to **None,** as shown in Figure 12-30.

Figure 12-30: Setting the DataBindings property of EmpId ComboBox to **None**

Step 2-8: Run the form. Select an id from the EmpId ComboBox and verify that the record for the selected id is displayed.

How does it find the record when there is no code to do it? The secret is that the ComboBox is getting the EmpIds from the Employee data table that also is bound to other controls on the form. So, when you select an EmpId from the ComboBox, you are changing the current record in the Employee data table, which in turn changes the data in all bound controls. This won't happen if you bind the ComboBox to one data table and the other controls to a different data table.

It's time to practice! Do Exercise 12-1.

Review Question

12.8   The data from a table named Customer is displayed on a form in Details View. The controls on the form were created by dragging and dropping the data table on the form after changing CustomerId field to a ComboBox. The purpose of the ComboBox is to let the user select a CustomerId from a list of all CustomerIds and display the record for the selected id. What properties of the ComboBox need to be changed from the default setting? What should be the setting for those properties?

12.7 Selecting a Group of Records Using the Binding Source

When a table contains a large number of records, it is important to be able to look at a group of records, like employees from a department. To understand ways to limit the records displayed on a form, let's take another look at the objects (components) that are used in moving data from the database to the form. As shown in Figure 12-31, the table adapter controls what is displayed in the data table in memory. Similarly, the binding source controls what is displayed on the form from the data table.

Figure 12-31: Relationship between database components

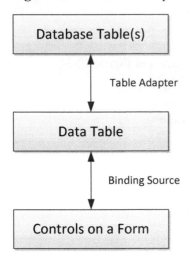

Figure 12-31 suggests that there are two ways you can select the records to be displayed on a form:

1. Use the binding source to control what is displayed in the controls on the form. For example, you may load the entire Employee database table into the Employee data table, but filter what is displayed in the controls on a form. You use the Filter property of the binding source to do it.

2. Use the table adapter to select a group of records from the database and load it into the data table. To do this, you specify the criterion to select records using the SQL language.

In this section, we look at selecting records using the binding source.

To select records using the binding source, assign the criteria (for selecting records) to the Filter property of the binding source. The general syntax is
    BindingSource.Filter = "criterion";
The criterion could be any Boolean expression like
    "LastName = 'Hope' AND DeptNo = 2"

Here are examples of complete statements:
```
employeeBindingSource.Filter = "LastName = 'Hope'";
employeeBindingSource.Filter = "Salary > 40000";
employeeBindingSource.Filter = "DateOfBirth > #01-01-1990#";
employeeBindingSource.Filter = "FullTime = True";
```

Note that the string data are enclosed in a pair of apostrophes, and Date type is enclosed in # signs. If the field is of numeric or Binary type, then do not enclose the data in apostrophes.

The whole criterion must be enclosed in quotes.

Filtering Records Using the Last Name from a TextBox

Rather than hard-coding the value of the field (like 'Hope' and 40000), you can make the criterion more flexible by getting the value from a TextBox or ComboBox.

Consider the first example:
```
employeeBindingSource.Filter = "LastName = 'Hope'";
```

How does the criterion look if the last name is taken from the TextBox, txtLastName?
You need to concatenate three different strings:
    Everything before the last name Hope: LastName ='
    The last name from the TextBox: txtLastName.Text
    The last apostrophe: '

The concatenated string looks as follows:
```
"LastName = '" + txtLastName.Text + "'"
```

The last part of the criterion, "'", is an apostrophe enclosed in quotes.

The complete statement would be
```
employeeBindingSource.Filter = "LastName = '" + txtLastName.Text + "'";
```

Or, you may split the criterion into four parts, as in
```
employeeBindingSource.Filter = "LastName = " + "'"+ txtLastName.Text + "'"
```

Similarly, if the Salary comes from a TextBox, txtSalary, the statement would look as follows:
```
employeeBindingSource.Filter = "Salary > " + txtSalary.Text;
```

Step 2-9: Add a TextBox named txtLastName and Button named btnFilterByLastName to the form, as shown in Figure 12-32.

Figure 12-32: The EmployeeSearch form that filters records using a last name

Add the code from Figure 12-33 to the Click event handler of the button.

Figure 12-33: Code to filter employee records using a last name

```
37    private void btnFilterByLastName_Click(object sender, EventArgs e)
38    {
39        employeeBindingSource.Filter = "LastName = '" + txtLastName.Text + "'";
40        // Or, use the wild card character "*" for to specify incomplete strings
41        employeeBindingSource.Filter = "LastName LIKE '" + txtLastName.Text + "*'";
42    }
```

Step 2-10: Run the form. Enter the name Hope into the TextBox and click the button. As the navigation bar shows (1 of 2), only the two records with the last name Hope are available on the form.

Next, enter "M" to display all employees whose last name starts with "M." Verify the result.

Filtering Records Using the Last Name Selected from a ComboBox

Rather than typing the last name into a TextBox, it would be easier to select a last name from a list of all valid unique last names. You could display all last names in the ComboBox by setting the Data Source to employeeBindingSource and Display Member to LastName. However, this would display the same last name multiple times if there are multiple records with that last name. So, we will create a new data table called LastNames that contains only unique last names, no duplicate names.

Adding a New Data Table and Corresponding Table Adapter

To add a new data table in the DataSet window, you go through many of the same steps you followed when you added the data source. Creating a data table also creates the table adapter for the table.

Step 2-11: Open the DataSet window by double clicking on the file HRDataSet.xsd in Solution Explorer. You will see the Employee data table and its table adapter.

Step 2-12: Right click anywhere in the window and select *Add*, *TableAdaper* (not Data Table). You will be asked to choose the connection. Because you are using the same connection, click *Next*. You will be asked to choose the Command Type.

Step 2-13: Keep the default, *Use SQL Statements*. Click *Next*. You will be asked to enter a SQL statement.

Step 2-14: Click *Query Builder* instead of typing in the query. Select Employee table from the *Add Table* window and click *Add*. Close *Add Table* window.

Step 2-15: Check LastName field, and click OK.
You will see the following SQL to select last names:
```
SELECT      LastName
FROM        Employee
```

Step 2-16: Change the SQL by inserting the key word DISTINCT before LastName, as follows:
```
SELECT      DISTINCT LastName
FROM        Employee
```

Step 2-17: Click *Advanced Options*. Uncheck the CheckBox for Insert, Update and Delete statements because the Last Names data table is used only for displaying data. Click OK.

Step 2-18: Make sure that *Fill a Data Table* is checked. Click *Next*. Click *Finish*. You will see the new data table with the name DataTable1.

Step 2-19: Right click the table name Employee1 and rename it to LastNames, as shown in Figure 12-34. Rename Employee1TableAdapter to LastNamesTableAdapter.

Figure 12-34: Renaming the data table, Employee1

So, now you have a data table named LastNames that contains only unique last names from Employee database table. You also have the table adapter that gets the data from the database.

Now you can display the last names in a ComboBox by setting its Data Source and Display Member properties. Because no control on the form is yet bound to LastNames table, no binding source was created for LastNames. So, the DataSource is to be set to the LastNames table, which in turn will create the binding source.

Step 2-20: Add a ComboBox by the name cboLastNames. Note that the purpose of this ComboBox is to provide a list of last names, not to display the last name from the current record. So, you do not set the DataBindings property to LastName. Do not delete the txtLastName TextBox.

Step 2-21: Set the Data Source property of the ComboBox to LastNames table by selecting *Other Data Source, Project Data Sourcc, HRDataSet, LastNames*. The DataSource property now shows the newly created binding source that has the name LastNamesBindingSource, as shown in Figure 12-35.

Figure 12-35: Setting the Data Source property of the ComboBox to LastNames

Set the Display Member property to LastName, as shown in Figure 12-36.

Figure 12-36: Display Member property of ComboBox

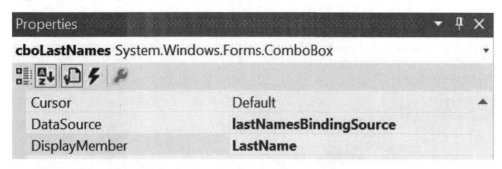

The ValueMember Property of ComboBox

You can display data from one field in a ComboBox, but use the corresponding data from another field to find a record. For example, to filter Employee records using department number, you may display department names (from Department table) in the ComboBox to make selection easier, but use the corresponding DeptNo to filter records. In this case, you would specify department name for DisplayMember and DeptNo for ValueMember property. The criterion for filtering would use the **SelectedValue** property that has the value (DeptNo), not the Text property (department name) that has the text displayed in the ComboBox.

Because using the ValueMember property generally gives more flexibility, we will use it in this example for demonstration, though it is not necessary.

Step 2-22: Set the ValueMember property of the ComboBox to LastName field.

Step 2-23: Run the form, and make sure the ComboBox displays all unique last names.

Next, write the code to filter records using the selected name, and run this code automatically whenever the user selects a name from the ComboBox. What event of the ComboBox should you use?

SelectedIndexChanged versus SelectionChangeCommitted Events

Both events are fired when the item selected in a ComboBox changes. However, there are some differences that are important for a data bound ComboBox.

SelectedIndexChanged event also is fired when SelectedIndex is set programmatically, like when you set the SelectedIndex to -1 to reset the ComboBox, and when the DataSource property of the ComboBox is set at runtime. Executing the code to filter records when these two events take place can have some unintended consequences.

But the **SelectionChangeCommitted** is not fired under the above two conditions. It is raised only when a user selects an item from the ComboBox.

Neither event is fired when items are added through the **Items** property of the ComboBox. So, in general, it is better to use the SelectionChangeCommitted event for data bound ComboBoxes. The code to filter Employee records using the selected name is shown in Figure 12-37.

Figure 12-37: Code to filter employee records

```
44      private void cboLastNames_SelectionChangeCommitted(object sender, EventArgs e)
45      {
46          employeeBindingSource.Filter = "LastName = '" + cboLastNames.SelectedValue + "'";
47      }
```

Note that you should use the **SelectedValue** (not the Text property) of the ComboBox, which gives the value of the field specified by the ValueMember property. The Text property is not changed at the time this event is raised, because this event is raised before the SelectedIndexChanged event; so, you cannot use the Text property in this event handler.

Step 2-24: Create the SelectionChangeCommitted event handler for cboLastNames, and add the code from Figure 12-37. Run the form and test the code.

Cancelling a Filter

To cancel a filter specified using the BindingSource.Filter property, set the Filter to a null string.

Step 2-25: Add a button named btnCancelFilter, and add the code from Figure 12-38 to the Click event handler. Run the form and test the code.

Figure 12-38: Code to cancel filter

```
48
49      private void btnCancelFilter_Click(object sender, EventArgs e)
50      {
51          employeeBindingSource.Filter = "";
52      }
```

It's time to practice! Do Exercise 12-2

Selecting a List from One Table to Filter Another Table

Suppose you want to select a department number from a ComboBox and filter employee records to view only employees from that department. The HR database includes a second table named department, as shown in Figure 12-39.

Figure 12-39: Department table

DeptNo	DeptName	DeptPhone
1	Accounting	920-424-1414
2	Finance	920-424-3377
3	Marketing	920-424-2442
4	Purchasing	920-424-0405
5	Human Resources	920-424-1166
6	Information Technology	920-424-3020

A list of unique department numbers for display in the ComboBox can be obtained from the Employee table by altering the SQL, or it can be obtained from the Department table. Because the Employee table is significantly larger than the Department table, it would be more efficient to get such a list from the Department table. Getting department numbers from the Department table also gives you the option to display department names in the ComboBox but use department numbers to filter records.

To get the department numbers (and names, if needed), you need to first create a data table and table adapter for Department table using the same steps that you followed to create the LastNames data table.

It's time to practice! Do Exercise 12-3.

Review Question

12.9 The data from a table named Employee is displayed on a form in Details View in TextBoxes. The controls on the form were created by dragging and dropping the data table on the form.

What would be the effect of executing the following statement on the set of records that the user can see on the form?
employeeBindingSource.Filter = "CountryId = 22";

What would be the effect of executing the above statement on the records stored in the Employee data table in memory?

12.8 Selecting Records Using the Table Adapter

In the previous section, you selected records using the Filter property of the binding source that controls what is transferred between a data table and the form. Another way to select records is to use the table adapter to change the SQL language that it uses to transfer data from the database to the dataset. This is fundamentally different from using the binding source.

Binding Source versus Table Adapter to Select Records

When you use binding source to filter records, typically the entire data from a table(s) is loaded from the database into a data table in memory. The binding source is used to help the user select different groups of data to work with. After the data are loaded into the data table, any change made to the database is not reflected in the data that the user sees. So, the user may not be seeing the latest data. Further, large tables consume a large amount of memory. A benefit of loading the entire table into memory and then filtering records using the binding source is that the database needs to be accessed only once in the beginning, thus reducing the network traffic.

When you use the table adapter, every time the user wants a group of data, it is loaded from the database into a data table and displayed on a form. So, each time a group of data is loaded, it will include the latest changes. The memory usage is lower because only selected records are held in data tables. This also means that the database is accessed every time the user requests a group of records.

Next, we look at the use of the table adapter to select records.

Fill Method of Table Adapter

So far, the auto-generated code on all forms used the Fill method of the table adapter to load data from the database to the Employee data table. The fill method uses the following SQL query:

```
SELECT EmpId, LastName, FirstName, DateOfBirth, Phone, DeptNo, CountryId, Salary,FullTime
FROM dbo.Employee
```

Note: To see this SQL, double click HRDataSet.xsd and click on *Fill, GetData()*, under *EmployeeTable Adapter*. You will see the above SQL in the CommandText property of the Fill method.

Creating a Parameter Query

To load only a subset of the database table, you need to create a new Fill method with a modified form of the standard SQL query, to include a criterion with a parameter for the last name to be specified, as follows:

```
SELECT EmpId, LastName, FirstName, DateOfBirth, Phone, DeptNo, CountryId, Salary,FullTime
FROM dbo.Employee
WHERE LastName = @LastName
```

The @ symbol indicates that **LastName** is a parameter whose value should be provided when you run the SQL.

For Access databases, use the question mark (?) in place of @ symbol, as in ?LastName.
For Oracle databases, use a colon (:), as in :LastName.

To **add a new query** (and a Fill method that uses the query), you right click the table adapter in the dataset schema window.

Tutorial 3: Parameter Query to Select Records

The parameter query, discussed above, is created in a new form, EmployeeSearch_ParmQry.

Step 3-1: Add a new form named **EmployeeSearch_ParmQry**.

Step 3-2: Choose Details mode for Employee data table, and ComboBox style for EmpId field. Drag and drop Employee table on the form.

Step 3-3: Open dataset Schema window by double clicking on HRDataSet.xsd in Solution Explorer.

Right click on employeetable adapter and select *AddQuery*.

Make sure that "Use SQL Statement" is selected. Click *Next*.

Make sure that "Select which returns rows" is selected. Click *Next*.

Note: If the option to "Select which returns rows" is grayed out, then you did not right click the Employee Table Adapter to create the query. Press *Cancel* and right click the Employee Table Adapter to add a query.

Change the SQL by adding the criterion as follows:

```
SELECT EmpId, LastName, FirstName, DateOfBirth, Phone, DeptNo, CountryId, Salary,
FullTime
FROM    dbo.Employee
WHERE LastName   = @LastName
```

Click *Next*.

Change the name of the method to **FillByLastName**. Click *Next*. Click *Finish*.

You will see the new method FillByLastName added to employeeTableAdapter, as shown in Figure 12-40.

Figure 12-40: The Parameter query for EmployeeTableAdapter

Using a Parameter Query

When you use a parameter query, you need to provide a value for the parameter(s). You will use data from a TextBox, txtLastName, for FillByLastName query as follows:

```
employeeTableAdapter.FillByLastName(hRDataSet.Employee, txtLastName.Text);
```

Step 3-4: Add a TextBox named txtLastName and a Button named btnFillByLastName with the caption "Select Records," as shown in Figure 12-41.

Figure 12-41: The EmployeeSearch_ParmQry form

Step 3-5: Comment out the statement in the Load event handler, which uses the standard Fill method, as shown in Figure 12-42. Add the code from Figure 12-42 to the Click event handler of the button.

Figure 12-42: Load event handler for the EmployeeSearch_ParmQry form

```
28          private void EmployeeSearch_ParmQry_Load(object sender, EventArgs e)
29          {
30              // TODO: This line of code loads data into the 'hRDataSet.Employee' table.
31              //this.employeeTableAdapter.Fill(this.hRDataSet.Employee);
32          }
33          private void btnSelectRecords_Click(object sender, EventArgs e)
34          {
35              employeeTableAdapter.FillByLastName(hRDataSet.Employee, txtLastName.Text);
36          }
```

Step 3-6: Run the form and test the code.

Step 3-7: To give the user an option to display all records, add a button, btnAllRecords, with the caption "Display All." Add the code to use the standard query, shown in Figure 12-43. Run the form and test the code.

Figure 12-43: Standard Fill method to display all records

```
37
38          private void btnAllRecords_Click(object sender, EventArgs e)
39          {
40              this.employeeTableAdapter.Fill(this.hRDataSet.Employee);
41          }
```

It's time to practice! Do Exercise 12-4.

Review Question

12.10   The data from a table named Employee are displayed on a form in Details View in TextBoxes. The controls on the form were created by dragging and dropping the data table on the form. A parameter query named FillByLastName was created using the following SQL:

SELECT EmpId, LastName, FirstName, DateOfBirth, Phone, DeptNo, CountryId, Salary, FullTime
FROM   dbo.Employee
WHERE DeptNo = @DeptNo

What would be the effect of executing the following statement on the Employee data table?
employeeTableAdapter.FillByLastName(hRDataSet.Employee, 11)

12.9 Untyped Datasets: Displaying Records

As discussed earlier in this chapter, the relationship between database tables and controls on a form can be represented by the illustration in Figure 12-43.

Figure 12-43: Relationship between database tables and controls on a form

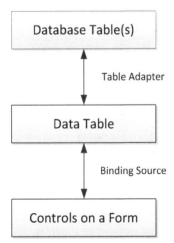

In the examples presented in previous sections, the data tables and other data access objects (data adapter and binding source), which are required to link database tables to controls on a form, are generated by the wizard at design time. The data types of the fields in the data table are determined at design time based on the data type of the corresponding fields in the database table. So, data tables created at design time contain fields that are of the same type as the fields in the database table; therefore, they are called **typed data tables**. The datasets that contain such tables are called **typed datasets**.

Typed datasets make it easy to check mismatch of data types between fields in the database table and the data that the application tries to store in those fields. However, generating typed datasets and other objects, like the data adapter, at design time makes it difficult to reuse the code in other forms or projects because the code is tied to objects within a form.

Bound versus Unbound Controls

Previous sections also used **data bound controls** that are easily generated by the wizard when you drag and drop the data table on the form. Data bound controls are bound to specific fields in the data table through the binding source. So, data on the form is linked tightly to the fields in the data table. The current row is displayed automatically in the controls. However, bound controls provide less flexibility in working with the data and validating it, compared to **unbound controls**, which are not directly bound to any field.

Creating objects at runtime using code, rather than at design time using the wizard, and displaying data in unbound controls require more code. However, this approach makes the code more flexible, efficient and reusable.

In this section, you learn how to develop applications that use code to create the dataset, table adapter and navigation bar at runtime, and display data in unbound controls. The datasets that are created at runtime using code are called **untyped datasets** because the data tables do not contain fields whose types are defined.

Tutorial 4: Display Records Using Untyped Datasets

This tutorial develops a form to display employee records using **untyped datasets** and **unbound controls**. You work with Employee table from HR2.mdf database file found in the DataFiles folder. You will develop a simple form to display employee records, as shown in Figure 12-44.

Figure 12-44: The Employee_Display form

Step 4-1: Open the project, Ch12_Databases_UntypedDatasets. Open the form, Employee_Display. This form contains all necessary controls, but you need to add the code.

Note that the controls were not generated by dragging and dropping data tables; they were manually added to the form. Therefore, they are **unbound** controls that are not bound to fields in the data table, as indicated by the DataBindings property of the controls.

Displaying Records from a Table

Displaying records from a table involves the following steps:

1. Create the data table, data adapter and binding source objects.
2. Fill data from the database table into the data table using the Fill method of the data adapter.
3. Link the binding source to the data table. The binding source has methods/properties to move between rows, get the position/count of rows and select rows using the Filter property.
4. Copy data from the current row of the binding source to the controls on the form.

These are the same steps that were taken to create typed datasets using the wizard, which generated the necessary code. In this section, you need to develop the code because you are not using the wizard.

Creating the Data Table(s)

A data table object is created from the DataTable class using the standard syntax for creating objects:
    DataTable *variableName* = new DataTable()
where *variableName* is any valid variable name.

For example, the following statement creates a data table object for the Employee table, and references it using the variable, dtEmployee.
```
DataTable dtEmployee = new DataTable();
```

Creating the Data Adapter

A data adapter is created from the DataAdapter class:
```
DataAdapter variableName = new DataAdapter(connectionString, Sql)
```

The first parameter, *connectionString*, provides the necessary information to connect to the database. To connect to a local database, the syntax of the connection string is
```
@"Data Source = (LocalDB)\MSSQLLocalDB; AttachDbFilename = path\fileName";
```

Here is an example:
```
@"Data Source = (LocalDB)\MSSQLLocalDB; AttachDbFilename = C:\VC\HR2.mdf";
```

If the database is to be accessed from a server, the Data Source would provide the server address. The structure of the connection string may vary with the version of SQL Server you are using. The connection string may include additional information on security, user id and password.

The second parameter, **Sql**, represents the SQL statement to select data. The syntax to select data from a single table is
```
SELECT fieldNames FROM tableName WHERE criterion
```

In the above statement, *fieldNames* represents the list of field names separated by commas, *tableName* represents the table name, and *criterion* represents the criterion, if any, to select records. If all fields are to be selected, you may replace field names by an asterisk.

For example, to select all fields from all records of the Employee table, use the following SQL statement:
```
SELECT * FROM Employee
```

The WHERE clause is discussed briefly later in this section.

Thus, the following code creates a data adapter to access all Employee records, and references it using the variable, daEmployee.
```
string connString = @"Data Source = (LocalDB)\MSSQLLocalDB; AttachDbFilename =
                      C\VC\HR2.mdf";
string empSQLStr = "SELECT * FROM Employee";
SqlDataAdapter daEmployee = new SqlDataAdapter(empSQLStr, connString);
```

Creating the Binding Source

The binding source is created from the BindingSource class:
```
BindingSource variableName = new BindingSource;
```

Here is an example that creates a binding source, which is referenced by the variable, bsEmployee:
```
BindingSource bsEmployee = new BindingSource();
```

Filling the Data Table

Filling the data table consists of loading data from the database table (or from a view that might have data from multiple tables) to the corresponding data table. The following statement fills the data table, dtEmployee, using the table adapter daEmployee:
```
daEmployee.Fill(dtEmployee);
```

The data adapter knows what data should be loaded into the data table, because it contains the necessary SQL statement that was specified when the data adapter was created.

Linking the Binding Source to the Data Table

To link a binding source to a data table, the **DataSource** property of the binding source is set to the data table, as in
```
bsEmployee.DataSource = dtEmployee;
```

The complete code to create the create the data access objects, fill the data table and link the data table to the binding source is shown in Figure 12-45.

Figure 12-45: Code to create the data access objects

```csharp
10      using System.Data.SqlClient;    // add this using directive
11
12      namespace Ch12_Databases_UntypedDatasets
13      {
14          public partial class Employee_Display : Form
15          {
16              public Employee_Display()
17              {
18                  InitializeComponent();
19              }
20              DataTable dtEmployee = new DataTable(); // creates the data table
21              SqlDataAdapter daEmployee;  // declares a reference variable for the data adapter;
22              BindingSource bsEmployee = new BindingSource(); // creates the binding source
23
24              private void Employee_Display_Load(object sender, EventArgs e)
25              {
26                  // Create the data adapter
27                  string connString = @"Data Source = (LocalDB)\MSSQLLocalDB; AttachDbFilename =
28                                        C:\VC\HR2.mdf";
29                  string empSQLStr = "SELECT * FROM Employee";
30                  daEmployee = new SqlDataAdapter(empSQLStr, connString);
31                  // Fill the data table, dtEmployere
32                  daEmployee.Fill(dtEmployee);
33                  // Link binding source to the data table
34                  bsEmployee.DataSource = dtEmployee;
```

Line 10: The using directive specifies the namespace, SqlClient, which contains the data access objects, like data table and data adapter.

Lines 20–22: The reference variables for the three objects are declared at the class level so that they can be used in multiple methods. The data table and binding source are created here, but the data adapter is created within the Load event handler of the form (lines 26–30) because it requires creating the connection and SQL strings.

Lines 27–30: The data adapter is created using the connection string and SQL statement, as discussed earlier.

Lines 31–34: The Fill method of the data adapter is used to fill the data table, and the DataSource property of the binding source is set to the data table.

In the next three steps of the tutorial, you add the above code to your form, and copy the database.

Step 4-2: Add the using directive shown in line 10, Figure 12-45, to the code window.

Step 4-3: Copy the database file, HR2.mdf, to a local folder outside the project folder.
Add the code from lines 20–22 at the class level.
Add the code from lines 26–30 to the Load event handler of the form to create the data adapter.
Make sure you specify the correct path for your database file.

Step 4-4: Add the code from lines 31–34 to fill the data table and link the binding source.

Copying Data from the Data Table into Controls on the Form

Because there is no concept of current row in a data table, the binding source is used to navigate through the rows and display records on the form.

The data in a binding source is just a view of the data from the data table. The view could consist of the entire table, or a subset of the data selected using the Filter property of the binding source. That is, the binding source helps you to look at the data table through different windows. A view also could present the data sorted in a certain order.

The data types of the binding source rows and the data table rows are not the same. The data viewed through a binding source is of **DataView** type, and each row of the binding source is of **DataRowView** type. But each row of a DataTable is of **DataRow** type.

The data from different columns of the current row of the data source is displayed in controls on the form using two steps:

1. Store the entire row in a DataRowView object.
2. Copy data from each column of the DataRowView into a control on the form.

Accessing a Binding Source Row

You access a row using its index. For example, bsEmployee[2] gets the third row in the binding source.

The index of the current row in the binding source is given by the **Position** property of the binding source, as in
```
int currentRow = bsEmployee.Position;
```
The index is used in the following statement to get the entire current row. The row is cast to DataRowView type and its reference is assigned to the variable, drvEmployee, which is of DataRowView type.
```
DataRowView drvEmployee = (DataRowView) bsEmployee[currentRow];
```
Alternate method: If the binding source doesn't filter rows, and therefore includes all rows from the data table, you may use the index of the current row to get the row from the **Rows** collection of the data table, as follows:
```
DataRow drEmployee = dtEmployee.Rows[currentRow];
```
This doesn't work when the data displayed on the form are filtered by the binding source because the first row of the binding source, for example, may not be the first row in the data table.

Copying Data from DataRowView to Controls on a Form

You can access individual fields of the DataRowView using the field name or the index of the field. For example, you may get the employee id from the DataRowView, drvEmployee, using one of the following two options:
```
drvEmployee["EmpId"].ToString();
```
or
```
drvEmployee[0].ToString();
```

Because records are to be displayed repeatedly when you move from one row to another, you will create a separate method named DisplayCurrentRow to copy data from DataRowView to controls.

Figure 12-46 shows the code within the method to store the current row in a DataRowView, drvEmployee, and to copy data from the DataRowView to controls on the form.

Figure 12-46: Displaying data from a data table in controls on a form

```
45      private void DisplayCurrentRow()
46      {
47          // store the current row of the data table in a DataRowView
48          int currentRow = bsEmployee.Position;
49          DataRowView drvEmployee = (DataRowView) bsEmployee[currentRow];
50
51          // copy data from the DataRowView to controls on the form:
52          txtEmpId.Text = drvEmployee["EmpId"].ToString();
53                  // Or, use the index as in, drvEmployee[0].ToString();
54          txtFirstName.Text = drvEmployee["FirstName"].ToString();
55          txtLastName.Text = drvEmployee["LastName"].ToString();
56          txtDOB.Text = drvEmployee["DateofBirth"].ToString();
57          txtPhone.Text = drvEmployee["Phone"].ToString();
58          txtCountryId.Text = drvEmployee["CountryId"].ToString();
59          txtDepNo.Text = drvEmployee["DeptNo"].ToString();
60          txtSalary.Text = drvEmployee["Salary"].ToString();
61          chkFullTime.Checked = bool.Parse(drvEmployee["FullTime"].ToString());
62          txtPosition.Text = drvEmployee["Position"].ToString();
63      }
```

Line 52: The data from the EmpId column of the DataRowView, drvEmployee, is displayed in the TextBox, txtEmpId.

Lines 54–62: Data from the remaining columns are displayed in the controls on the form.

Step 4-5: Add the code from Figure 12-46 to display the current row.
Add the code from Figure 12-47 to call DisplayCurrentRow method.
Run the project and test the code. The form should display the first record.

Figure 12-47: Statement to call DisplayCurrentRow method

```
35
36                  // call DisplayCurrentRow method
37                  DisplayCurrentRow();
```

Navigating through the Rows

You use the MoveFirst, MovePrevious, MoveNext and MoveLast methods of the binding source to move between rows.

Step 4-6: Add code to the Click event handlers of the navigation buttons, as shown in Figure 12-48.

Figure 12-48: Code to move between rows

```
64      private void btnFirst_Click(object sender, EventArgs e)
65      {
66          bsEmployee.MoveFirst();
67      }
68      private void btnLast_Click(object sender, EventArgs e)
69      {
70          bsEmployee.MoveLast();
71      }
72      private void btnPrevious_Click(object sender, EventArgs e)
73      {
74          bsEmployee.MovePrevious();
75      }
76      private void btnNext_Click(object sender, EventArgs e)
77      {
78          bsEmployee.MoveNext();
79      }
```

PositionChanged Event of the Binding Source

To display the current row, the DisplayCurrentRow method needs to be called every time you move to a new row. Rather than calling this method from each Click event handler shown in Figure 12-48, we will call this method whenever the position of the current row changes. To do this, you will create a method named bsEmployee_PositionChanged and link (subscribe) this method to the PositionChanged event of the binding source.

The PositionChanged event is fired whenever the position of the current row changes—that is, whenever you move to a new row. The PositionChanged method that called DisplayCurrentRow, and displays the current row number, is shown in Figure 12-49.

Figure 12-49: PositionChanged method

```
81      private void bsEmployee_PositionChanged(object sender, System.EventArgs e)
82      {
83          // display data & row number whenever a the current record position changes
84          if (bsEmployee.Count > 0)
85          {
86              DisplayCurrentRow();
87              lblRecordNo.Text = (bsEmployee.Position + 1) + " of " + bsEmployee.Count;
88          }
89      }
```

Lines 84 uses the Count property to check whether the binding source is empty.

Line 87 uses the Position and Count properties to display the row number of the current row and the number of rows in the set.

Next, you need to link (subscribe) the above method to the PositionChanged event of the binding source, as shown in line 43 in Figure 12-50.

Figure 12-50: Linking PositionChanged event to bsEmployee_PositionChanged method

```
38
39          // Display row number and count of rows
40          lblRecordNo.Text = (bsEmployee.Position + 1) + " of " + bsEmployee.Count;
41
42          // Link (subscribe) PositionChanged event to bsEmployee_PositionChanged method
43          bsEmployee.PositionChanged += bsEmployee_PositionChanged;
```

Step 4-7: Add the code from Figure 12-49 to create bsEmployee_PositionChanged method.
Add the code from line 43 in Figure 12-50 to link the method to the PositionChanged event of the binding source.
Add the statement from line 40 to display the row number and row count, the first time data are displayed on the form.

Step 4-8: Run the project and test the code.

The following is the complete code for the Employee_Display form:

```csharp
10   using System.Data.SqlClient;    // add this using directive
11
12   namespace Ch12_Databases_UntypedDatasets
13   {
14       public partial class Employee_Display : Form
15       {
16           public Employee_Display()
17           {
18               InitializeComponent();
19           }
20           DataTable dtEmployee = new DataTable(); // creates data table
21           SqlDataAdapter daEmployee; //declares a ref. variable for data adapter;
22           BindingSource bsEmployee = new BindingSource(); //creates binding source
23
24           private void Employee_Display_Load(object sender, EventArgs e)
25           {
26               // Create the data adapter
27               string connString = @"Data Source = (LocalDB)\MSSQLLocalDB; AttachDbFilename =
28                                     C:\VC\HR2.mdf";
29               string empSQLStr = "SELECT * FROM Employee";
30               daEmployee = new SqlDataAdapter(empSQLStr, connString);
31               // Fill the data table, dtEmployere
32               daEmployee.Fill(dtEmployee);
33               // Link binding source to the data table
34               bsEmployee.DataSource = dtEmployee;
35
36               // call DisplayCurrentRow method to display the current row
37               DisplayCurrentRow();
38
39               // Display row number and count of rows
40               lblRecordNo.Text = (bsEmployee.Position + 1) + " of " + bsEmployee.Count;
41
42               // Link (subscribe) PositionChanged event to bsEmployee_PositionChanged method
43               bsEmployee.PositionChanged += bsEmployee_PositionChanged;
44           }
```

```csharp
45      private void DisplayCurrentRow()
46      {
47          // store the current row of the data table in a DataRowView
48          int currentRow = bsEmployee.Position;
49          DataRowView drvEmployee = (DataRowView) bsEmployee[currentRow];
50
51          // copy data from the DataRowView to controls on the form:
52          txtEmpId.Text = drvEmployee["EmpId"].ToString();
53                  // Or, use the index as in, drvEmployee[0].ToString();
54          txtFirstName.Text = drvEmployee["FirstName"].ToString();
55          txtLastName.Text = drvEmployee["LastName"].ToString();
56          txtDOB.Text = drvEmployee["DateofBirth"].ToString();
57          txtPhone.Text = drvEmployee["Phone"].ToString();
58          txtCountryId.Text = drvEmployee["CountryId"].ToString();
59          txtDepNo.Text = drvEmployee["DeptNo"].ToString();
60          txtSalary.Text = drvEmployee["Salary"].ToString();
61          chkFullTime.Checked = bool.Parse(drvEmployee["FullTime"].ToString());
62          txtPosition.Text = drvEmployee["Position"].ToString();
63      }
64      private void btnFirst_Click(object sender, EventArgs e)
65      {
66          bsEmployee.MoveFirst();
67      }
68      private void btnLast_Click(object sender, EventArgs e)
69      {
70          bsEmployee.MoveLast();
71      }
72      private void btnPrevious_Click(object sender, EventArgs e)
73      {
74          bsEmployee.MovePrevious();
75      }
76      private void btnNext_Click(object sender, EventArgs e)
77      {
78          bsEmployee.MoveNext();
79      }
80
81      private void bsEmployee_PositionChanged(object sender, System.EventArgs e)
82      {
83          // display data & row number whenever a the current record position changes
84          if (bsEmployee.Count > 0)
85          {
86              DisplayCurrentRow();
87              lblRecordNo.Text = (bsEmployee.Position + 1) + " of " + bsEmployee.Count;
88          }
89      }
```

Review Questions

12.11   To create a table adapter object using code, you need to specify the _____ and _____.

12.12   What are the benefits of creating data access objects (data adapter, data table and binding source) using code at runtime, rather than at design time using the wizard?

12.13   How does an untyped data table differ from a typed data table?

12.14   What is the difference between a data bound control and an unbound control?

12.15   An application creates the data access objects (data adapter, data table and binding source) using code at runtime. What object and method do you use to load the data from the database table to the untyped data table?

12.16   How do you link a binding source to a data table?

12.17   List three major functions of a binding source.

12.18   An application creates the data access objects (data adapter, data table and binding source) using code at runtime. After loading data from a database table to an untyped data table, what are the major steps involved in displaying the current row in controls on the form?

12.10 Untyped Datasets: Selecting Records

Sections 12.7 and 12.8 discussed selecting records using the typed dataset, binding source and data adapter objects that were created at design time by the wizard. In this section, you will learn how to select records from **untyped** datasets using the binding source and data adapter objects that are created by code at runtime.

Consider the form, Employee_Select, shown in Figure 12-51. The form uses the Employee and Department tables from the HR2 database. The purpose of the form is to let you select employee records using two different methods: (1) by filtering records from the data table, and (2) by changing the SQL to select records from the database table.

Figure 12-51: The Employee_Select form

The form has two ComboBoxes on the right side; both let the user select a department number from a list of department numbers from the Department table.

The department number selected from the first ComboBox (labelled "Fill By Dept") provides the value for a parameter query that selects employee records from the database and loads them into the data table. The department number selected from the second ComboBox (labeled "Filter By Dept") provides the value to filter rows from the data table using the binding source.

The "Display All" button selects and displays all records.

The code to display Employee records is the same as that in the Employee_Display form discussed previously. Additional code is needed to create the data table and the data adapter for the Department table so that a list of DeptNo's can be added to the two ComboBoxes.

Figure 12-52 shows the class-level code that includes the code to create data access objects for the Department table. Note that the Department table doesn't need a binding source, because the only use of the table is to add DeptNo's to the ComboBox, which can be done directly from the data table.

Figure 12-52: Creating data access objects for Employee and Department tables

```
21              // data table, data adapter and binding source for Employee table
22              DataTable dtEmployee = new DataTable();
23              SqlDataAdapter daEmployee;
24              BindingSource bsEmployee = new BindingSource();
25
26              // data table and data adapter for Department table
27              DataTable dtDept = new DataTable();
28              SqlDataAdapter daDept;
```

Figure 12-53 shows the code to display Employee records on the form. This is the same as in the Employee_Display form.

Figure 12-53: Code to display Employee records

```
29              private void Employee_Select_Load(object sender, EventArgs e)
30              {
31                  // Fill employee data table, dtEmployee
32                  string connString = @"Data Source = (LocalDB)\MSSQLLocalDB; AttachDbFilename =
33                                                      C:\VC\HR2.mdf";
34                  string empSQLStr = "SELECT * FROM Employee";
35                  daEmployee = new SqlDataAdapter(empSQLStr, connString);
36                  daEmployee.Fill(dtEmployee);
37
38                  // Display employee data
39                  bsEmployee.DataSource = dtEmployee; // links binding source to the data table
40                  DisplayCurrentRow();
41                  lblRecordNo.Text = (bsEmployee.Position + 1) + " of " + bsEmployee.Count;
42                  bsEmployee.PositionChanged += bsEmployee_PositionChanged;
```

Adding DeptNo's to ComboBoxes

In Section 12.6, which used typed datasets to display values in a ComboBox from a data table column, you set the following two properties of the ComboBox at design time:

Set the DataSource property of the ComboBox to the binding source.
Set the DisplayMember property to the column name.

In this form, you will set an additional property of the binding source:

> Set the ValueMember property to the column name.

If an application uses the SelectedValue of the ComboBox, the ValueMember must be set to a column name. The ValueMember could be set to a column different from the DisplayMember. For example, the Display Member could be the DeptName, and the ValueMember could be the DeptNo. That means the DeptNames are displayed in the ComboBox, but when the user selects a DeptName, the corresponding DeptNo is available through the SelectedValue property of the ComboBox.

Figure 12-54 shows the code to fill the data table, dtDept, and to add the list of DeptNo's from dtDept to the two ComboBoxes. Note that, as stated earlier, the Department table doesn't need a binding source; so, the DataSouce for the ComboBox is set to the data table.

Figure 12-54: Code to display DeptNo's in the ComboBoxes

```
45              // Fill department data table, dtDept
46              string DeptSQLStr = "SELECT DeptNo, DeptName FROM Department";
47              daDept = new SqlDataAdapter(DeptSQLStr, connString);
48              daDept.Fill(dtDept);
49
50              //Display DeptNo's from dtDept, in the ComboBox, cboFillByDept:
51              cboFillByDept.DataSource = dtDept;
52              cboFillByDept.DisplayMember = "DEPTNO";
53              cboFillByDept.ValueMember = "DEPTNO";
54
55              //Display DeptNo's from dtDept, in the ComboBox, cboFilterByDept
56              cboFilterByDept.DataSource = dtDept;
57              cboFilterByDept.DisplayMember = "DEPTNO";
58              cboFilterByDept.ValueMember = "DEPTNO";
```

Tutorial 5: Select Records Using Untyped Datasets

This tutorial develops the Employee_Select form shown in Figure 12-51.

Step 5-1: Open the project Ch12_Databases_UntypedDatasets. Open the form, Employee_Select.
Currently, the form has all the controls, and the code to display Employee records and to navigate between them.
Make sure that the connection string in line 33 points to the right folder.
Run the form to make sure that the employee records are displayed.

Next, you will add the code to display DeptNo's in the ComboBoxes, and to select Employee records.

Step 5-2: Open the code window and add the missing class-level code from Figure 12-52 to create the data access objects for Department.

Step 5-3: Add the missing code from Figure 12-54 to the Load event handler of the form to display DeptNo's in the ComboBoxes.

Step 5-4: Run the form and test the code. The ComboBoxes should display the DeptNo's.

Filtering Rows Using the Binding Source

Filtering involves using the Filter property of the binding source to select rows from the data table for display in controls on the form. Section 12.7 discussed filtering using data objects created at design time using the wizard. The process is the same for untyped data tables. To refresh your memory, the syntax of the statement to filter rows from the data table is

```
bindingSource.Filter = "criterion";
```

where *bindingSource* is the name of the binding source object, and *criterion* is any Boolean expression. Examples of the statement for string and numeric fields can be found in Section 12.7.

In the Employee_Search form, the criterion is that the employee's DeptNo should be equal to the value selected in the cboFilterByDept ComboBox. Figure 12-55 shows the code that specifies the criterion.

Because we want the selected rows to be displayed whenever the user selects a DeptNo from the ComboBox, the code is placed in the SelectionChangeCommitted event handler of the ComboBox. The SelectionChangeCommitted event is fired when the user selects an item from the ComboBox.

As discussed in Section 12.7, the SelectedIndexChanged event is less suitable because it is fired under other conditions, too.

Figure 12-55: Filtering using the binding source

```
61     private void cboFilterByDept_SelectionChangeCommitted(object sender, EventArgs e)
62     {
63         bsEmployee.Filter = "DeptNo = " + cboFilterByDept.SelectedValue.ToString();
64         if (bsEmployee.Count > 0)
65         {
66             lblRecordNo.Text = (bsEmployee.Position + 1) + " of " + bsEmployee.Count;
67             DisplayCurrentRow();
68         }
```

Step 5-5: Generate the SelectionChangeCommitted event handler for cboFilterByDept ComboBox. (Select cboFilterByDept. In the properties window, click on the **events** button [lightning bolt]. Double click the SelectionChangeCommitted event).

Add the code from Figure 12-55 to the event handler to filter records.
Run the form and test the code by selecting the value 3 from the cboFilterByDept. You should be able to see only two employee records.

Selecting Records from the Database Table using the Table Adapter

Selecting records from a database table using the table adapter involves changing the SQL statements to specify the criterion. The selected records are loaded into a data table. As discussed in Section 12.8, this is different from filtering, which selects rows from the data table for display on the form.

In Section 12.8, where the data access objects were created at design time, you used the wizard to create a separate Fill method that includes a criterion in its SQL. Thus, the table adapter had two different Fill methods: one to select all records, and another to select specified records.

The method you use in this section for untyped dataset is different. You change the original SQL by assigning the new SQL to the data adapter's **SelectCommand** property (which is of SqlCommand type, and is generated when you create the data adapter). This is done by assigning the new SQL to the CommandText property of the SelectCommand, as follows:

dataAdapterName.SelectCommand.CommandText = "*sqlString*";

where *dataAdapterName* represents the name of the data adapter, and *sqlString* represents the SQL statement. Figure 12-56 shows the code to select records for the DeptNo value selected in the ComboBox.

Figure 12-56: Using the table adapter to select records

```
72    private void cboFillByDept_SelectionChangeCommitted(object sender, EventArgs e)
73    {
74        //Specify the SQL
75        daEmployee.SelectCommand.CommandText = "SELECT * FROM Employee where DeptNo = " +
76                                               cboFillByDept.SelectedValue.ToString();
77        //Clear and fill the data table
78        dtEmployee.Clear();
79        daEmployee.Fill(dtEmployee);
80    }
```

Alternatively, you may use a **parameter** (a variable name preceded by the "@" sign) in place of the value of DeptNo in the SQL, and provide the value in a separate statement, as follows:

```
daEmployee.SelectCommand.CommandText = "SELECT * FROM Employee WHERE DeptNo = @deptNo";
daEmployee.SelectCommand.Parameters.AddWithValue("@deptNo",
                                                  cboFillByDept.SelectedValue.ToString());
```

The parameter, @deptNo, within the SQL, is replaced by the value, cboFillByDept.SelectedValue.ToString(), specified in the AddWithValue method in the second statement.

Step 5-6: Add the code from Figure 12-56 to select records using the table adapter.
Run the form and test the code by selecting the value 3 from the cboFillByDept ComboBox. You should be able to see only two employee records.

Selecting All Records

To select all records, you change the SQL so that no criterion is included, as follows:
daEmployee.SelectCommand.CommandText = "SELECT * FROM Employee";

In addition, you cancel the filter by setting the Filter property of the binding source to null. The complete code to display all records is shown in Figure 12-57.

Figure 12-57: Displaying all records

```csharp
82  private void btnDisplayAll_Click(object sender, EventArgs e)
83  {
84      dtEmployee.Clear();
85      daEmployee.SelectCommand.CommandText = "SELECT * FROM Employee";
86      daEmployee.Fill(dtEmployee);
87
88      bsEmployee.Filter = ""; // remove the filter
89
90      lblRecordNo.Text = (bsEmployee.Position + 1) + " of " + bsEmployee.Count;
91      DisplayCurrentRow();
92  }
```

Step 5-7: Add the statements from Figure 12-57 to the Click event handler of btnDisplayAll.
Run the form and test the code by clicking the DisplayAll button after filtering records, and after selecting records using the data adapter.

The complete code for Employee_Select form is shown on the following pages:

The complete code for Employee_Select form

```csharp
21            // data table, data adapter and binding source for Employee table
22            DataTable dtEmployee = new DataTable();
23            SqlDataAdapter daEmployee;
24            BindingSource bsEmployee = new BindingSource();
25
26            // data table and data adapter for Department table
27            DataTable dtDept = new DataTable();
28            SqlDataAdapter daDept;
29
30    private void Employee_Select_Load(object sender, EventArgs e)
31    {
32        // Fill employee data table, dtEmployee
33        string connString = @"Data Source = (LocalDB)\MSSQLLocalDB; AttachDbFilename =
34                              C:\VC\HR2.mdf";
35        string empSQLStr = "SELECT * FROM Employee";
36        daEmployee = new SqlDataAdapter(empSQLStr, connString);
37        daEmployee.Fill(dtEmployee);
38
39        // Display employee data
40        bsEmployee.DataSource = dtEmployee; // links binding source to the data table
41        DisplayCurrentRow();
42        lblRecordNo.Text = (bsEmployee.Position + 1) + " of " + bsEmployee.Count;
43        bsEmployee.PositionChanged += bsEmployee_PositionChanged;
44
45        // Fill department data table, dtDept
46        string DeptSQLStr = "SELECT DeptNo, DeptName FROM Department";
47        daDept = new SqlDataAdapter(DeptSQLStr, connString);
48        daDept.Fill(dtDept);
49
50        //Display DeptNo's from dtDept in the ComboBox, cboFillByDept:
51        cboFillByDept.DataSource = dtDept;
52        cboFillByDept.DisplayMember = "DEPTNO";
53        cboFillByDept.ValueMember = "DEPTNO";
54
55        //Display DeptNo's from dtDept in the ComboBox, cboFilterByDept:
56        cboFilterByDept.DataSource = dtDept;
57        cboFilterByDept.DisplayMember = "DEPTNO";
58        cboFilterByDept.ValueMember = "DEPTNO";
59    }
60
61    private void cboFilterByDept_SelectionChangeCommitted(object sender, EventArgs e)
62    {
63        bsEmployee.Filter = "DeptNo = " + cboFilterByDept.SelectedValue.ToString();
64        if (bsEmployee.Count > 0)
65        {
66            lblRecordNo.Text = (bsEmployee.Position + 1) + " of " + bsEmployee.Count;
67            DisplayCurrentRow();
68        }
69
70    }
```

```csharp
        private void cboFillByDept_SelectionChangeCommitted(object sender, EventArgs e)
        {
            //Specify the SQL
            daEmployee.SelectCommand.CommandText = "SELECT * FROM Employee where DeptNo = " +
                                        cboFillByDept.SelectedValue.ToString();
            //Clear and fill the data table
            dtEmployee.Clear();
            daEmployee.Fill(dtEmployee);
        }

    private void btnDisplayAll_Click(object sender, EventArgs e)
    {
        dtEmployee.Clear();
        daEmployee.SelectCommand.CommandText = "SELECT * FROM Employee";
        daEmployee.Fill(dtEmployee);

        bsEmployee.Filter = ""; // remove the filter

        lblRecordNo.Text = (bsEmployee.Position + 1) + "  of  " + bsEmployee.Count;
        DisplayCurrentRow();
    }

    private void DisplayCurrentRow()
    {
        // store the current row of the data table in a DataRow
        int currentRow = bsEmployee.Position;
        DataRowView drvEmployee = (DataRowView)bsEmployee[currentRow];

        // copy data from the datarow to controls on the form:
        txtEmpId.Text = drvEmployee["EmpId"].ToString();
        // Or, use the index as in, drvEmployee[0].ToString();
        txtFirstName.Text = drvEmployee["FirstName"].ToString();
        txtLastName.Text = drvEmployee["LastName"].ToString();
        txtDOB.Text = drvEmployee["DateofBirth"].ToString();
        txtPhone.Text = drvEmployee["Phone"].ToString();
        txtCountryId.Text = drvEmployee["CountryId"].ToString();
        txtDepNo.Text = drvEmployee["DeptNo"].ToString();
        txtSalary.Text = drvEmployee["Salary"].ToString();
        chkFullTime.Checked = bool.Parse(drvEmployee["FullTime"].ToString());
        txtPosition.Text = drvEmployee["Position"].ToString();
    }
```

```
113
114     private void btnFirst_Click(object sender, EventArgs e)
115     {
116         bsEmployee.MoveFirst();
117     }
118     private void btnPrevious_Click(object sender, EventArgs e)
119     {
120         bsEmployee.MovePrevious();
121     }
122     private void btnLast_Click(object sender, EventArgs e)
123     {
124         bsEmployee.MoveLast();
125     }
126     private void btnNext_Click(object sender, EventArgs e)
127     {
128         bsEmployee.MoveNext();
129     }
130
131     private void bsEmployee_PositionChanged(object sender, System.EventArgs e)
132     {
133         // display row number and data, whenever a the current record position changes
134         if (bsEmployee.Count > 0)
135         {
136             lblRecordNo.Text = (bsEmployee.Position + 1) + " of " + bsEmployee.Count;
137             DisplayCurrentRow();
138         }
139     }
```

Review Question

12.19 A set of **unbound** textboxes is used to display records from the database table, Student (Id, Name, Major, Advisor, Phone), using the following objects:
        data adapter   - daStudent
        data table   - dtStudent
        binding source - bsStudent
The binding source is linked to the data table using appropriate statements in the code.
Two different code segments are shown below:
```
// code segment #1
daStudent.SelectCommand.CommandText = "SELECT * FROM Student WHERE Major = 'IS' "
daStudent.Fill(dtStudent);
DisplayCurrentRow();  // a method that displays current row, in the TextBoxes

// code segment #2
daStudent.Fill(dtStudent);
bsStudent.Filter = "Major = 'IS' "
DisplayCurrentRow();  // a method that displays current row, in the TextBoxes
```

How is the effect of executing code segment #1 different from the effect of executing code segment #2?

12.11 Untyped Datasets: Add/Edit/Delete Records

In this section, you will learn how to add, edit and delete Employee records using untyped dataset, data adapter and binding source that are created at runtime using code. Compared to using the Add, Save and Delete buttons on the navigation bar generated by the wizard, this method gives you more control on validating the data and controlling user actions.

As discussed in the previous section, the SelectCommand property of the data adapter is generated when you create the data adapter, and the SQL to select records was assigned to the SelectCommand. To add, edit and delete records, you need to generate the **InsertCommand**, **UpdateCommand** and **DeleteCommand** properties for the data adapter using the SqlCommandBuilder.
The following statement builds these commands for the data adapter, daEmployee:

```
SqlCommandBuilder commandBuilder = new SqlCommandBuilder(daEmployee)
```

where *commandBuilder* is any valid variable name.

The generated commands will have the necessary SQL. For example, the UpdateCommand will have the SQL to update the database table.

You will use the form shown in Figure 12-58 to learn how to develop the code to add, edit and delete records using untyped data tables.

Figure 12-58: The form to add, edit and delete Employee records

A difference in the user interface between the current version of the form and previous versions is the use of RadioButtons, rather than a TextBox, to make it easy to enter or change an employee's job position. In addition, a ComboBox is used to let the user select a DeptNo from a ComboBox to make it easy to enter a DeptNo while adding a new record or changing an existing record.

The **Add** button just clears the controls on the form so that you can enter the data for a new record and save the data.

The **Cancel** button displays the current row from the binding source so that any changes you made in the controls are removed.

The **Save** button saves new records as well as changes to the current record.

Tutorial 6: Add/Edit/Delete Using Untyped Datasets

This tutorial develops the code for the form Employee_AddEdDel.

Step 6-1: Open the project Ch12_Databases_UntypedDatasets. Open the form, Employee_AddEdDel.
This form includes all controls and the code to display Employee records without the Position field.
Open the code window and view the code.
Make sure that the connection string in line 38 specifies the right directory.
You will add the remaining code, including the code to add, save and delete records, and cancel changes made to the data.

Currently, the Load event handler includes the code to display employee records. In addition, the event handler includes the code to fill the data table, dtDept. You need to add the code to display DeptNo's in the ComboBox from the data table. The Department table doesn't need a binding source, because the only use of the table is to add DeptNo's to the ComboBox, which can be done directly from the data table.

Step 6-2: Add the following code to the Load event handler to display DeptNo's in the ComboBox.
Run the form, and make sure the ComboBox displays the DeptNo's.

```
52          //Display DeptNo's from data table, in cboDept:
53          cboDeptNo.DataSource = dtDept;
54          cboDeptNo.DisplayMember = "DEPTNO";
55          cboDeptNo.ValueMember = "DEPTNO";
```

Displaying Employee's Position in RadioButtons

As in previous forms, data from the current row is first stored in a DataRowView, and then copied from the DataRowView to controls on the form. This is done in the Display_Records method. The code to be added to display the Position field in RadioButtons is shown in Figure 12-59.

Figure 12-59: Code to display employee's position in RadioButtons

```
81              switch (drvEmployee["Position"].ToString())
82              {
83                  case "VP":
84                      rdbVP.Checked = true;
85                      break;
86                  case "Manager":
87                      rdbMgr.Checked = true;
88                      break;
89                  case "Staff":
90                      rdbStaff.Checked = true;
91                      break;
92              }
```

Step 6-3: Add the code from Figure 12-59 to the Display_Records method.
Run the form and make sure the Position data are displayed in RadioButtons.

Adding New Records

Adding new records involves clearing the controls, entering the data, and saving the data. Figure 12-60 shows the Click event handler of the Add button, and the ClearControls method that it calls.

The event handler also sets the Boolean variable adding. This flag is used to determine whether the data to be saved is a new record or an existing record.

The ClearControls clears all the TextBoxes using a foreach loop to access each control from the Controls collection.

Figure 12-60: Clearing controls

```
95          private void btnAdd_Click(object sender, EventArgs e)
96          {
97              ClearControls();    // clear all controls
98              adding = true;
99          }

214         private void ClearControls()
215         {
216             foreach (Control control in Controls)
217             {
218                 if (control is TextBox)
219                     control.Text = null;    //or, control.Text = "";
220             }
221             cboDeptNo.Text = null;
222             rdbVP.Checked = false;
223             rdbMgr.Checked = false;
224             rdbStaff.Checked = false;
225             chkFullTime.Checked = false;
226         }
```

Step 6-4: Add the code from Figure 12-60 to the Click event handler of the Add button. The Boolean variable, adding, is already declared at the class level.

Saving Records

Saving a record is a two-step process: First, save the data from the controls into the **data table**, and then save the data from the data table to the **database table** using Update method of the table adapter.

The same **Save** button is used to save changes to existing records and to add new records. Figure 12-61 shows the event handler. Two methods, CopyToDataRowView and CopyToDataRow, that are called from the event handler (lines 109 and 123) are shown in Figure 12-62 and Figure 12-63, respectively.

The Boolean variable *adding* is used to check whether the data in the controls is a new record. Saving new records into the **data table** involves the following steps:

- Create a blank DataRow (line 108).
- Copy data from the controls to the DataRow by calling the method, CopyToDataRow (line 109). Note that the data table is a collection of rows of the type, DataRow.
- Add the DataRow to the data table (line 111).

Saving changes to an existing record into the **data table** also involves three steps:

- Create a variable of **DataRowView** type that references the current row of the binding source (line 121). The binding source rows are of DataRowView type.
- Copy data from the controls to the DataRowView by calling CopyToDataRowView (line 123).
- Apply the changes in the DataRowView to the data table using the **EndEdit** method of the binding source (line 124).

Figure 12-61: Saving data

```
101     private void btnSave_Click(object sender, EventArgs e)
102     {
103         int noOfRecords = 0;
104         // Save data to the data table:
105         if (adding == true)    // if it is a new record
106         {
107             // Create a blank DataRow by calling the NewRow() method of data table.
108             drEmployee = dtEmployee.NewRow();
109             CopyToDataRow();    // copy all data, into DataRow
110             // Add the new DataRow to the Rows collection of the data table
111             dtEmployee.Rows.Add(drEmployee);
112
113             lblRecordNo.Text = (bsEmployee.Position + 1) + " of " + bsEmployee.Count;
114             adding = false;
115         }
116         else // it's an existing record
117         {
118             // Create a DataRowView that references the current row of
119             // the binding source:
120             int currentRow = bsEmployee.Position;
121             drvEmployee = (DataRowView)bsEmployee[currentRow];
122
123             CopyToDataRowView();  // Copy data from controls the DataRowView:
124             bsEmployee.EndEdit();   //Save the DataRowView into data table
125         }
126         // Update the database table
127         noOfRecords = daEmployee.Update(dtEmployee); // Save to database table
128         MessageBox.Show("Changes saved to database");
129     }
```

The EndEdit Method of the Binding Source

The EndEdit method (line 124) applies the changes in the binding source rows to the data table it is linked to. When a binding source row is updated using unbound controls, as in the current example, it is important to copy the data from the controls into the binding source row, and call the **EndEdit** method to save the changes in the binding source row to the data table.

The Update Method of the Data Adapter

The Update method of the data adapter saves the changes in the data table to the database table, as coded in line 127:

```
noOfRecords = daEmployee.Update(dtEmployee);
```

The method returns the number of records updated.

Step 6-5: Add the code from Figure 12-61 to the Click event handler of btnSave. (The methods, CopyToDataRowView and CopyToDataRow, discussed below, are already created.)

The CopyToDataRowView Method

Saving an edited record involves copying data from all controls, except the primary key, into a DataRowView by calling the method, CopyToDataRowView, shown in Figure 12-62. The process of copying data from the controls into a DataRowView essentially is the opposite of displaying data from a DataRowView. The value of Position column is determined based on which RadioButton is checked (lines 141–147).

Figure 12-62: Copying data from controls to DataRowView

```
131     private void CopyToDataRowView()
132     {
133         // Copy data from controls into columns in the DataRowView
134         drvEmployee["LastName"] = txtLastName.Text;
135         drvEmployee["FirstName"] = txtFirstName.Text;
136         drvEmployee["DateOfBirth"] = txtDOB.Text;
137         drvEmployee["Phone"] = txtPhone.Text;
138         drvEmployee["CountryId"] = txtCountryId.Text;
139         drvEmployee["DeptNo"] = cboDeptNo.Text;
140         drvEmployee["Salary"] = txtSalary.Text;
141         drvEmployee["FullTime"] = chkFullTime.Checked;
142         string empPosition = null; //empPosition represents the job title
143         if (rdbVP.Checked == true)
144             empPosition = "VP";
145         else if (rdbMgr.Checked)
146             empPosition = "Manager";
147         else if (rdbStaff.Checked)
148             empPosition = "Staff";
149         drvEmployee["Position"] = empPosition;
150     }
```

Saving a new record involves copying data from all controls, including the primary key, to a DataRow by calling the method, CopyToDataRow, shown in Figure 12-63.

Figure 12-63: Copying data from controls to DataRow

```csharp
152     private void CopyToDataRow()
153     {
154         // Copy data from controls into columns of the DataRow
155         drEmployee["EmpId"] = txtEmpId.Text;
156         drEmployee["LastName"] = txtLastName.Text;
157         drEmployee["FirstName"] = txtFirstName.Text;
158         drEmployee["DateofBirth"] = txtDOB.Text;
159         drEmployee["CountryId"] = txtCountryId.Text;
160         drEmployee["DeptNo"] = cboDeptNo.Text;
161         drvEmployee["Salary"] = txtSalary.Text;
162         drEmployee["FullTime"] = chkFullTime.Checked;
163         string empPosition = null;
164         if (rdbVP.Checked == true)
165             empPosition = "VP";
166         else if (rdbMgr.Checked)
167             empPosition = "Manager";
168         else if (rdbStaff.Checked)
169             empPosition = "Staff";
170         drEmployee["Position"] = empPosition;
171     }
```

Deleting Records

Deleting a record from the database involves two steps:

1. Remove it from the data table using the RemoveCurrent method of the binding source.
2. Update the database table using the Update method of the data adapter.

Figure 12-64 shows the code. The code uses a MessageBox to confirm deletion of the record. The MessageBox returns a value of DialogResult type, which is an enumeration. The value, which depends on the button selected by the user, is cast to short type, and stored in the variable, response. The record is deleted if the value of response is the number corresponding to the *Yes* button.

Step 6-6: Add the code from Figure 12-64 to the Click event handler of btnDelete. Run the form.

Figure 12-64: Deleting a record

```csharp
173         private void btnDelete_Click(object sender, EventArgs e)
174         {
175             short response = 0;
176             response = (short)MessageBox.Show("Delete Record? ", "Confirm ",
177                     MessageBoxButtons.YesNo, MessageBoxIcon.Question);
178             if (response == (short)System.Windows.Forms.DialogResult.Yes)
179             {
180                 //Remove current record from dataset table
181                 bsEmployee.RemoveCurrent();
182
183                 //Update the database table with the dataset table
184                 this.daEmployee.Update(dtEmployee);
185                 MessageBox.Show("Record deleted from the database");
186             }
187         }
```

The following is the complete code for the Employee_AddEdDel form:

```csharp
22        DataTable dtEmployee = new DataTable();
23        SqlDataAdapter daEmployee;
24        BindingSource bsEmployee = new BindingSource();
25
26        // data table, data adapter and binding source Department table:
27        DataTable dtDept = new DataTable();
28        SqlDataAdapter daDept;
29        BindingSource bsDept = new BindingSource();
30
31        bool adding = false;
32        DataRowView drvEmployee;
33        DataRow drEmployee;
34
35        private void Employee_AddEdDel_Load(object sender, EventArgs e)
36        {
37            // Fill employee data table, dtEmployee
38            string connString = @"Data Source =
39                    (LocalDB)\MSSQLLocalDB; AttachDbFilename = C:\VC\HR2.mdf";
40            string empSQLStr = "SELECT * FROM Employee";
41            daEmployee = new SqlDataAdapter(empSQLStr, connString);
42            //Build UpdateCommand, DeleteCommand, and InserCommand
43            SqlCommandBuilder commandBuilder = new SqlCommandBuilder(daEmployee);
44
45            daEmployee.Fill(dtEmployee);
46
47            // Fill department data table, dtDept
48            string DeptSQLStr = "SELECT DeptNo, DeptName FROM Department";
49            daDept = new SqlDataAdapter(DeptSQLStr, connString); //Create data adapter
50            daDept.Fill(dtDept);
51
52            //Display DeptNo's from Department table, in cboDept:
53            cboDeptNo.DataSource = dtDept;   // DeptNo's are obtained directly from data table
54            cboDeptNo.DisplayMember = "DEPTNO";
55            cboDeptNo.ValueMember = "DEPTNO";
56
57            // Display employee data
58            bsEmployee.DataSource = dtEmployee; // links the binding source to the data table
59            DisplayCurrentRow();
60            lblRecordNo.Text = (bsEmployee.Position + 1) + " of " + bsEmployee.Count;
61            bsEmployee.PositionChanged += bsEmployee_PositionChanged;
62        }
```

```csharp
63
64      private void DisplayCurrentRow()
65      {
66          // Copy the current row into a DataRowView:
67          int currentRow = bsEmployee.Position;
68          DataRowView drvEmployee = (DataRowView)bsEmployee[currentRow];
69
70          // Copy data from the datarowview to controls on the form:
71          txtEmpId.Text = drvEmployee["EmpId"].ToString();
72          txtLastName.Text = drvEmployee["LastName"].ToString();
73          txtFirstName.Text = drvEmployee["FirstName"].ToString();
74          txtDOB.Text = drvEmployee["DateofBirth"].ToString();
75          txtPhone.Text = drvEmployee["Phone"].ToString();
76          txtCountryId.Text = drvEmployee["CountryId"].ToString();
77          txtPhone.Text = drvEmployee["Phone"].ToString();
78          cboDeptNo.Text = drvEmployee["DeptNo"].ToString();
79          txtSalary.Text = drvEmployee["Salary"].ToString();
80          chkFullTime.Checked = bool.Parse(drvEmployee["FullTime"].ToString());
81          switch (drvEmployee["Position"].ToString())
82          {
83              case "VP":
84                  rdbVP.Checked = true;
85                  break;
86              case "Manager":
87                  rdbMgr.Checked = true;
88                  break;
89              case "Staff":
90                  rdbStaff.Checked = true;
91                  break;
92          }
93      }
94
95      private void btnAdd_Click(object sender, EventArgs e)
96      {
97          ClearControls();    // clear all controls
98          adding = true;
99      }
```

```csharp
private void btnSave_Click(object sender, EventArgs e)
{
    int noOfRecords = 0;
        // Save data to the data table:
        if (adding == true)      // if it is a new record
        {
            // Create a blank DataRow by calling the NewRow() method of data table.
            drEmployee = dtEmployee.NewRow();
            CopyToDataRow();    // copy all data, into DataRow
            // Add the new DataRow to the Rows collection of the data table
            dtEmployee.Rows.Add(drEmployee);

            lblRecordNo.Text = (bsEmployee.Position + 1) + "  of  " + bsEmployee.Count;
            adding = false;
        }
        else // it's an existing record
        {
            // Create a DataRowView that references the current row of
            // the binding source:
            int currentRow = bsEmployee.Position;
            drvEmployee = (DataRowView)bsEmployee[currentRow];

            CopyToDataRowView(); //copy data from controls into columns of DataRowView:
            bsEmployee.EndEdit();    //Save the DataRowView into data table
        }
        // Update the database table
        noOfRecords = daEmployee.Update(dtEmployee); // Save to database table
        MessageBox.Show("Changes saved to database");
}

private void CopyToDataRowView()
{
    // Copy data from controls into columns in the DataRowView
    drvEmployee["LastName"] = txtLastName.Text;
    drvEmployee["FirstName"] = txtFirstName.Text;
    drvEmployee["DateOfBirth"] = txtDOB.Text;
    drvEmployee["Phone"] = txtPhone.Text;
    drvEmployee["CountryId"] = txtCountryId.Text;
    drvEmployee["DeptNo"] = cboDeptNo.Text;
    drvEmployee["Salary"] = txtSalary.Text;
    drvEmployee["FullTime"] = chkFullTime.Checked;
    string empPosition = null; //empPosition represents the job title
    if (rdbVP.Checked == true)
        empPosition = "VP";
    else if (rdbMgr.Checked)
        empPosition = "Manager";
    else if (rdbStaff.Checked)
        empPosition = "Staff";
    drvEmployee["Position"] = empPosition;
}
```

```csharp
        private void CopyToDataRow()
        {
            // Copy data from controls into columns of the DataRow
            drEmployee["EmpId"] = txtEmpId.Text;
            drEmployee["LastName"] = txtLastName.Text;
            drEmployee["FirstName"] = txtFirstName.Text;
            drEmployee["DateofBirth"] = txtDOB.Text;
            drEmployee["CountryId"] = txtCountryId.Text;
            drEmployee["DeptNo"] = cboDeptNo.Text;
            drvEmployee["Salary"] = txtSalary.Text;
            drEmployee["FullTime"] = chkFullTime.Checked;
            string empPosition = null;
            if (rdbVP.Checked == true)
                empPosition = "VP";
            else if (rdbMgr.Checked)
                empPosition = "Manager";
            else if (rdbStaff.Checked)
                empPosition = "Staff";
            drEmployee["Position"] = empPosition;
        }

        private void btnDelete_Click(object sender, EventArgs e)
        {
            short response = 0;
            response = (short)MessageBox.Show("Delete Record? ", "Confirm ",
                        MessageBoxButtons.YesNo, MessageBoxIcon.Question);
            if (response == (short)System.Windows.Forms.DialogResult.Yes)
            {
                //Remove current record from dataset table
                bsEmployee.RemoveCurrent();

                //Update the database table with the dataset table
                this.daEmployee.Update(dtEmployee);
                MessageBox.Show("Record deleted from the database");
            }
        }

        private void btnFirst_Click(object sender, EventArgs e)
        {
            bsEmployee.MoveFirst();
        }
        private void btnNext_Click(object sender, EventArgs e)
        {
            bsEmployee.MoveNext();
        }
        private void btnPrevious_Click(object sender, EventArgs e)
        {
            bsEmployee.MovePrevious();
        }
        private void btnLast_Click(object sender, EventArgs e)
        {
            bsEmployee.MoveLast();
        }
        private void bsEmployee_PositionChanged(object sender, System.EventArgs e)
        {
            if (bsEmployee.Count > 0)
            {
                lblRecordNo.Text = (bsEmployee.Position + 1) + " of " + bsEmployee.Count;
                DisplayCurrentRow();
            }
        }
```

```
213
214    private void ClearControls()
215    {
216        foreach (Control control in Controls)
217        {
218            if (control is TextBox)
219                control.Text = null;     //or, control.Text = "";
220        }
221        cboDeptNo.Text = null;
222        rdbVP.Checked = false;
223        rdbMgr.Checked = false;
224        rdbStaff.Checked = false;
225        chkFullTime.Checked = false;
226    }
227
228    private void btnCancel_Click(object sender, EventArgs e)
229    {
230        DisplayCurrentRow();
231        adding = false;
232    }
```

Review Questions

12.20   What does the EndEdit method of binding source do?

12.21   True or false: The Rows of a data table are of DataRowView type.

12.22   What does the Update method of the data adapter do?

12.12 Command Object and DataReader

In the previous sections, you used the data adapter in combination with the data table to retrieve, update and delete records from a database. The data adapter has various SqlCommand objects like SelectCommand and UpdateCommand to do these tasks. In this section, you will create and use SqlCommand objects independent of data adapter and data table, and use the objects to execute SQL statements.

The data table was used in previous applications to keep the data in the memory while you work with the data. The data table, in combination with the data adapter and binding source, provides a convenient way to navigate through the records, and update, delete and add records. However, data tables take up memory. Further, updates to the database after filling a data table are not reflected in the data table because the connection to the database server is closed as soon as the data are loaded into the data table.

A **DataReader** (SqlDataReader, OdbcDataReader, or OracleDataReader) provides an efficient way to sequentially access data retrieved by an SQL statement. As the SQL executes, each row retrieved by the SQL is stored temporarily in the client's buffer, accessed sequentially by the Read method of the DataReader, and then removed from the buffer. So, DataReader can help reduce memory usage and improve performance. But, you cannot move back to previous records; it's a forward-only access, and the DataReader does not have methods to update, delete or add records. The connection is kept open for exclusive use by the DataReader until the DataReader is closed.

To understand the use of SqlCommand objects and DataReaders, you will create the form, Employee_Reader1, shown in Figure 12-65, that performs three different tasks:

1. Find the phone number of an employee by specifying the EmpId. This involves using the SqlCommand object to execute an SQL to find a single value.
2. Find the last name, first name and salary by specifying the EmpId. This involves using the SqlCommand object to execute an SQL, and using a DataReader to access the row retrieved by the SQL.
3. Update the salary.

Figure 12-65: Use of SqlCommand and SqlDataReader

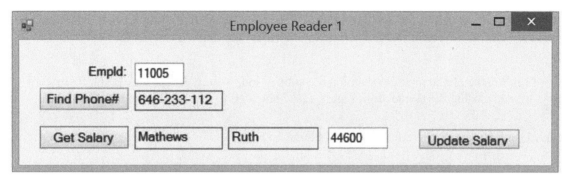

Creating the SqlCommand Object

The SqlCommand object, cmdEmployee, is created from the SqlCommand class, as follows:
   SqlCommand cmdEmployee = new SqlCommand()

where cmdEmployee could be any valid variable name. Because the Command object is used in multiple methods on this form, it is created at the class level.

You need to add the using directive, *using System.Data.SqlClient,* to specify the namespace that contains the class.

After you create the Command object, you need to specify the connection string and the SQL. To specify the connection, you create the SqlConnection object and assign it to the Connection property, as shown in line 26 of Figure 12-66. The connection string then is stored in the ConnectionString property of the Connection object.

Figure 12-66: Specifying the connection information

```
20      // Create a SqlCommmand object
21      SqlCommand cmdEmployee = new SqlCommand();
22
23      private void Employee_Reader1_Load(object sender, EventArgs e)
24      {
25          // Create a Connection object and assign it to the Connection property:
26          cmdEmployee.Connection = new SqlConnection();
27          // Specify the connection string for the Connection object
28          cmdEmployee.Connection.ConnectionString = @"Data Source = (LocalDB)\MSSQLLocalDB;
29                                      AttachDbFilename = C:\VC\HR2.mdf";
30      }
```

Next, you specify the SQL to find the phone number for a specified EmpId. The SQL is stored in the CommandText property of the SqlCommand object, as follows:

```
cmdEmployee.CommandText = "SELECT Phone FROM Employee WHERE EmpId = " + txtEmpId.Text;
```

EmpId is a numeric field; so, the value of EmpId is not enclosed in apostrophes, as in the case of string data. If EmpId is of string type, its value should be enclosed within apostrophes, as follows:

```
"SELECT Phone FROM Employee WHERE EmpId = '" + txtEmpId.Text + "'";
```

ExecuteScalar method of SqlCommand

The SqlCommand object has different methods for different types of SQL statements. An SQL with a single value in the result set can be executed using the ExecuteScalar method of the Command object, which returns the value. If the result set has more than one value, the method returns the first column of the first row.

The process of finding the phone includes opening the connection, calling the ExecuteScalar method to execute the SQL, and closing the connection. Figure 12-67 shows the code to find the phone number.

Figure 12-67: Using ExecuteScalar method to find phone number

```
32          private void btnPhone_Click(object sender, EventArgs e)
33          {
34              // Specify the SQL for the Command object
35              cmdEmployee.CommandText = "SELECT Phone FROM Employee WHERE EmpId = "
36                                       + txtEmpId.Text;
37              cmdEmployee.Connection.Open();  // open the connection
38              // Call the ExecuteScalar method to execute the SQL, and
39              // store the returned value in the variable, phone
40              string phone = cmdEmployee.ExecuteScalar().ToString();
41
42              lblPhone.Text = phone;
43              cmdEmployee.Connection.Close();
44          }
```

Line 40 shows the statement to call the ExecuteScalar method that returns the result (phone number) from executing the SQL specified in line 35. The returned value is stored in the variable, phone, and displayed in the Label, lblPhone.

Tutorial 7: SqlCommand and SqlDataReader to Work with Tables

This tutorial develops the Employee_Reader1 form, shown in Figure 12-65, and another form named Employee_Reader2 that retrieves and displays multiple Employee records.

Step 7-1: Open the project, Ch12_Databases_DataReader. Open the form, Employee_Reader1.

Add the code from Figure 12-66 to create the SqlCommand object and specify the connection information. Make sure that you specify the correct path for the database file.
Add the code from Figure 12-67 to specify the SQL and display the phone number.
Run the form and test the code using EmpId 11005.

ExecuteReader Method of SqlCommand

The ExecuteReader method executes an SQL, creates and populates a DataReader with the result set, and returns the DataReader that lets you access the rows, sequentially. So, this method is useful when the SQL result set includes multiple values from one or more rows. In the current form, you will use this method to execute an SQL that retrieves a single row containing the LastName, FirstName and Phone for a specified EmpId.

The DataReader returned by ExecuteReader may be assigned to a variable of DataReader type, as in the following statement:
```
SqlDataReader drEmployeeReader = cmdEmployee.ExecuteReader();
```

where drEmployeeReader could be any valid variable name.

Read Method of the DataReader

The rows in the result set of an SQL are accessed using the Read method of the DataReader, as follows:
    drEmployeeReader.Read();

The Read method advances the reader to the next row, thus letting you access the values within the row. Initially, the position of the Reader is before the first record. To read multiple rows, you use a loop.

Figure 12-68 shows the code that uses the Read method and performs the following tasks:
- Specifies the SQL
- Calls the ExecuteReader method
- Calls the Read method of DataReader
- Displays column values in Labels

Figure 12-68: Code to display column values

```
46      private void btnGetSalary_Click(object sender, EventArgs e)
47      {
48          // Assign the SQL to the CommandText property
49          cmdEmployee.CommandText = "SELECT LastName, FirstName, Salary FROM Employee WHERE EmpId = "
50                                                      + txtEmpId.Text;
51          cmdEmployee.Connection.Open();
52          // Call the ExecuteReader method to execute the SQL.
53           SqlDataReader drEmployee = cmdEmployee.ExecuteReader();
54
55          // Move the DataReader position to the first (and only) row.
56          drEmployee.Read();
57
58          // Get column values from the current row of the DataReader, and display in Labels:
59          lblLastName.Text = drEmployee["LastName"].ToString();
60          lblFirstName.Text = drEmployee["FirstName"].ToString();
61          txtSalary.Text = drEmployee["Salary"].ToString();
```

Line 53 shows the statement that calls the ExecuteReader method and assigns the DataReader returned by the method to the variable, drEmployeeReader, which is of SqlDataReader type.

The statement in **Line 56** advances the DataReader position to the first and only row of the DataReader.

If the result set of the SQL has multiple rows, you will call the Read method inside a loop to access the rows, as discussed later in the Employee_Reader2 form.

Getting Column Values from a Row Using the DataReader

You can get values from a column of the current row by specifying the name of the column, as shown in lines 59–61 in Figure 12-68. The data retrieved using this method is object type, which is converted to string type for display in Labels.

Step 7-2: Add the code from Figure 12-68 to display the name and phone number.
Run the form, and test the code.

Alternate Methods to Get the Column Values

You also may use the ordinal (the position) of the column, like an index, in place of the column name to get the value. Line 65 in Figure 12-69 uses the ordinal, 2, of the Salary column. The ordinal would be 0 for the first column, 1 for the second column, and so on. Salary is the third column in the result set of the SQL, so its ordinal is 2. In general, it is more efficient to retrieve data using the ordinals. However, this is prone to errors due to using the wrong ordinals.

Figure 12-69: Alternate methods to get column values

```
62
63          // Alternate methods to get the salary
64          // Use the ordinal, 2, in place of the column name, Salary
65          txtSalary.Text = drEmployee[2].ToString(); // Salary is the 3rd column of the row
66          // or,
67          Int32 salaryOrdinal = drEmployee.GetOrdinal("Salary");
68          txtSalary.Text = drEmployee[salaryOrdinal].ToString();
69
70          // Get Salary in its original type, double
71          double salary = drEmployee.GetDouble(2);
72          txtSalary.Text = salary.ToString();
73
74          drEmployee.Close();
75          cmdEmployee.Connection.Close();
76      }
```

To use meaningful names for ordinals, you may get the ordinal using the column names, store it in a variable, and use the variable name in place of the literal values. The statement in line 67 uses the GetOrdinal method to get the ordinal, and stores it in a variable. Line 68 uses the variable name to get the salary. This method would help to minimize errors due to using literal values like 1 or 2 for ordinals.

Regardless of the type of fields in the database table, the column values retrieved using the above methods are object types. So, they need to be converted to appropriate types. For example, if Salary is to be used in calculations, it must be converted to a numeric type.

An alternative is to get values in their original types using methods like GetDouble, GetDateTime and GetString, shown in Table 12-1. This approach minimizes the need to convert data after it is retrieved. For example, if Salary needs to be used in calculations, it can be retrieved as Double, which is its type in the database table, using the following statement shown in line 71:

```
double salary = drEmployee.GetDouble(2);
```
This approach, however, requires knowledge of the exact type of the fields, when data are retrieved.

Step 7-3: Comment out line 61, and add the code from line 65 in Figure 12-69 to display salary using the ordinal. Run the form to test the code.

Comment out line 65, and add the code from lines 67 and 68, which use the GetOrdinal method. Run the form to test the code.

Comment out lines 67 and 68, and add the code from lines 71 and 72, which use the GetDouble method. Uncomment lines 74 and 75. Run the form to test the code.

The DataReader has several other methods and properties that help to work with it. Some commonly used methods/properties are shown in Table 12-1.

Table 12-1: Methods and Properties of SqlDataReader

Method/property name	Description	Example
Close	Closes the DataReader object	drEmployee.Close
Dispose	Releases resources used	drEmployee.Dispose
GetBoolean, GetDateTime, GetDecimal, GetDouble, GetFloat, GetInt32, GetString	Gets the value in its original type	drEmployee.GetString(0) drEmployee.GetDouble(2)
GetOrdinal(String)	Gets the ordinal	drEmployee.GetOrdinal["EmpId"]
GetSchemaTable	Returns a data table that contains field attributes	drEmployee.GetSchemaTable
GetValue	Gets the value as object	drEmployee.GetValue(0)
GetValues	Gets the values in the row as an array of objects	object[] colValues = new object[10]; drEmployee.GetValues(colValues);
Read	Advances to next row	drEmployee.Read()
Properties		
FieldCount	Gets the number of columns	int count = drEmployee.FieldCount
HasRows	Indicates whether the DataReader contains any rows	if (drEmployee.HasRows)

ExecuteNonQuery Method of SqlCommand

Previous sections used the ExecuteScalar method and ExecuteReader method of the SqlCommand to retrieve data from tables. The SqlCommand has a third method, **ExecuteNonQuery**, to execute queries that do not retrieve any data, but can be used to update, insert and delete records, and to do other database operations like create tables. For update, insert and delete, the method returns the number of records affected. For most other operations, the return value is -1.

You will use the ExecuteNonQuery to update the salary when the user enters the new salary into the TextBox, txtSalary, and clicks the Update Salary button on the Employee_Reader1 form presented earlier in Figure 12-65. The code is shown in Figure 12-70.

Figure 12-70: Use of ExecuteNonQuery to update record

```
77
78      private void btnUpdate_Click(object sender, EventArgs e)
79      {
80          cmdEmployee.CommandText = "UPDATE Employee SET Salary = " + txtSalary.Text
81                                      + " WHERE EmpId = " + txtEmpId.Text;
82          cmdEmployee.Connection.Open();
83          int count = cmdEmployee.ExecuteNonQuery();
84          MessageBox.Show(count + " record updated");
85          cmdEmployee.Connection.Close();
86      }
87  }
```

Line 80 stores the SQL statement in the CommandText property of the SqlCommand.

Line 83 calls the ExecuteNonQuery method to execute the SQL. The method returns the number of records updated, which is stored in the variable, count, although it is not used in this form.

You may use the same method to delete records and add new records.

Step 7-4: Add the code from Figure 12-70 to update the salary.

Accessing Multiple Rows Using a DataReader

Next, we will look at how to access multiple rows of an SQL result set. The code is essentially the same as that used to display a single row, except that we use a loop to call the Read method multiple times.

Consider an application that lists all employee records for a DeptNo selected from a ComboBox, as illustrated in the form Employee_Reader2, shown in Figure 12-71. A ListView is used to display the data.

Figure 12-71: Accessing multiple records using DataReader

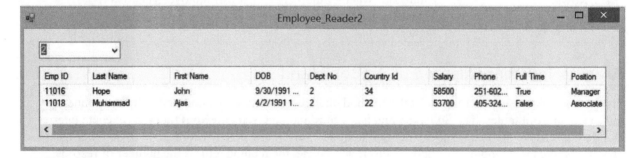

This application involves the following steps:

- Get DeptNo's from the Department table and add them to the ComboBox.
- Get the Employee records for the selected DeptNo.
- Store each record in a String array, and add it to the ListView.

Looping through Multiple Rows

This application involves accessing multiple rows of DeptNo's and Employee data. This is done by repeatedly calling the Read method of the DataReader in a while loop, as illustrated in Figure 12-72.

Figure 12-72: Using the while loop to access multiple rows

```
while (drEmployee.Read() == true)
{
        (statements to access the row)
}
```

The Read method in the while loop advances the reader to the next row, each time through the loop. The loop terminates when there are no more rows in the DataReader, as identified by the value, *false*, returned by the DataReader.

Getting DeptNo's from Department Table

As before, you will use the SqlCommand object and SqlDataReader objects to retrieve DeptNo's. Figure 12-73 shows the code to retrieve the data into a DataReader, and add them to the ComboBox.

You will use the same SqlCommand object, cmdEmployee, to retrieve Department and Employee data by changing the SQL assigned to the CommandText property.

Figure 12-73: Retrieving DeptNo's and adding to a ComboBox

```
19    // Create a SqlCommmand object
20    SqlCommand cmdEmployee = new SqlCommand();
21
22    private void Employee_Reader2_Load(object sender, EventArgs e)
23    {
24        // Create a Connection object and specify the connection string
25        cmdEmployee.Connection = new SqlConnection();
26        cmdEmployee.Connection.ConnectionString = @"Data Source = (LocalDB)\MSSQLLocalDB;
27                                                    AttachDbFilename = C:\VC\HR2.mdf";
28        // Add DeptNo's from the Department table, to the Combobox
29        cmdEmployee.CommandText = "SELECT DeptNo FROM Department"; // specify SQL
30        cmdEmployee.Connection.Open();
31        // Call the ExecuteReader method to execute the SQL.
32        SqlDataReader drEmployee = cmdEmployee.ExecuteReader();
33
34        while (drEmployee.Read() == true)  // loop through the rows
35        {
36            cboDeptNo.Items.Add(drEmployee["DeptNo"]);  // add DeptNo to the ComboBox
37        }
38        cmdEmployee.Connection.Close();
39    }
```

Line 32 executes the SQL, and creates the DataReader, drEmployee, that contains multiple rows, each containing a single DeptNo.

The while loop in **line 34** processes the Read method repeatedly to access the rows.

Line 36 adds each DeptNo to the ComboBox.

Step 7-5: Open the form Employee_Reader2. Add the missing code from Figure 12-73. Run the form. Make sure that the ComboBox displays the DeptNo's.

Displaying Employee Records in a ListView

Displaying the data in a ListView involves the following steps:

- Retrieve Employee records, and access each row using a while loop.
- Use the **GetValues** method to get the entire current row, and store it in an object array.
- Convert the object array to a String array.
- Create a ListViewItem using the String array.
- Add the ListViewItem to the ListView.

Figure 12-74 shows the code to implement these steps.

Figure 12-74: Displaying Employee data in a ListView

```
41    private void cboDeptNo_SelectedIndexChanged(object sender, EventArgs e)
42    {
43        cmdEmployee.CommandText = "SELECT * FROM Employee WHERE DeptNo = "
44                                  + cboDeptNo.SelectedItem;
45        cmdEmployee.Connection.Open();
46        SqlDataReader drEmployee = cmdEmployee.ExecuteReader();
47
48        // display data in a ListView by storing the current row in an array
49        lvwEmployee.Items.Clear();
50        while (drEmployee.Read() == true)   // loop through the rows
51        {
52            int colCount = drEmployee.FieldCount;  // get number of columns
53            object[] colValuesObj = new object[colCount];  // create an object array
54
55            // Get values from the current row, and store in the array, colValuesObj
56            drEmployee.GetValues(colValuesObj);
57            // Convert objects to strings and store in the String array, colValues
58            String[] colValues = Array.ConvertAll(colValuesObj, element => element.ToString());
59
60            // Create a ListViewItem using the array, colValues
61            ListViewItem lviEmp = new ListViewItem(colValues);
62            lvwEmployee.Items.Add(lviEmp);   // add the ListViewItem to ListView:
63        }
64        cmdEmployee.Connection.Close();
65    }
```

Let's look at the steps in more detail.

Retrieve Employee records, and access each row using a while loop

Lines 43–46 retrieve Employee records for the selected DeptNo, and create the DataReader, as before.

Line 50 uses the Read method in a while loop to access the rows.

Use the GetValues method to get the entire current row, and store it in an object array

Lines 52 and 53 use the **FieldCount** property, which gets the number of columns, to create the object array.
```
int colCount = drEmployee.FieldCount;
object[] colValuesObj = new object[colCount];
```

Line 56 gets the values, and stores them in the object array, colValuesObj. Note that GetValues returns the set of values as an object array.
```
drEmployee.GetValues(colValuesObj);
```

Convert the object array to a String array

Line 58 uses the Array.ConvertAll method to convert the values from object to String:

```
String[] colValues = Array.ConvertAll(colValuesObj, element => element.ToString());
```

The **Array.ConvertAll** method converts the elements of an array of one type to another type, and returns an array of the new type.

> The first argument, colValuesObj, is the input array.
> The second argument "element => element.ToString()" can be thought of as a function with element as the input, and element.ToString() as the output. The word, element, in the second argument could be any word or notation.

Create a ListViewItem using the String array

Line 61 creates the ListViewItem, as discussed in Chapter 10:
```
ListViewItem lviEmp = new ListViewItem(colValues);
```

Add the ListViewItem to the ListView

Line 62 adds the ListViewItem to the ListView, lvwEmp:
```
lvwEmployee.Items.Add(lviEmp);
```

Accessing Individual Column Values

If it is necessary to work with individual column values rather than displaying data in a ListView, the values may be accessed directly from the DataReader using the column name or the ordinal, as shown in the following code:

```
while (drEmployee.Read() == true)
    {
        double salary = (double) drEmployee["Salary"];
        // or,
        Int32 salaryOrdinal = drEmployee.GetOrdinal("Salary");
        double salary = drEmployee.GetDouble(salaryOrdinal);
        // (statements that use salary go here)
    }
```

You also may store an entire row into an array, and access the individual columns from the array:

```
while (drEmployee.Read() == true)   // loop through the rows
{
        int colCount = drEmployee.FieldCount;  // get number of columns
        object[] colValuesObj = new object[colCount];   // create an object array

        // Get values from the current row, and store in the array, colValuesObj
        drEmployee.GetValues(colValuesObj);

        Int32 salaryOrdinal = drEmployee.GetOrdinal("Salary");
        double salary = (double) colValuesObj[salaryOrdinal];
        // (statements that use salary go here)
}
```

Step 7-6: Add the code from Figure 12-74. Run the form to test the code.

Review Questions

12.23 What is the function of a DataReader?

12.24 True or false: When you use a data table and data adapter to access records from a database table, the connection to the database table is maintained until the data table is closed.

12.25 True or false: A DataReader provides an efficient way to move forward and backward through the records retrieved by an SQL statement.

12.26 An SqlCommand object is to be used to execute an SQL statement that retrieves the salary for an employee, by specifying the employee's id. What method of the Command object should you use?

12.27 An SqlCommand object is to be used to execute an SQL statement that retrieves an employee's id, name and phone number using the employee's id. What method of the Command object should you use?

12.28 An SqlCommand object is to be used to execute an SQL statement that updates an employee's salary. What method of the Command object should you use?

12.29 What does the Read method of the SqlDataReader do?

Exercises

Exercise 12-1

Create a new project named Exercise12-1 and develop a form named DeptSearch that lets the user display records from Department table, one at a time using typed data table and data bound controls. Use a ComboBox for DeptNo field and TextBoxes for DeptName and DeptPhone. Use the ComboBox to let the user select a DeptNo from a list of all DeptNo's and display the corresponding record.

Exercise 12-2

Add a ComboBox named cboCountries to EmployeeSearch form within the project Ch12_Databases. Change the Text property of the ComboBox to "Filter By Country." The ComboBox should display a list of all unique CountryId's. When the user selects a CountryId from the ComboBox, filter the employee records using the selected CountryId. That is, the records displayed on the form should be limited to those for the selected CountryId.

Exercise 12-3

Add a ComboBox named cboDeptNo to EmployeeSearch form within the project Ch12_Databases. Change the Text property of the ComboBox to "Filter By Dept." The ComboBox should display a list of all DeptNo's from the Department table. When the user selects a DeptNo from the ComboBox, filter the employee records using the selected DeptNo; that is, the records displayed on the form should be limited to those for the selected DeptNo.

Next, modify the form so that the ComboBox displays a list of all department names instead of department numbers. When the user selects a name, use the corresponding DeptNo (ValueMember) to filter records.

Exercise 12-4

Create a new project named Exercise12_4 and develop a form named EmpSearch that lets the user display records from Employee table, one at a time, in TextBoxes. Use typed data table and data bound controls. Add a ComboBox named cboDeptNo to the form. Change the Text property of the ComboBox to "Fill By Dept." The ComboBox should display a list of all DeptNo's from the Department table. When the user selects a DeptNo from the ComboBox, load employee records for the selected DeptNo from the Employee database table to the Employee data table so that the user will be able to view only the selected records.

Exercise 12-5

Create a new project named Exercises12_Untyped, and develop a form named DeptDisplay that lets the user display records from Department table, one at a time using untyped data table and unbound controls. Allow the user to move between records. Use TextBoxes for all fields.

Exercise 12-6

Add a ComboBox named cboCountries to Employee_Select form within the project Ch12_Databases_UntypedDatasets. Change the Text property of the ComboBox to "Filter By Country." Create a data table, load the unique CountryId's into the table and display the CountryId's in the ComboBox. When the user selects a CountryId from the ComboBox, filter the employee records using the selected CountryId; that is, the records displayed on the form should be limited to those for the selected CountryId.

Exercise 12-7

Create a new project named Exercises12_Untyped, and develop a form named EmployeeSelect that lets the user display records from Employee table, one at a time, in TextBoxes. Use typed data table and data bound controls. Add a ComboBox named cboDeptNo to the form. Change the Text property of the ComboBox to "Fill By Dept." The ComboBox should display a list of all DeptNo's from the Department table. When the user selects a DeptNo from the ComboBox, load employee records for the selected DeptNo from the Employee database table to the data table, dtEmployee, so that the user will be able to view only the selected records.

Exercise 12-8

Create a new project named Exercises12_Untyped, and develop a form named DeptDAddEdDel that lets the user display records from Department table, one at a time using untyped data table and unbound controls. Allow the user to add, edit and delete records. Use TextBoxes for all fields.

Exercise 12-9

Open the project, Employee_DataReader. Open the form, Employee_Reader1. Add a button to this form so that the user can delete the currently displayed record by clicking the button.

Exercise 12-10

Create a new project named Exercises12_DataReader, and develop a form named DeptSelect that uses SqlCommand and SqlDataReader objects to let the user do the following tasks:

1. Type in a DeptNo into a TextBox and display the department's name in a Label.

2. Select a DeptNo from a ComboBox. Display the department's name and phone number in a ListBox.

Rentals Assignment 3

The main goal of this assignment is to give you experience in working with database tables, text files, DataGridView and ListView.

This assignment is a modified version of assignment 2. A major change is the use of database tables (instead of sequential files) to store information. You are to develop an application for "Ace Auto Rentals," which is a small business that rents automobiles.

The **Rentals** database used in this assignment can be found in Tutorial_Starts/DataFiles folder. This database includes two tables: the Rental table that stores information on car rentals, and AutoTypes that contains the daily rental rates for different types of automobiles. Copy the database file to a local folder.

The following is a sample of the records in the **Rental** table:

RentalId	CustNumber	CustName	RentalDate	NumOfDays	AutoTypeId	DiscCategory	TotalCost
1001	10138	Josh Philip	11/5/2015	2	3	Business	81
1002	10150	Mini George	10/10/2015	1	2	Favorite	48
1003	10144	Jeffrey Mathews	12/22/2015	3	1	State	179.5
1004	10130	Cathy Arnold	12/21/2015	2	2	Standard	120
1005	10135	Gina Rodriguez	11/22/2015	5	4	Standard	200
1006	10126	Faith Abraham	11/23/2015	4	3	Favorite	112
1007	10115	Chris Cheng	12/22/2015	4	2	Standard	80
1008	10124	Sashi Nair	12/20/2015	2	6	Business	45

The following table shows the field types of Rentals. The data types char() and varchar() are text data.

Name	Data Type	Allow Nulls
RentalId	int	☐
CustNumber	char(5)	☑
CustName	varchar(40)	☑
RentalDate	date	☑
NumOfDays	int	☑
AutoTypeId	int	☑
DiscCategory	varchar(15)	☑
TotalCost	float	☑

The **AutoTypes** table looks as follows:

AutoTypeId	Description	DailyRate
1	SUV	70
2	mini-van	60
3	luxury	45
4	full-size	40
5	mid-size	35
6	compact	25

Part 1

Create a form named DisplayRentals that lets the user display **Rental** records, one at a time, as shown below.

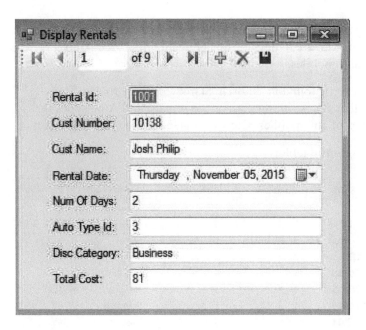

Note: If the version of SQL Server you use is incompatible with the database, when you create the data source, you will be asked whether you want to update the database. Answer "Yes."

Again, make sure that you answer **"No"** when the DataSource ConfigurationWizard asks you whether you want to copy the database to the project folder. (If you answer "Yes," the changes you make to the table will be lost the next time you run the project.)

The primary key, RentalId, is an auto-increment field. To view the fields and their properties, right click Rental table in Server Explorer window, and select *Open Table Definition*. The T-SQL window displays the SQL that creates the table. The key word **IDENTITY(1,1)** specifies that RentalId is an auto-increment field that increments by 1, starting with 1.

To view the data tables, open RentalDataSet.xsd. Select RentalId field and view its properties. Make sure that the auto-increment properties are set as follows:

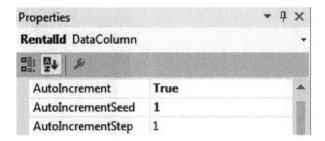

Part 2

Create a second form named EnterRentals that lets the user enter new records into the Rental table, as shown below.

Use a DataGridView to display records from the table AutoTypes (in Rentals database), and to let the user select an AutoTypeId. When the user clicks the "Insert Auto Type" button, copy the selected AutoTypeId to the corresponding TextBox in the new record.

Similarly, use a ListView to display records from the DiscCategories.txt file, shown below, and to let the user select a discount category. Keep DiscCategories file in the Bin/Debug folder of the project. When the user clicks the "Insert Disc Type" button, the selected discount category is copied to the corresponding TextBox in the new record.

```
Normal, 0
State, 15
Business, 10
Favorite, 20
```

When the user clicks "**Compute Cost**" button, the total cost is computed and inserted into the TextBox for Total Cost in the new record.
    Total cost: number of days * Daily Rate *(1- discPercent/100))

To compute total cost, use the daily rate from the selected row in the DataGridView (see EmpDirectory form in Ch12_Database). Similarly, use the discount percent from the selected row in the ListView (see ListViewApp form in Ch10_GUIadditionalControls).

Part 3

Create a third form, **Rentals**, that serves as a switchboard with buttons and **menus** that allow the user to go to EnterRentals and DisplayRentals forms. Make Rentals the startup form.

Help Desk Assignment 3

The main goal of this assignment is to give you experience in working with database tables, text files, DataGridView and ListView.

In this assignment, you will develop a modified version of the system specified in Help Desk Assignment 2. A major change is the use of database tables (instead of sequential files) to store information.

To refresh your memory, the system maintains information on support provided to employees by the IT help desk in an organization. The help desk provides over-the-phone and on-site assistance with technical problems to employees from different departments at multiple locations. The department of the employee is charged for the services.

The charge for service consists of a fixed charge and a variable charge. The fixed charge is $10, and the standard variable charge is the product of duration of service in minutes and charge per minute. The charge/min varies with the area in which the support is provided, as specified later.

Typically, request for services are put in a standard queue, and services are provided in the order in which requests are received. However, services that can wait may be put in a low priority queue to receive a 25% reduction in service charge. Urgent services that are put in a high priority queue are charged 25% over the standard service charge. Services that are put in the standard queue don't have any additional charge.

The **Helpdesk** database used in this assignment can be found in Tutorial_Starts/DataFiles folder. This database includes two tables: the **Service** table that stores information on services, and the **ServiceArea** table that contains the charge/min for different service areas.

Copy the database file to a local folder.

The following is a sample of the records in the **Service** table:

ServiceId	EmpId	FirstName	LastName	StartTime	Duration	SupportAreaId	PriorityCode	TotalCharge
1001	010256	Ann	Abraham	9/20/2019 2:32:00 PM	25	1	standard	35
1002	010128	Amy	Luke	11/5/2019 8:42:00 AM	6	4	standard	17.5
1003	010583	Terry	Chang	9/9/2019 1:13:00 PM	64	2	high	126
1004	010173	David	Bowie	10/8/2019 8:05:00 AM	15	1	low	20
1005	010242	Carlos	Lopez	9/23/2019 9:08:00 AM	12	3	standard	28
1006	010234	Murna	Gee	12/3/2019 9:44:00 AM	24	5	low	36.8
1007	010445	Ravi	Menon	11/5/2019 9:01:00 AM	10	2	standard	22.5

The following table shows the field types of Service table. The data types char() and varchar() are text data.

Name	Data Type	Allow Nulls
ServiceId (PK)	int	☐
EmpId	char(6)	☑
FirstName	varchar(20)	☐
LastName	varchar(30)	☑
StartTime	datetime	☑
Duration	int	☑
SupportAreaId	int	☑
PriorityCode	varchar(10)	☑
TotalCharge	float	☐

The **ServiceArea** table looks as follows:

ServiceAreaId	Description	ChargePerMt
1	computer system	1
2	ERP	1.25
3	BI	1.5
4	cloud	1.25
5	data warehouse	1.5

Note that the field ServiceAreaId in this table represents the same data item as the field **SupportAreaId** in the Service table.

Part 1

Create a project named **HDassignment3**, and develop a form named **DisplayService** that lets the user display **Service** records, one at a time, as shown below.

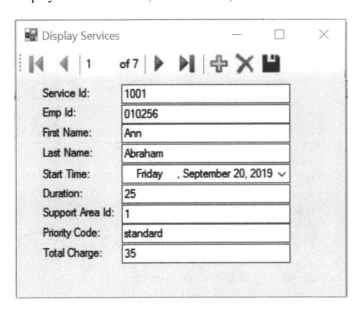

Note: If the version of SQL Server you use is incompatible with the database, when you create the data source, you will be asked whether you want to update the database. Answer "Yes."

Again, make sure that you answer **"No"** when the DataSource ConfigurationWizard asks you whether you want to copy the database to the project folder. (If you answer "Yes," the changes you make to the table will be lost the next time you run the project.)

The primary key, ServiceId, is an auto-increment field. To view the fields and their properties, right click Service table in the Server Explorer window, and select *Open Table Definition*. The T-SQL window displays the SQL that creates the table.

To view the data tables, open ServiceDataSet.xsd. Select ServiceId field and view its properties. Make sure that the auto-increment properties are set as follows:

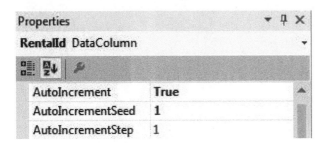

Part 2

Create a second form named **EnterService** that allows the user to enter new records into the **Service** table and to edit and delete existing records, as shown below.

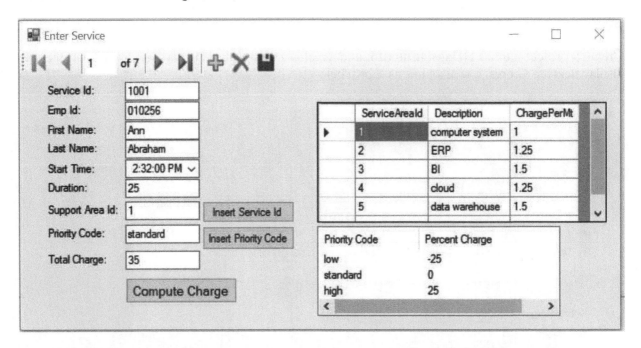

Use a DataGridView to display records from the ServiceArea table so that the user can select a ServiceAreaId and insert the Id into a new Service record or an existing one to change it. When the user clicks the "Insert Service Type" button, copy the selected ServiceTypeId to the corresponding TextBox (labeled "Support Area Id").

Similarly, use a ListView to display records from the **PriorityCode.txt** file, shown below, to let the user select a priority code and insert it into a Service record. Create the PriorityCode.txt file and keep it in the Bin/Debug folder of the project. When the user clicks the "Insert Priority Code" button, the selected priority code is copied to the corresponding TextBox labeled "Priority Code."

```
low, -25
standard, 0
high, 25
```

When the user clicks the "**Compute Charge**" button, the total charge is computed and inserted into the TextBox for Total Charge in the Service record.

Total charge: (fixed charge + duration in mts * charge/mt) *(1 + (Priority percent charge/100))

To compute the total charge, use the charge/min from the selected row in the DataGridView (see the EmpDirectory form in Ch12_Database). Similarly, use the priority percent from the selected row in the ListView (see the ListViewApp form in Ch10_GUIadditionalControls).

Part 3

Create a third form named **Helpdesk** that serves as a switchboard with buttons and **menus** that allow the user to go to **EnterService** and **DisplayService** forms. Make **Helpdesk** the startup form.

The factors that determine your grade include the following:

Use meaningful and easily understood names for forms, controls and variables used in the code. Use standard naming conventions. Do not use names like form1 and button1.

Make sure the code is easy to understand. Use comments to specify what each part of the code does. Avoid unnecessary duplication of code.

Use the C# features you learned effectively.

Chapter 13

Object-Oriented Programming

Most professional programming is done in languages based on the object-oriented paradigm. C# is no exception and is in fact one of the languages with the strongest support for proper object-oriented programming. Throughout the previous chapters of this book, you have actually made frequent use of many object-oriented principles and techniques. For instance, every user interface element, such as Button, TextBox and Label, are all objects. In this chapter, you will discover what object-oriented programming (OOP) is, how you can create and work with your own objects, and how to develop applications using OOP.

One of the primary reasons object-oriented programming is so popular in professional programming is that if you follow proper object-oriented principles, it leads you in the direction of a well-designed system. However, care still has to be taken in how this is achieved. The goal is to create a set of classes that can be reused in different contexts. For instance, you will see an Employee class throughout this chapter that could be used in a Payroll system, a Sales system, or a Performance Evaluation system. If it is used for multiple purposes, once it has been put into production, it cannot be changed without considering the effect on all systems.

Learning Objectives

After studying this chapter, you should be able to:

- Describe the differences between classes and objects.
- Define objects in terms of their identity, state and behavior.
- Create class UML diagrams to show key features of classes.
- Create object UML diagrams to show the state of an object-oriented system.
- Describe how encapsulation leads to better systems design.
- Create classes in C#.
- Use constructors to create new objects.
- Describe how properties expose object state.
- Describe how method calls can be used to pass messages between objects.
- Create a system with multiple forms where data can be passed between the forms using object-oriented principles.

Topics

13.1	Introduction to Objects and Classes	13.4	Properties
13.2	Classes versus Objects	13.5	Calling Methods (Sending Messages to Objects)
13.3	Information Hiding (Encapsulation)		

13.1 Introduction to Objects and Classes

Object-oriented programmers generally distinguish between the problem domain and the application domain. The problem domain involves the parts of the real world that the computer system is working with and solving problems for. For instance, the problem domain for a payroll system would contain the employees of the company, the actual hours worked by those employees, and all the rules governing how salaries, taxes, and other deductions are calculated and paid out. The application domain, on the other hand, is the actual payroll computer system and its users. The users will work with representations of the problem domain (real world) in order to solve the problems the system is intended to solve.

With object-oriented programming, we start by creating representations of the problem domain inside the application. The problem domain typically contains multiple entities like employee and payroll. The instances of entities we want to keep track of in the problem domain are represented in the application domain by *objects*. So, in the payroll system, each employee becomes an object, every pay period becomes an object, every pay check becomes an object, and so on. The set of objects that represent the instances of an entity is described in computer code with a construct called a *class*. A class is like a blueprint that can be used as a template to create many instances (objects) that have the same properties. In this way, a class represents an entity of the real-world problem addressed by the application. For example, a payroll application for an organization may use an Employee class to represent employees of the organization.

13.2 Classes versus Objects

With object-oriented programming, you start by creating classes that represent real-world entities. A class consists of code that describes a group of data items that represent the attributes of the entity and methods that represent the behavior of the entity. The name, birth date and address of the employee are examples of the data items of the Employee class. The data items in a class are accessed using an interface of publicly available methods and properties of the class. Behaviors are the activities or functions of an entity. Updating the hourly rate for an employee would be a behavior of the employee entity, which may be implemented by a method called PayRaise within the Employee class.

Figure 13-1 shows the Employee class in Unified Modeling Language (UML) notation. In UML, a class is represented by a box divided into three sections. The top portion contains the class name, the middle portion contains the attributes or properties and the lower portion contains the methods. The plusses and minuses signify whether the element is public or private, respectively. Public elements make up the public interface that other classes in the system can access, whereas private elements can only be accessed from within the class itself. The notation for the methods is similar to C# but also slightly different. The name of the method is given first, followed by parentheses that list the parameters that the method accepts, as well as the data type for each parameter. If the method returns a value, the data type for the return value is given after the parameters. In this example, the PayRaiseAmount method is public and takes a single parameter of the type decimal. It doesn't return anything.

Figure 13-1: Employee class in UML notation

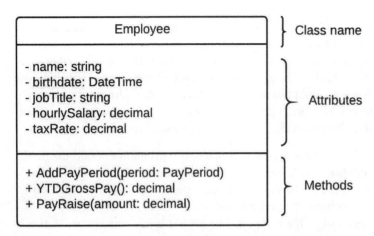

The Employee class describes the attributes and behavior of the employees of an organization. To represent an individual employee like John Smith, an application creates an instance of the Employee class in memory, called an object. So, the object is an *instantiation* of the class.

The class is an abstraction of the real-world entity written in computer code; thus, the same Employee class can be used to create multiple objects, each object representing a different employee. You can think of a class as the blueprint used to describe how to build a house, and each specific house built based on that blueprint is an object. Just like with a house, even though the objects are based on the same blueprint, once they are instantiated, they will diverge in some ways. For objects, the values of all their variables will be different.

Objects in C# are characterized by three general concepts:

- **Identity**: Just like every employee is distinct from every other employee, so too is every object in the computer system distinct from every other object. Once an object has been created, we can distinguish it from all other objects in the system. This is similar to the concept of a primary key in a database, but object-oriented systems automatically implement an identity mechanism.
- **State**: The state of an object is the set of values of the attributes that we care about regarding that object. For employees, we would likely care about things like their name, birth date, address, job title, and pay rate, whereas we are not likely to be concerned about their hair color. Each object has specific values for the things we care about. So, we might have two employees with these two states:

Table 13-1: Examples of objects

Attributes	Employee 1	Employee 2
Name	John Smith	Rebecca Jones
Birthdate	12/10/1993	10/5/1994
Address	200 Main St.	100 Elm St.
Job Title	Network Engineer	Software Developer
Hourly Pay	$35	$45

It's important to realize that the state of an object changes over time. In fact, anytime the value of an attribute changes, the state of the object has changed. So, for instance, if Rebecca gets a pay raise to $47 per hour, the state of the Employee 2 object has changed.

- **Behavior:** Each object has specific behavior that is also modeled in the system. In the problem domain, we might have employees punch in for work, punch out, get their salary paid out, get a pay raise and so on. In the application domain, behavior is implemented as methods that can be called on an object. The method code specifies what action happens when the method is called on a particular object.

The concepts of class and object are often confused and described in overlapping terms, but they are two distinct concepts that are important to keep separated. Classes are described in code and are used as the blueprints to instantiate (create) the objects. Each object in an application represents an instance of a real-world entity. There would only be one Employee class, but many Employee objects (one for each actual employee in the organization).

Review Questions

13.1 Identify several objects in your world from the following classes:
- Book
- Car
- Account
- Student
- Professor

13.2 For each of the classes above, identify a few attributes and behaviors that might be relevant to represent in an information system.

13.3 In your own words, describe the difference between class and object.

13.3 Information Hiding (Encapsulation)

One of the defining principles of object-oriented programming is that of Information Hiding (sometimes also referred to as encapsulation). The idea is that the way the data is presented by an object to other parts of the system is independent of how it is actually stored in the object.

This distinction provides several advantages. First, it allows for a simple and consistent internal representation of data in an object. For example, the total time worked by an employee could be stored in minutes, but it could be presented in the public interface by a method that returns fractional hours. Second, it protects the state of the object from being changed in inappropriate ways. For example, if the time worked by an employee is really represented by successively punching in and out, it would be inappropriate to be able to change the total time worked directly; it should only be changed through the transactions. Lastly, it also allows for restricting how the data inside an object can be accessed. Some attributes of an object should not be changed from outside of that object. As an example, consider a pay rate for an employee that must fall within certain bounds. If the pay rate could be changed directly, it could not be guaranteed to stay within its bounds.

To achieve this, each object provides a private implementation of data and a public interface, and only the public interface is available to other parts of the system. The implementation is thus "hidden" or "encapsulated" inside the object. In Figure 13-1, the attributes and methods marked with a plus represent the public interface, whereas the ones with a minus are private. A UML class diagram might omit some private implementations, and as we discuss later, attributes are implemented in C# using private fields that are exposed through public **properties** and methods.

As an example, consider the Employee object described above. We have specified that it has a Name attribute. How should that be actually stored in the object? A simple approach could be to store it in a single string. However, you could also split it in two and store first and last name separately. In that case, the attribute that provides the full name would be responsible for combining the first and last name and presenting it to its clients in that form. We could also do the opposite and define public access to both first and last name. If the name was stored internally in a single string, the first and last name attributes would be responsible for extracting the proper string and returning. Similarly, for hourly pay, which is defined as a whole number of dollars, we could store this as a decimal and provide public access as a rounded value and an integer.

When you design the system and decide on how to represent the data inside the objects, some representations are clearly better than others, so you have to carefully design both the public interface and the internal representation. It is much easier to combine a first and last name string into a full name than to attempt to create rules for extracting a first and last name from a full name.

As another example, consider the PayPeriod class shown in Figure 13-2. Each PayPeriod object contains data that allows for calculating the salary for an employee during that period. One of the items stored is the time that the employee worked during the pay period. But what is the best way to store this value—as whole minutes, fractional hours, or a combination of hours and minutes? This is obviously an important decision, but for the client classes that use the PayPeriod, it is largely irrelevant, as the programmer of the PayPeriod class could just provide methods for retrieving the time in all three formats and then have those methods do conversions as necessary.

Figure 13-2: PayPeriod class

```
┌─────────────────────────────────┐
│           PayPeriod             │
├─────────────────────────────────┤
│ - minutesWorked: int            │
│ - startDate: DateTime           │
│ - employee: Employee            │
│ + HoursWorked: double           │
│ + MinutesWorked: int            │
├─────────────────────────────────┤
│ + GrossPay(): decimal           │
│ + TaxWithheld(): decimal        │
│ + NetPay: decimal               │
│ + AddTime (minutes: int)        │
└─────────────────────────────────┘
```

There are several advantages to encapsulating the implementation like this:

- It allows for conversion between different formats or units—for example, between hours and minutes.
- It allows for having read-only and write-only attributes. There are many situations where it is useful to restrict the ability to change the attribute of a class directly. For instance, in an Account class you might have an attribute for the Balance on the account, but in most systems, the balance would only be changed by performing transactions on the account, and not by setting it directly. So, the Balance is better represented by a read-only attribute. Write-only attributes are rare in practice but would allow for the value to be changed but not read directly.
- The software becomes easier to maintain. By having a single internal representation and controlled access, there is only one place to make changes internally, and conversions can be done in a single spot as well.
- You will have fewer mistakes in the code due to unintended changes of values in an object. When access to the stored values in an object is properly protected, it is less likely to have unintended changes or end up with inconsistent values for an object.
- You can change the implementation of a class without affecting the public interface.

Review Questions

13.4　Imagine a Thermostat class that allows for a temperature to be set to specific values within minimum and maximum bounds. Discuss how encapsulation can be helpful for this class, as well as what problems might arise if data in the class were not encapsulated.

13.5　Draw a UML class diagram for the Thermostat class in the previous question.

Tutorial 1: Create an Employee Class

In this tutorial, you will create your first class and several objects from this class to be used in a payroll system. The class is the Employee class that was mentioned above. In the context of this simplified system, an employee will have the following attributes:

- Name—of type string
- Birthdate—of type DateTime
- Hourly salary—of type decimal
- Job title—of type string
- Tax rate—of type double (must be between 0 and 1)

Employee objects will have the following behavior:

- Given a number of hours worked during a pay period, calculate the gross salary, the tax amount and the net amount of money to be paid out.

Step 1-1: Start Visual Studio and create a new Windows Forms Application project called Payroll.

Step 1-2: Add the Employee class.

　　Right click the project in the Solution Explorer and select Add > Class…
　　Name the class Employee.cs and click Add.

Fields

When implementing classes, fields are variables at the class level that represent the attributes for the class and allow for storing the values of the attributes (the state) of an object. Fields should always be marked private. If you do not add an access modifier, C# will default to the most restricted access you could declare for that member. This would be private for fields, but it is recommended to still add the private key word to signal your intent clearly.

Step 1-3: Add fields to the Employee class.

Add the code shown in Figure 13-3 to the Employee class.

Figure 13-3: Fields of Employee class

```
 7   namespace Payroll
 8   {
 9       class Employee
10       {
11           private string name;
12           private DateTime birthday;
13           private decimal hourlySalary;
14           private string jobTitle;
15           private decimal taxRate;
16       }
17   }
```

Each of the fields declared here helps describe an Employee. Notice how they correspond to the attributes shown in Table 13-1. When you create different objects from the Employee class, each object will have a different set of values for the fields because each object represents a different employee.

Each field is declared using the private key word, which is a way to express that the variable cannot be directly accessed from outside the class and that details of how the data is stored is to be kept internal to the object. This will allow you to change the implementation without affecting the external usage of the class.

The Constructor

The constructor is a special method that is called when creating an object from a class. The constructor has two purposes: create the object and initialize the fields. Creating the object happens behind the scenes via the runtime environment when your program is running; you don't have to do anything special in the constructor code. However, you will have to write the code to initialize the fields, which you will see in the next step.

The constructor takes parameters that can be used to initialize the fields. In this case, there are parameters for each of the fields except the tax rate, which is set to a default value of 0.25 (all employees pay 25% tax by default).

In C#, the constructor follows a specific pattern:

- Constructors are given the same name as the class.
- Constructors do not return anything (not even void).

- Each field is initialized in the constructor. If you do not explicitly initialize a field, it will get a default value, depending on its data type (numbers become zero, strings become null, Booleans become false, etc.).

Step 1-4: Add the constructor to the Employee class.

Add the following code after the declaration of the fields, before the closing brace of the Employee class.

Figure 13-4: Employee constructor

```
16
17          public Employee(string name, DateTime birthday, decimal hourlySalary, string jobTitle)
18          {
19              this.name = name;
20              this.birthday = birthday;
21              this.hourlySalary = hourlySalary;
22              this.jobTitle = jobTitle;
23              taxRate = 0.25m;
24          }
25
```

The key word "this" is used to refer to the current object instance, and allows you to distinguish between fields and parameter values, as in line 19:

```
this.name = name;
```

At first glance, this line looks a little strange, but it is actually quite simple: this.name refers to the field name (line 11 in Figure 13-3), and the name on the right side of the equal sign refers to the parameter name. You can see this in Visual Studio by placing the cursor inside each of the two occurrences of name. When you click on the one on the right, the parameter is highlighted, and when you click on this.name, the field is highlighted. It isn't a requirement that the constructor parameters match the name of the fields, but it is fairly common that they do.

The ToString method

It is often useful to have an object be able to provide a brief one-line description of itself. This is usually done by implementing a method called ToString. There are technical reasons for the method to be called ToString that will be covered later, when we discuss the concept of *inheritance*. For now, just add this method to the Employee class.

Step 1-5: Add the ToString method to the Employee class.

Figure 13-5: ToString method

```
26          public override string ToString()
27          {
28              return string.Format("Name: {0}, Birthday: {1:d}, Hourly Salary: {2:c}, Job Title: {3}",
29                  name, birthday, hourlySalary, jobTitle);
30          }
```

As you can see, the method just returns a string that includes most of the fields. The meaning of the override key word in line 26 will also be discussed later during the coverage of inheritance.

Step 1-6: Create the user interface.

Rename Form1.cs in Solution Explorer to EmployeeForm.cs (and rename in the code as well when you're asked). Change the Text property of the form to Employees.

In order to demonstrate how objects are created, add a label named lblEmployee to the Form in the project.

Double click the background of the form (not the label) to switch to the code view of the form. It should look like Figure 13-6.

Figure 13-6: Initial code in the EmployeeForm class

```
11    namespace Payroll
12    {
13        public partial class EmployeeForm : Form
14        {
15            public EmployeeForm()
16            {
17                InitializeComponent();
18            }
19
20            private void EmployeeForm_Load(object sender, EventArgs e)
21            {
22
23            }
24        }
25    }
```

As you can see from line 13, the Form is in fact a class, called EmployeeForm. This means that when the program is running and the Form is being displayed, an EmployeeForm object is created. This class currently has two methods:

1. A constructor, specified in lines 15–18, that takes no parameters and calls the method InitializeComponent, which is created by Visual Studio to create all the user interface components in the form. If you want to see this auto-generated code, you can right click on the method name and select Go To Definition.

2. A method called EmployeeForm_Load, specified in lines 20–23, that takes two arguments. The EmployeeForm_Load method is called automatically by the system when the form is loaded. There are a number of other event handler methods that you can add to your form and insert code into to execute code at certain times in the life cycle of a form.

Creating an Object

When creating an object, you follow a specific pattern:

```
ClassName variableName = new ClassName();
```

This line declares a variable on the left side of the equal sign and on the right side calls the constructor for the class. The right side generates an object that is then assigned to the variable on the left side.

The key here is the use of the new key word in front of the call to the constructor. The new key word is used to indicate that an object is created from the specified class. Let's create our first Employee object.

Step 1-7: Create objects and display ToString.

Add the following code to the EmployeeForm_Load event handler:

```
DateTime birthday = new DateTime(1999, 5, 1);
Employee spongebob = new Employee("Spongebob", birthday, 10.5m, "Burger flipper");
lblEmployee.Text = spongebob.ToString();
```

This code actually creates two objects. First, a DateTime object is created for May 1, 1999. This is assigned to the variable birthday. Then, an Employee object is created using the birthday object and several other pieces of data. This is assigned to the spongebob variable. Finally, the ToString method is called on the spongebob object, and the result is assigned to the Text property of the label you added to the user interface above.

Run the program and you should see the description of the Employee object displayed on the form, as shown in Figure 13-7.

Figure 13-7: First object created

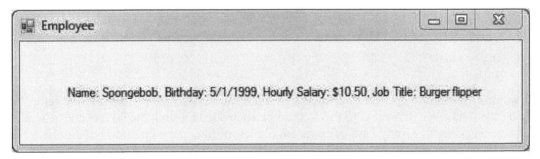

To make sure you understand how the object is created and displayed, add a Breakpoint on the second line of code in the EmployeeForm_Load event handler. Once the debugger stops on that line, examine the variables available at that point.
Next, step into the Employee constructor, and see how the values are initialized in the object. Once you return to the Form object, reexamine the variables. You should now see that the spongebob variable has values associated with each of the fields.

Continue stepping into the ToString method, and notice that you are back in the Employee class looking at the spongebob object.

It's time to practice! Do Step 1-8 and Exercise 13-1 at the end of the chapter.

Step 1-8: Add a second label and create a new Employee object with different values, and show the new object on the new label.

13.4 Properties

So far, you have seen fields and methods in a class. The fields contain the state of the object, and the methods implement the object's behavior. Because of encapsulation, the fields are always kept private so as not to allow other parts of the system to access the fields directly. However, in many cases we want to be able to expose the fields and be able to read and change them in a controlled manner. For instance, it is easy to imagine that you could create a form with controls for all the fields for the Employee class and allow a user to enter values into the controls and assign those values to the fields. But because the fields are private, this would not be possible. You can try right now in the program you created previously to see if you can assign a different value to one of the variables in the spongebob object after it has been created by adding this line after creating the object:

```
spongebob.name = "Spongebob Squarepants";
```

You will get a syntax error from trying this, because name is declared as private in the Employee class, so go ahead and comment out that line of code.

So, how can we allow the fields to remain private but still change their values? The C# language to a large extent was created by Microsoft as an improved version of Java. The best way to understand why C# has a concept of properties is to go back to the Java language.

In Java we would create public "getters" and "setters" to retrieve and change the field values. For the field, name, these would look like this:

```
public string getName()
{
    return this.name;      //return the value of the name field
}
public void setName(string value)
{
    this.name = value;     //assign a value to the name field
}
```

If a programmer wanted to not allow a client to change a field, he/she would simply not write the set method. By keeping the fields private, access to them could be controlled through the getters and setters. For instance, if it was necessary to do any kind of conversion, like splitting the name into last and first names, this could be done in the getters and setters.

If you implemented these methods, you could change the name with this line of code:

```
Spongebob.setName("Spongebob Squarepants");
```

However, one problem with this approach is that getters and setters are not a part of the Java language; they are merely a convention that most Java programmers follow. So, the C# language designers decided to introduce the concept of a **Property** that takes the place of the getters and setters in Java. Here is a property that provides the same functionality as the getters and setters above:

```
public string Name
{
    get
    {
        return this.name; //return the value of the name field
    }
    set
    {
        this.name = value; //assign a value to the name field
    }
}
```

This is a single construct called Name that includes a get section and a set section. The get simply returns the value in the named field. The set uses the reserved key word *value* to provide the value to be stored in the field. Just like with getters and setters in Java, you can add any valid C# code to the properties. You're not restricted to just assigning and returning values from fields. You could do conversions or have properties whose values are calculated and do not have an associated field.

Once the property has been added, you can change the name in this way:

```
spongebob.Name = "Spongebob Squarepants";
```

This is very similar to your first attempt at changing the name, but because the property is declared public, it will actually work.

How do you retrieve the name? You have seen many examples throughout this book of using properties, as they are very common with user interface controls and other parts of the .Net framework, so a single additional example here should suffice. Imagine a form that has a label called lblName and a TextBox named txtName:

```
lblName.Text = spongebob.Name;
spongebob.Name = txtName.Text;
```

In the first statement, the Name property of the spongebob object retrieves the value of the name field using the get section of the property. The value is assigned to the Text property of the Label.

In the second statement, the content of the textbox is retrieved using the Text property of the txtName object and assigned to the Name property of the spongebob object. When a value is assigned to a property, the set section of the property is used to assign the value to the field through the value key word in the set section of the property. So, the content of txtName is assigned to the name field of spongebob object.

The naming convention for properties is to change the first letter of the corresponding field to uppercase. If you wanted to have a read-only property, you would simply leave out the set section.

Automatic Properties

One final note about properties is that very often there is a lot of boilerplate code when creating a class:

- A private field
- A public property that simply gets and sets the values for the fields

When this is the case, the field and property code can be written using a simplified version, called an *Automatic Property*. Here's the Name property from earlier using the automatic property:

```
public string Name { get; set; }
```

This one line of code replaces the following private field and the property code:

```
private string name;
public string Name
{
    get
    {
        return this.name;
    }
    set
    {
        this.name = value;
    }
}
```

As you can see, Automatic properties are much shorter and easier to read. The usage remains the same as you saw above, except the field can no longer be accessed directly inside the property or in the rest of the class.

With an automatic property, the compiler will automatically create a backing field of the correct type and implement the standard get and set code you saw for the fully implemented property. So, the only difference is that it is much shorter to write and to read. When you use automatic properties, you have to remember that you cannot implement any conversion or other code in the get and set sections. You must also supply both get and set, so you cannot create automatic read-only or write-only properties.

Review Questions

13.6   What are the advantages and disadvantages of automatic properties?
13.7   Why should fields always be marked as private?
13.8   Why did the C# language designers choose to introduce properties instead of getters and setters like in Java?

Tutorial 2: User Interface and Properties

In this tutorial you will learn how to store multiple objects in a ListBox and retrieve them. This will help you understand how to work with references to objects. By the end of this tutorial, you will have a form where the user can enter information on an employee and click a button to create an Employee object. Multiple Employee objects will be created and displayed in a ListBox where the user can select one to be modified. Figure 13-8 shows the final user interface for Tutorial 2.

Figure 13-8: Final user interface for Tutorial 2

Step 2-1: Create a user interface form.

Continue working with the project from Tutorial 1. Move the Label you added earlier to the bottom of the Employee form and add controls to the EmployeeForm in the project so it looks like Figure 13-9.

Figure 13-9: Employees form

Name the controls on the right as follows (from top to bottom): txtName, dtpBirthday, txtJobTitle, txtPayRate, and btnSave.

Step 2-2: Code the button to create an object and display it in the label:

Double click the Save button to go to the code for the form.

Comment out the code you added to the Form_Load method and add the following code to the btnSave_Click method:

```
private void btnSave_Click(object sender, EventArgs e)
{
    string name = txtName.Text;
    DateTime birthday = dtpBirthday.Value;
    string title = txtJobTitle.Text;
    decimal hourlyRate = Decimal.Parse(txtPayRate.Text);
    Employee employee = new Employee(name, birthday, hourlyRate, title);
    lblEmployee.Text = "Employee: " + employee.ToString();
}
```

The code is fairly straightforward. First, all the values are read from the user interface; then an Employee object is created using the new key word. Finally, the lblEmployee's Text property is set to the result of calling the ToString method on the employee object.

Run the program and add values to the form fields. Click the Save button, and notice that the label at the bottom of the form is updated with information about the object, as shown in Figure 13-10.

Figure 13-10: Output from creating a single object

![Employee form showing Name: Spongebob, Birthday: Saturday, May 01, 1999, Job Title: Burger Flipper, Hourly Pay Rate: 10.75, with Save button and label text: "Employee: Name: Spongebob, Birthday: 5/1/1999, Hourly Salary: $10.75, Job title: Burger Flipper"]

Step 2-3: To help save multiple objects, add a ListBox and add objects to the ListBox with the ToString method.

Remove the lblEmployee label from the form and replace it with a ListBox named lstEmployees.

Now instead of adding each object to the label and losing them, we will add them to the ListBox and explore some ways to interact with them again later.

Remove the last line of code in the btnSave_Click method and replace it with this line of code:

```
lstEmployees.Items.Add(employee);
```

This line will add the employee object to the Items collection in the ListBox. Previously, when you have worked with ListBoxes, you added string objects to the Items collection. In this case, you have added an entire Employee object with several pieces of data included. However, when the ListBox is being displayed, it will automatically call the ToString method on each of the objects in its Items collection, and will then display the return value (string) from calling the method.

Run the code again, and you should see that the employee objects show up one after the next in the ListBox each time you click Save, as shown in Figure 13-8.

Step 2-4: To make it possible to change the values of the object, add properties to the Employee class.

So far, you can only add objects to the list, and there is no way to edit an existing object. To fix this, you will have to allow the user to select an item in the ListBox and then display the values of the fields for the object in the form. The behavior of the Save button must be changed to recognize whether an employee has been selected and then modify the selected object rather than create a new one. However, all the fields are private in the Employee class, so the first thing you need to do is to expose them as properties, so you can read and modify them as needed.

Comment out all the fields in the Employee class except taxRate and replace them with the following code:

```
//private string name;
public string Name { get; set; }
//private DateTime birthday;
public DateTime Birthday { get; set; }
//private decimal hourlySalary;
public decimal HourlySalary { get; set; }
//private string jobTitle;
public string JobTitle { get; set; }
private decimal taxRate;
public decimal TaxRate { get { return taxRate; } }
```

Now the code in the constructor and ToString is broken because the fields no longer exist. The properties are declared public and, as such, are always named with a starting uppercase letter. The fields were private and, as such, named with a lowercase initial letter. Because C# is case sensitive, the two are not interchangeable.

Because the code uses automatic property for all attributes, except taxRate, there is no explicit declaration of fields for those attributes. The taxRate field was left unchanged and a property added. We didn't use an automatic property for taxRate because we wanted to make it a read-only property so that the tax rate is not allowed to be changed.

Step 2-5: Fix the problems by changing the lowercase initial letter to uppercase for each of the errors in the constructor and ToString method (look for the squiggly red lines). See Figure 13-11 for the updated code. Run the form to test the code.

Figure 13-11: Updated constructor and ToString method in Employee

```
20    public Employee(string name, DateTime birthday, decimal hourlySalary, string jobTitle)
21    {
22        this.Name = name;
23        this.Birthday = birthday;
24        this.HourlySalary = hourlySalary;
25        this.JobTitle = jobTitle;
26        taxRate = 0.25m;
27    }
28
29    public override string ToString()
30    {
31        return string.Format("Name: {0}, Birthday: {1:d}, Hourly Salary: {2:c}, Job title: {3}",
32            Name, Birthday, HourlySalary, JobTitle);
33    }
```

Step 2-6: Recall the object from the ListBox and display in the user interface form.

Switch to the design view of the form and select the ListBox. Switch the Properties window to events by clicking the lightning bolt. Then find the SelectedIndexChanged event and double click in the field next to it (see Figure 13-12).

Figure 13-12: SelectedIndexChanged event for the ListBox

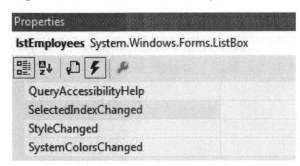

In the generated event handler, enter the following code:

```csharp
private void lstEmployees_SelectedIndexChanged(object sender, EventArgs e)
{
    Employee employee = (Employee)lstEmployees.SelectedItem;
    txtName.Text = employee.Name;
    txtPayRate.Text = employee.HourlySalary.ToString();
    dtpBirthday.Value = employee.Birthday;
    txtJobTitle.Text = employee.JobTitle;
}
```

The first statement in the event handler will get the item that the user selected in the ListBox (SelectedItem property). This item, which is of type object, is then cast to an Employee and assigned to the employee variable. The last four lines then change the user interface controls based on the values in the object. Notice how the properties in the Employee class are used to extract the values from the object.

Items in the ListBox are stored in a collection called Items. When you add an object to that collection, the ListBox will display it by calling the ToString method to display it in the ListBox. Since you have been storing Employee objects, you can get that entire object back and interact with it.

A common mistake when working with ListBoxes is to add a string describing the object, such as the name of the employee. Then, to retrieve it, you would have to get the index it was stored at to be used to find the object in some other collection. However, this approach is error prone because it assumes that the ListBox and the collection are kept in sync, with no objects added or deleted or any sorting applied to one but not the other. By saving the entire object to the ListBox, it doesn't matter what happens to the other collection.

Step 2-7: Save object values.

Once the object is displayed in the form fields, we need to be able to save any changes the user makes back into the object, rather than have it save as a new object when the Save button is clicked. To do this, you will use the properties again to save the values. However, first you have to find the right object to save into. In the previous step, the employee variable only has scope within the lstEmployees_SelectedIndexChanged method, so it cannot be accessed once that method finishes executing. However, the object is still selected in the ListBox, so you can check to see if an item is selected and then save the values into the selected item's object. Otherwise, if no object is selected, create a new object and add to the list.

The first step is to recall the object and update the properties. Revise the btnSave_Click method, as shown in Figure 13-13.

Figure 13-13: Making changes to object properties

```
37          private void btnSave_Click(object sender, EventArgs e)
38          {
39              //Grab values from the form.
40              string name = txtName.Text;
41              DateTime birthday = dtpBirthday.Value;
42              string title = txtJobTitle.Text;
43              decimal hourlyRate = Decimal.Parse(txtPayRate.Text);
44              Employee employee;
45              if (lstEmployees.SelectedIndex >= 0)
46              {
47                  //If an employee is selected in the ListBox, get reference
48                  //to that employee and update properties from the form.
49                  employee = (Employee)lstEmployees.SelectedItem;
50                  employee.Name = name;
51                  employee.Birthday = birthday;
52                  employee.JobTitle = title;
53                  employee.HourlySalary = hourlyRate;
54              }
55              else
56              {
57                  //If no employee is selected, create a new object based on
58                  //values in the form
59                  employee = new Employee(name, birthday, hourlyRate, title);
60                  lstEmployees.Items.Add(employee);  //Add new object to ListBox
61              }
62          }
```

Line 44 declares a variable of type Employee, which is populated in one of two ways—either by the employee selected in the ListBox (line 49) or as a new object (line 59).

The if statement in Line 45 checks whether an item is selected in the ListBox. If selected, the object is retrieved (line 49) and the object values are updated with the updated data (lines 50–53); if not, a new Employee object is created in line 59 and added as a new item to the ListBox in line 60.

If you run the program after making the changes in the above step, the ListBox doesn't behave as it should. When clicking Save after updating an existing employee, the corresponding employee information displayed in the ListBox doesn't update. So, if you select an employee that has been previously updated, the ListBox and the user interface fields won't show the same thing. There are a few ways to fix this, but the simplest is to keep a List (a collection) of the Employees as a field in the EmployeeForm class. Then every time a new Employee object is created, it is added to the List, and Employee objects in the List are updated without adding new objects. Every time the List is updated, you can clear the items out of the ListBox and refresh with the items in the List.

Step 2-8: Refresh the ListBox.

To refresh the ListBox, first create a private field called employees at the top of the EmployeeForm class and instantiate it in the EmployeeForm constructor as shown in line 20 of Figure 13-14.

Figure 13-14: List for storing Employee objects

```csharp
13  public partial class EmployeeForm : Form
14  {
15      private List<Employee> employees;
16
17      public EmployeeForm()
18      {
19          InitializeComponent();
20          employees = new List<Employee>();
21      }
```

Line 15 declares a List that can hold Employee objects, and line 20 creates the List object.

Next, update the code in btnSave_Click as shown in Figure 13-15.

Figure 13-15: Using a List to store Employee objects

```csharp
37      private void btnSave_Click(object sender, EventArgs e)
38      {
39          //Grab values from the form.
40          string name = txtName.Text;
41          DateTime birthday = dtpBirthday.Value;
42          string title = txtJobTitle.Text;
43          decimal hourlyRate = Decimal.Parse(txtPayRate.Text);
44          Employee employee;
45          if (lstEmployees.SelectedIndex >= 0)
46          {
47          //If an employee is selected in the ListBox, get reference
48          //to that employee and update properties from the form.
49              employee = (Employee)lstEmployees.SelectedItem;
50              employee.Name = name;
51              employee.Birthday = birthday;
52              employee.JobTitle = title;
53              employee.HourlySalary = hourlyRate;
54          }
55          else
56          {
57          //If no employee is selected, create a new object based on
58          //values in the form
59              employee = new Employee(name, birthday, hourlyRate, title);
60              //lstEmployees.Items.Add(employee); //Add new object to ListBox
61              employees.Add(employee); //Add new object to the collection
62          }
63      //Clear all items from the ListBox
64          lstEmployees.Items.Clear();
65      //Add all items from the List to the ListBox
66          lstEmployees.Items.AddRange(employees.ToArray());
67      }
```

Line 60 has been replaced by line 61, which adds the newly created Employee object to the List. Note that an object is added to the List only if it is a newly created object.

Lines 50–53 update the object in the selected item of the ListBox. It is important to note that if the selected object from the ListBox is updated, there is no need to add it to the List, because both the ListBox and List refer to the same set of objects, as explained below through Figure 13-16. When you update an object from the ListBox, the List that refers to the same object is also updated.

Line 64 clears out any existing items from the ListBox, and line 66 then adds all the Employee objects in the employees List to the ListBox. The AddRange method requires an array, but the List class conveniently provides a ToArray method that returns the contents of the List as an Array. Clearing the ListBox and Adding the entire List to the ListBox refreshes the display of the updated object in the ListBox.

Run the program again and you should be able to manipulate existing objects and add new ones.

You can refer to Figure 13-16 to help you understand which objects are in the system and define the relationships between them. Each box in this diagram represents an object in the system at a point in time (and represents the situation shown in Figure 13-8). The diagram starts with the object that represents the form itself. This object has a reference to the lstEmployees ListBox (among many other references to all the user interface elements), as well as to the employees List. The ListBox has a reference called Items to a ListBox.ObjectCollection, which in turn references the two Employee objects, just like the List does. What this shows is that when the objects are inserted into both the ListBox and the List, you can retrieve from either one and work on the object. Any changes made to the object will be reflected anywhere the object is referenced, since there is only a single copy of the object.

Figure 13-16: Object diagram showing objects in the system

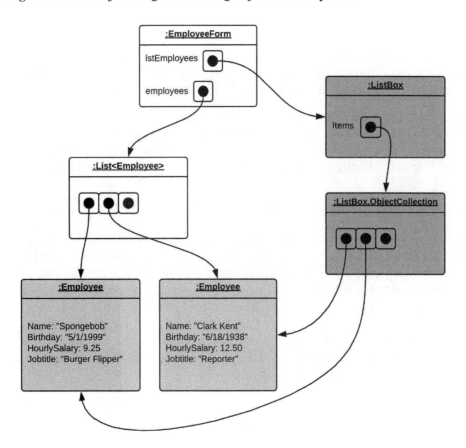

It's time to practice! Do Steps 2-9 and 2-10.

Step 2-9: Add the ability to delete an Employee from the list.

Step 2-10: Change the Employee class to store both a first and last name. Then change the rest of the code to reflect this change. (Add fields to the form and the code that interacts with the form and the Employee objects.)

13.5 Calling Methods (Sending Messages to Objects)

The primary way to interact with an object is to call a method on it or, said in another way, to send a message to that object. Imagine you have an Order object. You could then send messages to that object asking for the customer who placed the order, the total value of the order (calculated by the value of each of the related order line objects), the products in the order and so on. Because the order is an object that knows its own state, you can ask any question related to that state—or the state of related objects.

As an example, we could send a message to an Employee object to add another pay period in this way:

```
emp1.AddPayPeriod(payPeriod);
```

This line of code calls the method AddPayPeriod on an Employee object represented by the emp1 variable passing a parameter represented by the variable payPeriod. When calling a method on an object, you always start with a reference to the object followed by a dot, and then the method name and a pair of parentheses surrounding any parameter values.

Chapter 13: Object-Oriented Programming

If a method returns an object, you can then immediately call a method on that return value. This is sometimes called *dotting*. Here's an example:

```
emp1.ToString().ToUpper();
```

In this example, the ToString method is called on emp1. The resulting string is then used for the call to ToUpper, which transforms the resulting string to uppercase letters. There is no limit to how many times you can dot your way to making calls. However, it is sometimes advantageous to create intermediate variables, like this:

```
string empString = emp1.ToString();
string upper = empString.ToUpper();
```

While this code does take up more space, it can be easier to read, and it allows you to examine the value assigned to the intermediate result when debugging.

In the next tutorial you will see how you can call methods on objects and create methods in forms that can be called.

Tutorial 3: Calling Methods and Passing Data between Forms

When testing a program, have you found it tedious to have to enter a new employee every time you launch the program to test some new aspect of it? This can be fixed by hard coding a few Employee objects and passing them as parameters to the form to make them available every time you launch the program. It should be noted that hard coding data generally makes programs difficult to maintain. Here we use it to show how to pass data to a form.

This tutorial also illustrates how the Form class is just like any other class and can be modified to your liking using object-oriented techniques. We will illustrate this in the tutorial by having you modify the constructor for the Form class to accept parameters. This will allow you to select a value on one Form and pass that value to another form to be displayed. You will also see how to pass values back from a child form opened from a parent form. This can be done by creating a public method in the parent form and passing a reference to the parent object to the constructor of the child form.

Passing a List of Objects to a Form

A List of Employee objects can be passed to EmployeeForm by adding the List as a parameter to the constructor of the form.

Step 3-1: Modify Program.cs to create Employee objects, add them to a List and pass the List to EmployeeForm.

Open the Program.cs class and add the following lines of code at the beginning of the Main method:

```
List<Employee> employees = new List<Employee>();
DateTime birthday = new DateTime(1999, 5, 1);
employees.Add(new Employee("Spongebob", birthday, 10.5m, "Burger flipper"));
birthday = new DateTime(1995, 4, 18);
employees.Add(new Employee("Squidward Tentacles", birthday, 12.56m, "Cashier"));
```

There isn't much new here. The code creates a List to hold Employee objects and then creates two Employee objects and adds them to the List.

Next, modify the last line of the Main method like this:

```
Application.Run(new EmployeeForm(employees));
```

You have added the employees List as a parameter to the constructor of the EmployeeForm class. This will pass a reference to the List to the constructor so it can be used in the EmployeeForm. You will get a red squiggly line because the constructor hasn't yet been modified to accept a List. This will be fixed in the next step.

Step 3-2: Modify the EmployeeForm constructor to work with the Employee List.

Open EmployeeForm.cs to code view. Previously, you added a field named employees (see Figure 13-14), of type List<Employee>. Now you just need to assign the List<Employee> that is passed from Program.cs (when the EmployeeForm object is created) to the employees field in the EmployeeForm class. Figure 13-17 illustrates the changes that need to be made to the constructor.

Figure 13-17: Modified EmployeeForm constructor

```
13      public partial class EmployeeForm : Form
14      {
15          private List<Employee> employees;
16
17          public EmployeeForm(List<Employee> employees)
18          {
19              InitializeComponent();
20              if (employees == null)
21              {
22                  employees = new List<Employee>();
23              }
24              this.employees = employees;
25              lstEmployees.Items.AddRange(employees.ToArray());
26          }
```

This is all pretty simple. Notice that in line 17, a parameter is added to the constructor, so it matches up with what you did in Program.cs. Then, in line 20–23, we ensure that the List has been instantiated. Line 24 assigns the List, which is passed from Program.cs to the employees field. Finally, in line 25, the List is added to the ListBox using the AddRange method that you saw previously in Figure 13-15.

You may get a confusing error message that says something about inconsistent accessibility in the constructor. This is likely because the constructor in the form is public, whereas the Employee class doesn't have an access modifier (public/private). If no access modifier is specified, a class is given the access *protected*, which is in between public and private, and thus the class is less accessible than the form's constructor. Fix this, if necessary, by making the Employee class public.

Run the program and check that the two Employees show up in the form.

Creating a Child Form and Passing Data from Parent Form

Next, we will take a look at having multiple objects interact by introducing a PayPeriod class. The idea is that an Employee object would keep a list of PayPeriod objects, as shown in Figure 13-18.

Figure 13-18: Class diagram

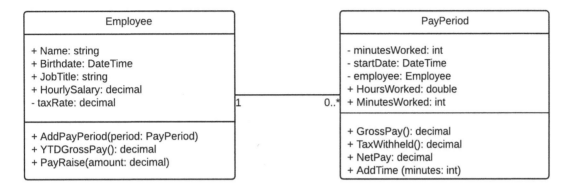

Figure 13-19 shows the code to create the PayPeriod class.

This class stores the total number of minutes worked in the pay period in the minutesWorked field. This field is modified by the AddTime method, which is called for each period worked to add the number of minutes to the pay period. The HoursWorked property (line 17) converts minutesWorked from integer minutes to decimal hours by dividing the minutes by 60. This way, the implementation is kept as minutes, but any client can request the time in fractional hours instead of minutes.

It's also worth noting that the properties MinutesWorked and HoursWorked are read-only. The actual time worked can only be changed by calling the AddTime method (lines 28–31). The reason for this restriction is to illustrate a situation where an employee would periodically add more time to their work record. It wouldn't be good if the amount of time could just be set to some arbitrary number. This approach is commonly used in transaction-oriented systems.

Figure 13-19: PayPeriod class

```
 9    public class PayPeriod
10    {
11        //Instance variables:
12            private int minutesWorked;
13            private DateTime startDate;
14            private Employee employee;
15        //Properties:
16            public int MinutesWorked { get { return minutesWorked; } }
17            public decimal HoursWorked { get { return minutesWorked / 60m; } }
18            public DateTime StartDate { get { return startDate; } }
19            public Employee Employee { get { return employee; } }
20        //Constructor:
21            public PayPeriod(int minutesWorked, DateTime startDate, Employee employee)
22            {
23                this.minutesWorked = minutesWorked;
24                this.startDate = startDate;
25                this.employee = employee;
26            }
27        //Methods:
28            public void AddTime(int minutes)
29            {
30                minutesWorked += minutes;
31            }
32
33            public decimal GrossPay()
34            {
35                return HoursWorked * employee.HourlySalary;
36            }
37
38            public decimal TaxWithholding()
39            {
40                return GrossPay() * employee.TaxRate;
41            }
42
43            public decimal NetPay()
44            {
45                return GrossPay() - TaxWithholding();
46            }
47    }
```

The properties StartDate and Employee (lines 18–19) are also read-only but cannot be changed once they are initialized in the constructor (lines 21–26).

The relationship between the two classes is implemented through the employee field (line 14). The data type for this variable is the Employee class. This means that each PayPeriod object has a variable that points to an Employee object. This allows each employee to have multiple pay periods.

Step 3-3: Add PayPeriod class.

Add the code shown in Figure 13-19 to create a PayPeriod class in the project.

Step 3-4: Modify Employee class to hold PayPeriod objects.

See Figure 13-20 for what to modify in the Employee class in the next steps.

First, add a private field called payPeriods, as shown in line 18. Initialize the variable in the constructor (line 27). This variable is a List that can hold multiple PayPeriod objects. This allows each Employee object to easily access all the employee's pay periods.

Figure 13-20: Employee class with List of PayPeriods

```csharp
 9    public class Employee
10    {
11        public string Name { get; set; }
12        public DateTime Birthday { get; set; }
13        public decimal HourlySalary { get; set; }
14        public string JobTitle { get; set; }
15        private decimal taxRate;
16        public decimal TaxRate { get { return taxRate; } }
17
18        private List<PayPeriod> payPeriods;
19
20        public Employee(string name, DateTime birthday, decimal hourlySalary, string jobTitle)
21        {
22            this.Name = name;
23            this.Birthday = birthday;
24            this.HourlySalary = hourlySalary;
25            this.JobTitle = jobTitle;
26            taxRate = 0.25m;
27            payPeriods = new List<PayPeriod>();
28        }
29
30        public void AddPayPeriod(PayPeriod payPeriod)
31        {
32            payPeriods.Add(payPeriod);
33        }
34
35        public override string ToString()
36        {
37            return string.Format("Name: {0}, Birthday: {1:d}, Hourly Salary: {2:c}, Job title: {3}",
38                Name, Birthday, HourlySalary, JobTitle);
39        }
```

Step 3-5: Add method to add PayPeriods to Employees.

Add the AddPayPeriod method as shown in lines 30–33. This method calls the Add method on the List object to add a new pay period.

Passing Data to Child Form

Next, we will take a look at how you can work with multiple forms as objects and pass data back and forth between them. Figure 13-21 shows a new form named PayPeriodForm that will be launched when the user clicks a button on the Employees form. The information about the selected Employee will be passed to the PayPeriodForm, where the pay periods can be added to the employee.

Figure 13-21: Employee and PayPeriodForm forms

Step 3-6: Create PayPeriod form.

Start by adding a new form named PayPeriodForm to the project.

Add the controls you see in Figure 13-21. The controls are named as follows: lblEmployee, dtpStartDate, txtHoursWorked, btnSave, and lstPayPeriods.

Launching a child form

To launch the PayPeriodForm from Employee form, you need to add a button to the Employee form and add the code shown in Figure 13-22 to the Click event handler of the button.

Figure 13-22: Launching the PayPeriodForm from the Employee form

```
76          private void btnPayPeriods_Click(object sender, EventArgs e)
77          {
78              Employee employee = (Employee)lstEmployees.SelectedItem;
79              if (employee != null)
80              {
81                  PayPeriodForm payPeriodForm = new PayPeriodForm(employee, this);
82                  payPeriodForm.Show();
83              }
84          }
```

The code will create a PayPeriodForm object in line 81 if an employee has been selected. Two parameters are passed to the constructor: references to the selected employee as well as a reference to the EmployeeForm object (this). Those two references will be used in the PayPeriodForm object, as you will see later. You have a syntax error in the call to the PayPeriodForm constructor in line 81, because by default the constructor doesn't take any parameters. You will fix this by modifying the PayPeriodForm constructor. Line 82 makes the Pay Period Form visible.

Step 3-7: Add a button to Employee Form, and add the code to launch the PayPeriodForm.

Switch to the EmployeeForm and add a button named btnPayPeriods with the text Pay Periods. To launch the PayPeriodForm, double click the button in the designer and add the code shown in Figure 13-22 to the generated method.

Step 3-8: In this step, you will add parameters to the constructor of PayPeriodForm, and add code to display the current employee's information in PayPeriodForm and save PayPeriod objects in the current employee object. In addition, you will add code to display all pay periods for the current employee in a ListBox on the PayPeriodForm.

Switch to the code for PayPeriodForm.cs and add private fields, and adjust the constructor as shown in Figure 13-23.

Figure 13-23: PayPeriodForm constructor

```
13      public partial class PayPeriodForm : Form
14      {
15          private Employee employee;
16          private EmployeeForm parent;
17
18          public PayPeriodForm(Employee employee, EmployeeForm parent)
19          {
20              InitializeComponent();
21              this.employee = employee;
22              this.parent = parent;
23              lblEmployee.Text = employee.ToString();
24          }
25      }
```

Next, switch to design view and double click the Save button. Then enter the code shown in Figure 13-24 to save a PayPeriod in the current employee object.

Figure 13-24: Saving a new PayPeriod

```
26          private void btnSave_Click(object sender, EventArgs e)
27          {
28              decimal hours;
29              if (Decimal.TryParse(txtHours.Text, out hours))
30              {
31                  DateTime startDate = dtpStartDate.Value;
32                  int minutesWorked = (int)Math.Round(hours * 60);
33                  PayPeriod pp = new PayPeriod(minutesWorked, startDate, employee);
34                  employee.AddPayPeriod(pp);
35
36                  updateListBox();
37              }
38          }
```

The method reads a fractional number of hours and converts to minutes (line 32) for storing in the PayPeriod object. Line 33 creates a PayPeriod object, and line 34 adds that object to the employee that the form is working on. Notice how you can call a method on an object and pass parameters to the method.

The call to updateListBox in line 36 is throwing an error because the method hasn't been created yet; this will happen in the next step.

Step 3-9: Update the ListBox.

> Hover over updateListBox() and click the lightbulb; then select the Generate method option. This will generate a new method. Replace the body of the method with the code shown in Figure 13-25.

Figure 13-25: Updating the ListBox

```
39
40    private void updateListBox() {
41        lstPayPeriods.Items.Clear();
42        foreach (PayPeriod pp in employee.PayPeriods)
43        {
44            string pString = string.Format("{0:d} // {1} hours // Gross: {2:c} // Tax: {3:c} // Net: {4:c}",
45                pp.StartDate, pp.HoursWorked, pp.GrossPay(), pp.TaxWithholding(), pp.NetPay());
46            lstPayPeriods.Items.Add(pString);
47        }
48    }
```

> The core of this method is a foreach loop that goes through all the PayPeriod objects in the employee and adds a string for the pay period to the ListBox. There is one problem, as you can see—the PayPeriods property doesn't yet exist in the Employee class, so switch to the Employee class and add the following read-only property:
>
> ```
> public List<PayPeriod> PayPeriods { get { return payPeriods; } }
> ```
>
> Run the program and select an employee; then add some pay periods to each of the employees.

It's time to practice! Do the following step:

Step 3-10: Modify the constructor in PayPeriodForm to display any existing pay periods for an employee when the form is first loaded.

Sending Messages to the Parent Form

You have seen how to pass data from a parent form to a child form as a parameter in the constructor. But what about the other way—how can you pass information back or call methods in the parent form if needed? Passing data back to the parent form is as simple as modifying a reference to an object on the child form. If the parent form also has a reference to the object, then the changes will be available on the main form without any further action. This is what you did in the previous part of the tutorial; every time you added a PayPeriod to an Employee on the child PayPeriodForm, that change was automatically reflected on the parent EmployeeForm because both the parent and child forms had references to the Employee object.

The following is an example of how to call a method on a parent form. Some of the work is already done. To launch the PayPeriodForm from Employee form, you added a button to the Employee form, and added the code shown in Figure 13-22 to the Click event handler of the button.

In Figure 13-22 and Figure 13-23, you set up a reference in the child form to the parent form. That reference can then be used to access any public method or property from the child form.

What you will do in this tutorial is calculate the average gross income for all employees year-to-date and then display the difference between the average and the income for the employee that is currently displayed on the PayPeriod form. To do this, you first need to add a method to the Employee class that will calculate the YTD gross income for one employee:

```
public decimal YTDGross()
{
    decimal result = 0m;
    foreach (PayPeriod p in payPeriods)
    {
        if (p.StartDate.Year == DateTime.Now.Year)
        {
            result += p.GrossPay();
        }
    }
    return result;
}
```

This method follows a fairly standard pattern for working with collections of objects. It uses a foreach loop to go through all the objects in the collection and then compares the year of the start date to the current year. If they match, the pay period's gross pay is added to a temporary variable, whose value is returned when the loop has gone through all the pay periods.

Step 3-11: Calculate the year to date gross income for an employee.

Add the code to create the YTDGross method shown above.

Next, call this method to compute the average gross income.

Step 3-12: Calculate average gross income.

Add the following method to EmployeeForm.cs:

```
public decimal YTDAverageGross()
{
    decimal sum = 0m;
    foreach (Employee e in employees)
    {
        sum += e.YTDGross();
    }
    return sum / (decimal)employees.Count;
}
```

This method follows a similar pattern by going through all the Employee objects in the employee collection and calling the method you just created to calculate the average gross income.

Step 3-13: Display averages on the PayPeriod form.

Add a Label below the ListBox on the PayPeriod form and name it lblCompareAverage. Leave the Text property blank.

Add lines 49–50 in Figure 13-26 to the updateListBox method in PayPeriodForm.cs.

Figure 13-26: Comparing the averages

```csharp
40      private void updateListBox()
41      {
42          lstPayPeriods.Items.Clear();
43          foreach (PayPeriod pp in employee.PayPeriods)
44          {
45              string pString = string.Format("{0:d} // {1} hours // Gross: {2:c} // Tax: {3:c} // Net: {4:c}",
46                  pp.StartDate, pp.HoursWorked, pp.GrossPay(), pp.TaxWithholding(), pp.NetPay());
47              lstPayPeriods.Items.Add(pString);
48          }
49          decimal average = parent.YTDAverageGross();
50          lblCompareAverage.Text = string.Format("{0:c}", employee.YTDGross() - average);
51      }
```

Line 49 makes a call on the parent object to the YTDAverageGross() method that you created in EmployeeForm. Line 50 then calls the YTDGross() method that you created in the Employee class and calculates the difference between the two values, which is then displayed on the form.

Run the program and observe how the average is calculated and updated as you switch between the forms, and add pay periods to each of the employees.

It's time to practice! Do exercises 13.8, 13.9, and 13.10 at the end of the chapter.

The following figures contain the full code for the project in Tutorial 3.

Figure 13-27: Program.cs

```csharp
static class Program
{
    /// <summary> The main entry point for the application. </summary>
    [STAThread]
    static void Main()
    {
        List<Employee> employees = new List<Employee>();
        DateTime birthday = new DateTime(1999, 5, 1);
        employees.Add(new Employee("Spongebob", birthday, 10.5m, "Burger flipper"));
        birthday = new DateTime(1995, 4, 18);
        employees.Add(new Employee("Squidward Tentacles", birthday, 12.56m, "Cashier"));

        Application.EnableVisualStyles();
        Application.SetCompatibleTextRenderingDefault(false);
        Application.Run(new EmployeeForm(employees));
    }
}
```

Figure 13-28: Employee.cs

```csharp
public class Employee
{
    public string Name { get; set; }
    public DateTime Birthday { get; set; }
    public decimal HourlySalary { get; set; }
    public string JobTitle { get; set; }
    private decimal taxRate;
    public decimal TaxRate { get { return taxRate; } }

    private List<PayPeriod> payPeriods;
    public List<PayPeriod> PayPeriods { get { return payPeriods; } }


    public Employee(string name, DateTime birthday, decimal hourlySalary, string jobTitle)
    {
        this.Name = name;
        this.Birthday = birthday;
        this.HourlySalary = hourlySalary;
        this.JobTitle = jobTitle;
        taxRate = 0.25m;
        payPeriods = new List<PayPeriod>();
    }

    public void AddPayPeriod(PayPeriod payPeriod)
    {
        payPeriods.Add(payPeriod);
    }

    public override string ToString()
    {
        return string.Format("Name: {0}, Birthday: {1:d}, Hourly Salary: {2:c}, Job Title: {3}",
            Name, Birthday, HourlySalary, JobTitle);
    }

    public decimal YTDGross()
    {
        decimal result = 0m;
        foreach (PayPeriod p in payPeriods)
        {
            if (p.StartDate.Year == DateTime.Now.Year)
            {
                result += p.GrossPay();
            }
        }
        return result;
    }

    public decimal YTDTax()
    {
        return YTDGross() * taxRate;
    }

    public decimal YTDNet()
    {
        return YTDGross() - YTDTax();
    }
}
```

Figure 13-29: EmployeeForm.cs

```csharp
public partial class EmployeeForm : Form
{
    private List<Employee> employees;

    public EmployeeForm(List<Employee> employees)
    {
        InitializeComponent();
        if (employees == null)
        {
            employees = new List<Employee>();
        }
        this.employees = employees;
        lstEmployees.Items.AddRange(employees.ToArray());
    }

    private void btnSave_Click(object sender, EventArgs e)
    {
        //Grab values from the form
        string name = txtName.Text;
        DateTime birthday = dtpBirthday.Value;
        string title = txtJobTitle.Text;
        decimal hourlyRate = Decimal.Parse(txtPayRate.Text);
        Employee employee;
        if (lstEmployees.SelectedIndex >= 0)
        {
            //If an employee is selected in the ListBox, get reference
            //to that employee and update the properties
            employee = (Employee)lstEmployees.SelectedItem;
            employee.Name = name;
            employee.Birthday = birthday;
            employee.JobTitle = title;
            employee.HourlySalary = hourlyRate;
        }
        else
        {
            //If no employee is selected, create a new Employee object
            //based on values in the form.
            employee = new Employee(name, birthday, hourlyRate, title);
            //lstEmployees.Items.Add(employee); //Add new objext to ListBox
            employees.Add(employee); //Add new object to collection
        }
        //Clear all items from the ListBox
        lstEmployees.Items.Clear();
        //Add all items from the employees collection to the ListBox
        lstEmployees.Items.AddRange(employees.ToArray());
    }

    private void lstEmployees_SelectedIndexChanged(object sender, EventArgs e)
    {
        Employee employee = (Employee)lstEmployees.SelectedItem;
        txtName.Text = employee.Name;
        txtPayRate.Text = employee.HourlySalary.ToString();
        dtpBirthday.Value = employee.Birthday;
        txtJobTitle.Text = employee.JobTitle;
    }

    private void btnPayPeriods_Click(object sender, EventArgs e)
    {
        Employee employee = (Employee)lstEmployees.SelectedItem;
        if (employee != null)
        {
            PayPeriodForm payPeriodForm = new PayPeriodForm(employee, this);
            payPeriodForm.Show();
        }
    }
}
```

```csharp
    public decimal YTDAverageGross()
    {
        decimal sum = 0m;
        foreach (Employee e in employees)
        {
            sum += e.YTDGross();
        }
        return sum / (decimal)employees.Count;
    }

}
```

Figure 13-30: PayPeriod.cs

```csharp
public class PayPeriod
{
    //Instance variables:
    private int minutesWorked;
    private DateTime startDate;
    private Employee employee;
    //Properties:
    public int MinutesWorked { get { return minutesWorked; } }
    public decimal HoursWorked { get { return minutesWorked / 60m; } }
    public DateTime StartDate { get { return startDate; } }
    public Employee Employee { get { return employee; } }
    //Constructor:
    public PayPeriod(int minutesWorked, DateTime startDate, Employee employee)
    {
        this.minutesWorked = minutesWorked;
        this.startDate = startDate;
        this.employee = employee;
    }
    //Methods:
    public void AddTime(int minutes)
    {
        minutesWorked += minutes;
    }

    public decimal GrossPay()
    {
        return HoursWorked * employee.HourlySalary;
    }

    public decimal TaxWithholding()
    {
        return GrossPay() * employee.TaxRate;
    }

    public decimal NetPay()
    {
        return GrossPay() - TaxWithholding();
    }
}
```

Figure 13-31: PayPeriodForm.cs

```csharp
public partial class PayPeriodForm : Form
{
    private Employee employee;
    private EmployeeForm parent;

    public PayPeriodForm(Employee employee, EmployeeForm parent)
    {
        InitializeComponent();
        this.employee = employee;
        this.parent = parent;
        lblEmployee.Text = employee.ToString();
        updateListBox();
    }

    private void btnSave_Click(object sender, EventArgs e)
    {
        Decimal hours;
        if (Decimal.TryParse(txtHoursWorked.Text, out hours))
        {
            DateTime startDate = dtpStartDate.Value;
            int minutesWorked = (int)Math.Round(hours * 60);
            PayPeriod pp = new PayPeriod(minutesWorked, startDate, employee);
            employee.AddPayPeriod(pp);
            updateListBox();
        }
    }

    private void updateListBox()
    {
        lstPayPeriods.Items.Clear();
        foreach (PayPeriod pp in employee.PayPeriods)
        {
            string pString = String.Format("{0:d} // {1} hours // Gross: {2:c} // Tax: {3:c} // Net: {4:c} // ",
                pp.StartDate, pp.HoursWorked, pp.GrossPay(), pp.TaxWithholding(), pp.NetPay());
            lstPayPeriods.Items.Add(pString);
        }
        decimal average = parent.YTDAverageGross();
        lblCompareAverage.Text = string.Format("{0:c}", employee.YTDGross() - average);
    }
}
```

We have covered a number of object-oriented principles and techniques in this chapter. As you have seen, it allows you to write very powerful and simple code. But we have only scratched the surface of what you can do with it—and have even left a good number of topics out in order to simplify the coverage and keep the page number reasonable in a single chapter. The next chapter contains one of the most significant topics in object-oriented programming: that of inheritance.

We encourage you to continue studying and using the object-oriented material you have learned here to further improve your coding skills.

Exercises

Exercise 13-1

Imagine you are creating a computer application to keep track of the work you need to do for your classes. Identify several classes that would be relevant to include in the application, and several objects for each class.

Exercise 13-2

Choose one of the classes from the previous exercise and identify several attributes and methods for that class.

Exercise 13-3

Create a new project that can track cars and the repairs on the cars. You will need two classes: Car and Repair. Start by creating a Car class with a few simple properties like Make, Model and Year. Add a form that can be used to create and modify Car objects and display them in a ListBox.

Exercise 13-4

Add a Repair class to the project. Repairs have a description and an amount for the repair (e.g., oil change, $30). Create a form that can be used to add a repair to a selected car.

Exercise 13-5

Add the ability to calculate the total amount for all repairs for a car, as well as simple statistics across cars, such as total for all cars, average repair amount, and average per car.

Exercise 13-6

Create a Book class that has the following attributes: Author, Title, Year published and the year of the most recent edition. Add code to ensure that the year published is higher than 1440, when Gutenberg invented the printing press, and no more than one higher than the current year. Also ensure that the year of the most recent edition isn't higher than the published year. Create several Book objects to demonstrate how this works.

Exercise 13-7

Add labels to the Employee Form to show the Gross Earnings, Tax Withholdings and Payment for the currently selected employee.

Exercise 13-8

Add more labels to show the difference in Gross Earnings, Tax Withholdings and Net Payment between the currently selected employee, and the averages for all employees.

Exercise 13-9

Add a method to the Employee form that calculates the number of employees with Gross Earnings above the average. Display this value on the form.

Exercise 13-10

Rewrite Programming Assignment 2, but use objects to represent an AutoType car and a Rental. You will need to create classes for each of the AutoType and Rental. When you read the data from the database, you pass the data from each record to the constructor of the class in order to create an object. To facilitate saving objects, you can have a method in the class that converts the data of an object to a proper form to be saved and returns it as a string. This string can then be saved in some other part of the program.

Chapter 14

Inheritance

When modeling the real world, we sometimes encounter entities that are specializations of some other entity. For example, in a banking application, all accounts behave in similar ways, but checking accounts, savings accounts and lines of credit all have subtle differences in what data is stored and how they behave. So, we could create a general account class that implements the general aspects of accounts, and then have specialized classes for each of the account types that implement the specific differences between the different types of accounts. We may also have different kinds of people in our system—clients and doctors, students and instructors, customers and employees and so on. In all these cases, one concept can be expressed as a more detailed version of another concept. We may have a generalized person concept that can be used as a template when modeling the more specialized types of people such as clients, students, customers.

This notion of a more specialized version of an existing concept is called *inheritance* in object-oriented programming, and is one of the foundational concepts that distinguishes object-oriented languages from traditional languages. Inheritance allows us to extend the functionality of an existing class by inheriting the features of that existing class. This allows us to modify some parts of a class and keep other parts intact. Conceptually, inheritance allows us to describe relationships between different classes. We can use inheritance to express that some entity is a specialized version of another entity.

Inheritance also plays a big role in the .Net framework classes. Even if there is no inheritance between the classes that make up your system, it is still vitally important to understand how it works and be able to apply the principles as you use the .Net framework for building user interfaces, storing data or working with collections.

Learning Objectives

After studying this chapter, you should be able to:

- Describe the concept of inheritance and how it can help improve code.
- Implement inheritance between classes.
- Use polymorphic method dispatch to execute the same function on objects of the same type.
- Discuss the differences and similarities of abstract classes and interfaces.
- Implement sorting of collections.
- Describe the use of access modifiers to implement encapsulation in C#.

Topics

14.1	Introduction to Inheritance	14.5	Polymorphism
14.2	Implementing Inheritance	14.6	Abstract Classes and Interfaces
14.3	Using Subclasses and Super classes	14.7	Using Interfaces to Sort Collections
14.4	Overriding Methods	14.8	Access Modifiers in C#

14.1 Introduction to Inheritance

In programming, we often need to modify already created code to add additional features or behaviors. For instance, having created a payroll system with an Employee class, we might recognize that there are significant differences in how salary is calculated between salaried and hourly employees. While we could add a variable to the class to hold the type of employee, the calculations would become rather complex, with many if statements checking on which type the employee is. A better approach is to use the concept of inheritance.

For instance, we can say that a Dog is a specialized version of Mammal, which itself is a specialized version of Animal. In this way, we create a *hierarchy* of classes (see Figure 14-1).

Figure 14-1: Inheritance hierarchy

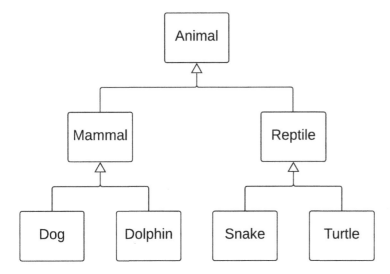

As you look at the relationship between two adjacent classes in the hierarchy, you can see the lower one is a more specialized version of the class above it—a Snake is a Reptile with additional characteristics that are not common to all Reptiles. Similarly, a Reptile is an Animal. Of course, it follows from this that Snakes are Animals. This relationship is often referred to as an *is-a* relationship (the lower class *is-a* higher class). We also say that the higher-level class is a *super class* of the lower-level class, which is a *subclass*. In C#, the super class is also called a *base class*. In UML diagramming, the inheritance relationship is shown with an empty triangle pointing to the super class.

The is-a relationship exists between classes, not between objects. It expresses that all Dolphins are Mammals and thus have all the properties and behaviors described by Mammal.

The alternative to the is-a relationship is the *has-a* relationship. This is also called *object composition* or aggregation, where one object is made up of other objects. For instance, we might say a car has an engine, four wheels, two or more doors, and so on. In other words, the car is made up of those components, and there would be a separate object for the engine, each wheel, and each of the doors. Object composition is implemented with a variable in one class, holding a reference to one or more objects of a different class as you saw with the relationship between Employee and PayPeriod in the previous section. In object composition, multiple objects exist to represent the relationship. For example, each Employee object would be related to multiple PayPeriod objects (one for each period that the employee gets paid).

However, with inheritance, only a single object is created, even when the object is of one of the subclasses. If a Dog object is created, there will not be a Mammal object or Animal object. However, the Dog will contain all the properties and methods implemented in the Mammal and Animal classes.

Review Questions

14.1 Indicate whether the following concepts represent inheritance. If so, which is the super class and which is the subclass?
- Oven and Kitchen
- Rectangle and Square
- Vehicle and Truck
- Batman and Superhero
- Student and Person
- Student and Employee
- Order and Customer
- Ferrari and Engine

14.2 Identify at least two subclasses for each of the following classes:
- Car
- Student
- Order
- Account
- Book
- Person

14.2 Implementing Inheritance

Returning to our example, consider the Employee class and two subclasses, as shown in Figure 14-2. This simplified hierarchy has two subclasses of Employee: HourlyEmployee and SalariedEmployee. We will use this hierarchy to calculate the gross pay (before any deductions, taxes, etc.) for each of the two types of employees. Hourly Employees' gross pay is calculated as the number of hours worked multiplied by the hourly pay rate, whereas the gross pay for Salaried Employees is calculated based on the annual salary and the number of days in the pay period. To keep things simple, we will use a two-week pay period and assume fifty-two weeks in a year.

Figure 14-2: Employee inheritance hierarchy

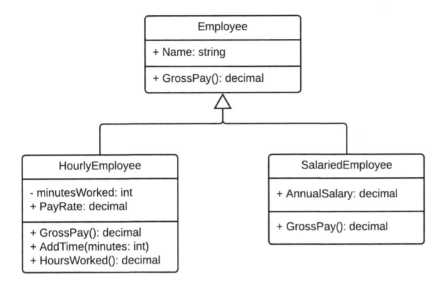

When implementing inheritance, you specify that the subclass inherits from the super class as shown in line 1 of Figure 14-3. The name of the super class is specified after the subclass name, separated by a colon.

Any constructor in a subclass must call a super class constructor. This is done in the signature line of the constructor, where you put a colon and then the method name *base* (line 11). Notice that parameters show that the constructor in Subclass takes a string (p1) and an integer (p2) as parameters. The integer parameter (p2) is then passed to the constructor in SuperClass, which would likely use it to initialize the corresponding field declared in SuperClass. We can thus infer from this that SuperClass has a constructor that takes a single integer. If you need to call a specific method in the super class, you can use the key word **base** in a similar way to the key word **this**, as shown in line 18.

Figure 14-6 shows a more specific example of a SalariedEmployee inheriting the characteristics an Employee class.

Figure 14-3: Implementing a subclass

```
 9      class Subclass: Superclass
10      {
11          public Subclass(string p1, int p2) : base(p2)
12          {
13      
14          }
15      
16          public void Method()
17          {
18              base.SuperMethod();
19          }
20      }
```

The next tutorial will show how to implement the employee Inheritance hierarchy.

Tutorial 1: Creating Subclasses

In this tutorial you will create a super class and several of its subclasses to give you some good experience with inheritance.

Step 1-1: Create a project and classes.

Start by creating a new Visual Studio Windows Forms project named Payroll-Inheritance. Once the project is created, add the following three classes by right clicking the project folder in Solution Explorer and selecting Add > Class: Employee, HourlyEmployee and SalariedEmployee.

Step 1-2: Implement Employee class.

Refer to Figure 14-4 for the initial implementation of the Employee class. This class is very simple, with an automatic property for name and a constructor that initializes the name property.

Figure 14-4: Employee class code

```
class Employee
{
    public string Name { get; set; }

    public Employee(string name)
    {
        this.Name = name;
    }
}
```

Step 1-3: Set up inheritance in SalariedEmployee.

Switch to SalariedEmployee class and implement the class as shown in Figure 14-5.

You'll note that this class does not have a property to store a name, but as you will see later, this is available through the Employee class. Line 9 specifies that Employee is the super class of this class. Note that the SalariedEmployee class doesn't have a property for name.

Line 12 has the call to the constructor in Employee, passing the name parameter on to the super class. Notice the difference between the two sets of parentheses in line 12: The first one declares two parameters—that is, it specifies that this constructor takes a string and a decimal. However, the second set contains a value passed to the constructor in Employee. Even though it looks a little strange with that "base" key word, this is just a regular call to the super class constructor. As such, we could have specified any string for the parameter passed to the super class, or even written one, like "Smith," directly in the code. It is common that, in addition to taking parameters for its own fields, a subclass constructor also takes all the parameters of its super class and simply passes them through.

Figure 14-5: SalariedEmployee code

```
 9      class SalariedEmployee : Employee
10      {
11          public decimal AnnualSalary { get; set; }
12          public SalariedEmployee(string name, decimal salary) : base(name)
13          {
14              this.AnnualSalary = salary;
15          }
16      }
```

Step 1-4: Instantiate objects.

Switch to the Program.cs and add the following line of code at the beginning of the Main method:
```
Employee emp1 = new Employee("Spongebob");
```

This line creates a generic Employee object, and assumes that Spongebob is neither salaried nor hourly (which may not really make sense in the real world—we'll discuss how to handle this later in the chapter).

Add the following line:
```
SalariedEmployee salEmp2 = new SalariedEmployee("Clark Kent", 45000);
```

This creates a SalariedEmployee object. It is important to realize that no Employee object is created by this line, even though SalariedEmployee is a subclass of Employee. So, each of the two lines will create exactly one object.

Add a breakpoint in the first of the two lines above and start the debugger (F5). Once you hit the breakpoint, click on Step Into (F11) and you should find that execution has moved to the Employee class. You can step through the program and notice that everything works as it normally does when creating an object. The string "Spongebob" gets assigned to the Name property inside the constructor.

Continue stepping until you get back to Main; then step into the constructor call for SalariedEmployee. In the constructor, execution will move to the end of line 12 as shown in Figure 14-6. Pay attention to the values in the name variables. They should be the same—namely, "Clark Kent" as was entered when the object was created in the Main method. If you expand *this* in the Locals window, you should also see that Name is defined in this object, even if it isn't in the code of SalariedEmployee. However, because it is in the super class, it shows up here. It's important to note that when a SalariedEmployee object is created, there isn't a corresponding Employee object. With inheritance, only one object is created, but this object contains all the properties of its super class.

Figure 14-6: Debugging SalariedEmployee

The "base" method call will call the constructor of the super class, so when you click Step Into again, you will find yourself in the Employee class where the Name property is set to Clark Kent. Expand *this* again, and note that even though you are in the Employee class, you can still see the AnnualSalary property, because you are really in the middle of creating a SalariedEmployee object, and not an Employee object. As you continue, you will return to the SalariedEmployee constructor where AnnualSalary is initialized to 45,000.

When you are back in the Main method, examine the two objects (emp1 and salEmp2), carefully noting that both the Employee object and the SalariedEmployee object contain a Name property but only SalariedEmployee contains an AnnualSalary property. The "base" call to the super class contructor does not create a separate object of the super class; you end up with a single object of the subclass that contains all the properties of the super class. Once you have examined the two objects, you can click the Stop debugging button.

Repeat the debugging until you are comfortable with the relationship between the classes and objects created. Feel free to create additional objects with different values to explore.

Step 1-5: Implement HourlyEmployee.

Now that you have a good understanding of how subclasses work, you will add a second subclass to see how two different subclasses inherit common elements from their super class but can add their own special elements.

Use the class diagram in Figure 14-2 as a guide in implementing the HourlyEmployee class (the code is in Figure 14-7).

Start by making the class a subclass of Employee (line 9). Then add the PayRate property and the private field, minutesWorked. These are the pieces of data needed to be stored for the HourlyEmployee objects, which are different from both Employee and SalariedEmployee.

Next, following the example of what you did with SalariedEmployee, create the constructor (lines 14–18), passing the name to the super class constructor and setting the property and field values.

Finally, add the two methods specific to the HourlyEmployee class. AddTime (lines 20–23) updates the minutesWorked field by adding the value of minutes parameter to the existing value of minutesWorked. HoursWorked (lines 25–28) converts the minutes worked and returns as fractional hours worked.

Figure 14-7: Code for HourlyEmployee class

```
 9      class HourlyEmployee : Employee
10      {
11          public decimal PayRate { get; set; }
12          private int minutesWorked;
13
14          public HourlyEmployee(string name, decimal payRate) : base(name)
15          {
16              this.PayRate = payRate;
17              this.minutesWorked = 0;
18          }
19
20          public void AddTime(int minutes)
21          {
22              this.minutesWorked += minutes;
23          }
24
25          public decimal HoursWorked()
26          {
27              return minutesWorked / 60.0m;
28          }
29      }
```

Step 1-6: Test HourlyEmployee class.

Open Program.cs and add the following three lines of code following the creation of the SalariedEmployee object that you added previously:

```
HourlyEmployee hourlyEmp3 = new HourlyEmployee("Batman", 40.0m);
hourlyEmp3.AddTime(30);
decimal hours = hourlyEmp3.HoursWorked();
```

The first line creates an HourlyEmployee object in the same way that you created a SalariedEmployee object previously. The second line calls the AddTime method on this object, and the last line calls the HoursWorked method on the object.

Remove your previous breakpoint; then place a new breakpoint on the first of these lines, and use the debugger to step into each of the method calls to ensure you understand which code gets called and the result. Note again that no Employee object is created and that the SalariedEmployee and HourlyEmployee objects are entirely independent of each other. Changing the state of the hourlyEmp3 object has no impact on the state of the salEmp2 object.

It's also important to note that you cannot call the AddTime and HoursWorked methods on the emp1 and salEmp2 objects, as those methods are declared in the HourlyEmployee class.

Try it by adding these lines to the Main method:

```
emp1.AddTime(60);
decimal h2 = emp1.HoursWorked();
salEmp2.AddTime(90);
decimal h3 = salEmp2.HoursWorked();
```

You will get red squiggly lines and error messages saying that Employee and SalariedEmployee don't contain a definition for the methods you're trying to call. This means that you cannot call methods declared in an object's subclasses or an object's sibling classes.

However, you *can* call methods that are declared in a super class (as long as they are declared public). Comment out the four lines you just added and add these lines instead:

```
string n1 = emp1.Name;
string n2 = salEmp2.Name;
string n3 = hourlyEmp3.Name;
```

You shouldn't get any syntax errors, and if you use the debugger and step through, you will see that n1, n2 and n3 get the Name values for each of the three objects (Spongebob, Clark Kent and Batman).

It's time to practice! Do the following steps.

Step 1-7: Create a Person class as a super class of Employee; then move the Name property to Person. Note that you should not need to make any changes outside of the Employee and Person classes to have the system function as it did before.

Step 1-8: Add a Customer class as a subclass of Person. Include a property for a Credit Limit. (Each customer may be able to buy up to their credit limit before having to pay.) Create several Customer objects and display their details on a form.

14.3 Using Subclasses and Super Classes

When you have super classes and subclasses, you can start doing some more flexible things with variables. For instance, as you have just seen, you can call on properties (and any other public methods) declared in a super class as if they were declared in the same class as the current object. However, this can be taken further. You can, in fact, declare a variable to be of a super class type but then assign that variable an object that is of a subclass type, like this:

```
Employee employee;
employee = new HourlyEmployee("Batman", 40.0m);
employee = new SalariedEmployee("Clark Kent", 45000);
```

Here you see a variable, employee, that is of the declared type Employee, but is assigned first an HourlyEmployee object and then a SalariedEmployee object. When you declare objects like that, you can call any method on the object that is available in the declared type of the variable. So, in the above example, you would be able to call on the Name property but not HoursWorked or AddTime—even when it's the HourlyEmployee object that is assigned to the variable. Thus, this approach is appropriate when you work with subclass objects and only need to use the properties and methods that are declared in the super class.

One common usage for this is to declare a generic collection to hold objects of a supertype and then at runtime be able to add a mix of subtype objects to the collection and call methods available in the super class on all or some of the objects in the collection. For instance, you might have a collection like this, containing the emp1, salEmp2 and hourlyEmp3 objects that were declared in the previous tutorial:

```
List<Employee> employees = new List<Employee>() { emp1, salEmp2, hourlyEmp3 };
```

With a collection like this, you can have a loop that goes through all the objects in the collection and takes some action for all of them. For instance, if we wanted to create a list of all the names of the employees, we could have a loop like this:

```
string names = "";
foreach(Employee e in employees)
{
    names += e.Name + " ";
}
```

This goes through all the items in the List and then accesses the Name property for each one. Since all the objects in the list are subclasses of Employee, which has the Name property, this will work.

14.4 Overriding Methods

Sometimes a subclass may need to modify the behavior of a method in a super class. This is called *overriding*. Overriding allows you to specify more specific behavior in a subclass. Often, you can specify at a super class level that a particular behavior is needed, but you cannot specify *how* this behavior will be implemented in a subclass. For instance, in the animal example earlier in the chapter, we could specify that all animals must have a Move behavior, but each kind of animal will have very different ways of moving, so the implementation of the Move behavior will be different in each class. For employees, you will see in the next tutorial that all employees will have a way to calculate how much money they get paid out for a pay period, but the calculation is done differently depending on whether someone is an hourly or salaried employee.

In order to override a method, you write the same method signature in the subclass that you had in the super class but add the key word *override*. It is common to override the ToString method, which is actually implemented in a class called Object—an implicit super class of every other class. So, in the Employee class, we might have this ToString method:

```
public override string ToString()
{
    return Name;
}
```

The ToString method in HourlyEmployee could look like this:

```
public override string ToString()
{
    return base.ToString() + " Pay Rate: " + PayRate;
}
```

Both of these methods override the version of the method implemented in the super class. It isn't necessary to have a version in the immediate super class. If Employee didn't implement ToString, HourlyEmployee could still implement it and would then override the Object version.

Note the call to base.ToString() in the HourlyEmployee version. This calls the method in the super class. So, in this case, the HourlyEmployee ToString method would return the Name along with the pay rate. The key word base works much like the key word this, except that base refers to the super class and this refers to the current object. The key word base is used in two different ways: as a way to call the super class constructor from the subclass constructor and as a way to access any method or property in the super class.

If the ToString method is implemented in both the Employee class and its subclasses, you might be confused as to which version of the method is actually called at runtime. This is especially confusing if a variable is declared of the supertype but you assign an object of a subtype. For instance, assume you have implemented ToString in Employee and HourlyEmployee, as discussed above, but not in SalariedEmployee. What will happen in the following code?

```
Employee employee;
string empStatus;
employee = new HourlyEmployee("Batman", 40.0m);
empStatus = employee.ToString();
employee = new SalariedEmployee("Clark Kent", 45000);
empStatus = employee.ToString();
```

Here we declare a variable employee of type Employee but then assign an object of type HourlyEmployee to that variable (which works because HourlyEmployee is a subclass of Employee). Then we call ToString on the employee variable and assign it to the string empStatus. After that, we assign a SalariedEmployee object and call ToString again. While the two lines with the call to ToString are identical, they will produce different outputs because the actual objects assigned to employee are different.

When you call an overridden method, the runtime environment will always look for the implementation of the method defined in the class in the inheritance hierarchy that is closest to the actual object at runtime. Thus, the first call to ToString will call the version in HourlyEmployee, but the second version will call the version in Employee because SalariedEmployee class doesn't override the ToString method.

Rules for Overriding Methods

When you create an overridden version of a method in a subclass, you have to follow a few rules:

- The methods must have the same signature—the same method name and the number and types of arguments must be the same.
- The return types have to be the same.
- The access modifier (public, private, protected) has to be the same.
- The overridden method must include the override key word.

Overriding versus Overloading

Overriding is often confused with overloading, as both are ways to create different implementations of a method with the same name. The big difference between the two is that overriding only occurs when a subclass implements the same method as a super class. Overloading, meanwhile, allows you to have methods with the same name in the same class. However, when overloading, you must give the methods different signatures—that is, the number and/or types of parameters for the methods must be different enough that the compiler can determine which one is intended to be called. With overloading, you can also provide different return types and can have different access modifiers.

Review Questions

14.3   If class A is a super class of class B, which is a super class of class C, which of the following are allowed?
- `A q = new A();`
- `A x = new B();`
- `B y = new A();`
- `C z = new A();`

14.4   Given the classes (A, B, C) in the previous question, explain why the following is allowed:
```
List<A> letters = new List<A>();
letters.Add(new A());
letters.Add(new B());
letters.Add(new C());
```

14.5   Given the classes (A, B, C) in the previous question, if class B implements a method P that is overridden in C, which version (in B or C) will be called in each of the following code snippets?
```
C c = new C();
c.P();

A a = new B();
a.P();

B b = new C();
b.P();
```

14.6   Given the classes (A, B, C) in the previous question, if class A implements a method Q that is overridden in C (but not in B), which version (in A or C) will be called in each of the following code snippets?
```
C c = new C();
c.Q();

A a = new B();
a.Q();

B b = new C();
b.Q();
```

Tutorial 2: Implementing the GrossPay Method

Let's return to the Employee example and the GrossPay method, which we discussed earlier. The GrossPay method will calculate the gross pay for a single two-week pay period for an employee. However, this calculation is done differently, depending on whether the employee is an hourly or salaried employee, or neither.

When we are done, we would like to be able to add both types of Employees to a collection and calculate the gross pay for all of them, regardless what type they are, with code that might look like this:

```
decimal totalGrossPay = 0m;
foreach (Employee emp in employees)
{
    totalGrossPay += emp.GrossPay();
}
```

We first have to determine which class(es) to implement the GrossPay method in. We have several options:

- In the Employee class only—With this option, we can't distinguish between the different kinds of employees in order to provide a different implementation of the method.
- In the HourlyEmployee and SalariedEmployee classes—This allows us to provide different implementations, but then the loop above wouldn't work and we can't have a collection with both types of employees.

Both of these options could be made to work, but the code becomes much less elegant and harder to maintain. A third option is

- Implement the method in Employee and each of the two subclasses—This allows for collections of all types of employees, and different implementations for each type of employees, and the loop above will work.

There is only one drawback to this last option: We can't provide a meaningful implementation for the method in the Employee class, since we don't have any information about salary, hours worked and so on in an object of type Employee. We could simply have this method return a predefined value (perhaps zero or a large negative value) as a signal that the method is undefined, but if the method is used in calculating statistics across many employees, this will likely not be noticed and the statistics will be off.

One solution to this dilemma is to declare that the method is *abstract*. When a method is abstract, no implementation is provided and only the method signature is included in the class. It would look like this for GrossPay in Employee:

```
public abstract decimal GrossPay();
```

Notice the key word abstract, indicating that the method is abstract, with the semicolon at the end of the method indicating there is no body to this method.

Step 2-1: Add GrossPay method in Employee class.

Go ahead and add the abstract GrossPay method as shown above to the Employee class. You will get a red squiggly line under the GrossPay method name. This is because once you have an abstract method in a class, the class itself must also be declared abstract.

When a class is abstract, you cannot create objects from this class. But in this case, that is exactly what we want. We were unable to provide an implementation of the GrossSalary method in the Employee class because we didn't have any data to allow us to make the calculation. This is a strong indication that we would never create Employee objects, but always either HourlyEmployee or SalariedEmployee objects—that is, we assume that every employee is salaried or hourly. So, we can't have a situation as we saw with Spongebob earlier (he is neither hourly or salaried)—employees have to be one or the other.

Step 2-2: Make Employee abstract.

Add the abstract key word to the class declaration line, like this:

```
abstract class Employee
```

Step 2-3: Add GrossPay to the SalariedEmployee class.

Switch to SalariedEmployee.cs and you will notice that this class now has an error. This is because when you inherit from an abstract class, which this class does, now that you made Employee abstract, you must either declare the subclass abstract or provide an implementation for all abstract methods.

To fix this, you implement GrossPay in SalariedEmployee by adding the following method anywhere in the class:

```
public override decimal GrossPay()
{
    return AnnualSalary / 52 * 2;
    //Gross pay is simply two weeks of annual salary (52 weeks a year)
}
```

Step 2-4: Similarly, you need to implement GrossPay in HourlyEmployee.

Add the following code to HourlyEmployee.cs:

```
public override decimal GrossPay()
{
        return HoursWorked() * PayRate;
}
```

If you build the system now, you may get some errors if you have created any Employee objects anywhere. Go through and comment all those lines out so the system builds. You might find some of these in the Main method in Program.cs.

Step 2-5: Test the inheritance structure.

To test the system, we will create a number of hourly and salaried employee objects, and then calculate the total and average gross pay across all the employees and output this to Labels on a Form. Start by adding the code in Figure 14-8 to the Main method in Program.cs to set up the data to be used for the test. Note that Form1.cs has been renamed to EmployeeInheritanceForm.cs.

Figure 14-8: Creating test data for inheritance

```csharp
15  static void Main()
16  {
17      Random r = new Random(); //Random number generator - to be used several times below
18      List<Employee> employees = new List<Employee>();
19      string[] employeeNames = {"Superman", "Green Lantern", "Batman", "Spiderman",
20                                "Thor", "Hal Jordan", "Wonder Woman", "Captain America"};
21      foreach (string name in employeeNames)
22      {
23          Employee emp = null;
24          int salOrHrl = r.Next(0, 2); //Random number between 0 and 1
25          if (salOrHrl == 0) //Create hourly employee
26          {
27              decimal payRate = (decimal)r.NextDouble() * 20 + 10; //Random number between 10 and 30
28              HourlyEmployee hourly = new HourlyEmployee(name, payRate);
29              hourly.AddTime(r.Next(30, 480)); //Add some random time for this employee
30              emp = hourly; //Assign hourly to the Employee variable that is added to the list later
31          }
32          else //create salaried employee
33          {
34              decimal salary = r.Next(40000, 80000); //Random number between 40,000 and 80,000
35              emp = new SalariedEmployee(name, salary);
36          }
37          employees.Add(emp); //Add the employee to the list
38      }
39
40      Application.EnableVisualStyles();
41      Application.SetCompatibleTextRenderingDefault(false);
42      Application.Run(new EmployeeInheritanceForm(employees));
43  }
```

In this code, you start by setting up a random number generator (line 17) that will be used to generate random numbers for test purposes.

The names for the employees are put into an array (lines 19–20), and a loop (lines 21–37) generates an employee object for each name. This object is declared in line 23 and added to the list in line 37. We use a random number between 0 and 1 (line 24) to determine if the employee will be hourly (if it's 0, line 25) or salaried (if it's 1, line 32). The random number method that produces integers is defined to not include the upper bound provided, so we need to specify the number to be between 0 and 2 (but excluding the upper bound).

For hourly employees (lines 27–30), the pay rate is determined as a random number between 10 and 30. In line 29, a random number of minutes is added to the employee to have worked. Because we need to call AddTime on the hourly employee, we can't use the variable emp, which is of type Employee. As discussed earlier, you can assign an object of a subclass to a variable of a super class type, but in that case, you can only use the methods and properties defined in the super class. So, we declare a variable of the type HourlyEmployee (line 28). Line 30 then assigns the HourlyEmployee to the Employee variable. This works because Employee is a super class of HourlyEmployee.

For salaried employees, we just need to determine a salary, which is determined as a random integer between 40,000 and 80,000.

Line 42 has been modified to call the EmployeeInheritanceForm constructor with the employees list as an argument. However, because the constructor hasn't been modified yet, you will get an error. This is fixed in the next step.

Step 2-6: Modify Form constructor.

Switch to EmployeeInheritanceForm.cs and add the code as shown in line 15, 17 and 19 in Figure 14-9 (you'll add the rest of the code later). Line 15 sets up a field to hold the list of employees that is passed into the constructor (line 17) and assigned in line 19. If you get an error message that List<Employee> is less accessible than the method EmployeeInheritanceForm, make the Employee class public.

Figure 14-9: Code in EmployeeInheritanceForm

```csharp
13      public partial class EmployeeInheritanceForm : Form
14      {
15          List<Employee> employees;
16
17          public EmployeeInheritanceForm(List<Employee> employees)
18          {
19              this.employees = employees;
20              InitializeComponent();
21
22              lblDetails.Text = employeeDetails();
23              lblNumEmployees.Text = employees.Count.ToString();
24              lblTotalGrossPay.Text = string.Format("{0:c}", totalGrossPay());
25              lblAverageGrossPay.Text = string.Format("{0:c}", averageGrossPay());
26          }
27
28          private decimal totalGrossPay()
29          {
30              decimal result = 0;
31              foreach (Employee e in employees)
32              {
33                  result += e.GrossPay();
34              }
35              return result;
36          }
37
38          private decimal averageGrossPay()
39          {
40              return totalGrossPay() / employees.Count;
41          }
42
43          private string employeeDetails()
44          {
45              string result = "";
46              foreach (Employee e in employees)
47              {
48                  result += string.Format("Name: {0}, Gross Pay: {1:c}\n", e.Name, e.GrossPay());
49              }
50              return result;
51          }
52      }
```

Step 2-7: Setup Form for output.

Open the Form in design view and add seven labels, as shown in Figure 14-10.

Figure 14-10: Form design for testing inheritance

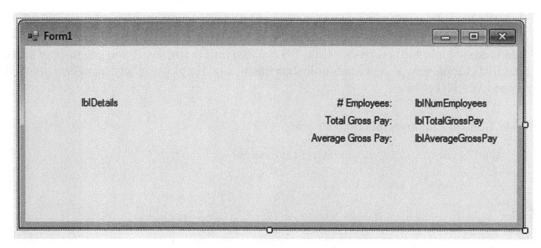

Step 2-8: Implement code for output.

Switch back to the code file and add the three methods shown in lines 28–51 and the rest of the constructor in lines 22–25 in Figure 14-8.

The most interesting thing here is the totalGrossPay method, as it illustrates how you can use inheritance to write very simple code that does something fairly powerful. The core of the method is a loop over the list of employees. On each Employee object, we call the method GrossPay, which returns the gross pay for that employee. However, remember that there is no implementation of the GrossPay method in the Employee class, so this will call the version in either SalariedEmployee or HourlyEmployee. Since the employees list contains both salaried and hourly employees, at runtime, the proper version of the method will be chosen based on the actual type of the object in the list. Note that because the Employee class is abstract, there could not be any plain Employee objects that aren't either salaried or hourly.

Step 2-9: Run program to test.

Run the program and you should see output similar to Figure 14-11. Notice that your salary numbers will be different since they are randomly generated.

It's time to practice! Do the following step:

Step 2-10: To ensure you understand what is happening, set a breakpoint in line 30 of Figure 14-9 and step through the code and into the GrossPay method. Notice that you will go into HourlyEmployee for some objects and SalariedEmployee for other objects.

Figure 14-11: Output from testing inheritance

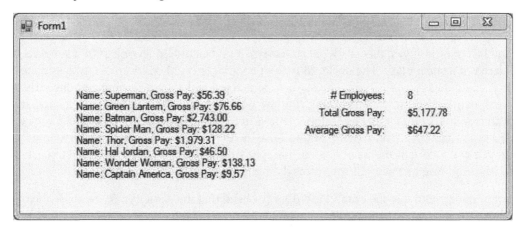

14.5 Polymorphism

We have been discussing the concept of *polymorphism*, which means "many forms." This works in several different ways here. First, an object can be considered to be of two different types, both the type that the variable holding the object was declared to be and the actual type of the object. The actual type must be either the same as the declared type, or a subclass of the declared type, as you saw when you created HourlyEmployee and SalariedEmployee objects and assigned them to an Employee variable.

You also used polymorphism when you defined a method in a super class and an overridden version in the subclass. You saw how the actual method called was the one in the actual type of the object. For example, in line 33 of Figure 14-9, the GrossPay method that was called for an hourly employee was the method in the HourlyEmployee class and the method that was called for a salaried employee was the one in the SalariedEmployee class, even though the declared type was Employee. The process used to determine the actual method called is called *dynamic dispatch*.

Polymorphism also allows you to create a collection to hold objects of a super class type and then add objects of that type and any subclass to it. You can then execute any method that is available on the super class type on the objects that are stored in the collection. You saw an example of this when calculating the total gross pay across all employees above. The loop in the totalGrossPay method in Figure 14-9 calls the GrossPay method in the Employee class, but at runtime, it executes either the GrossPay method in HourlyEmployee or the one in SalariedEmployee, depending on the actual type of the object.

To fully understand this concept, it helps to understand the difference between *static* and *dynamic* types. Consider the following line of code:

```
Car c1 = new Car();
```

What is the data type of the c1 variable?

It is clearly of the type Car, since the variable was declared of that type and an object of the Car class is created and assigned to the c1 variable.

However, if you consider the following line of code, where Vehicle is a super class of Car, things are less clear:

```
Vehicle v1 = new Car();
```

Again, a Car object is created and assigned to the variable v1. However, v1 is declared to be of the type Vehicle. So, it seems that v1 is both of the type Vehicle and the type Car. What gives? How can a variable be of two data types at the same time? Is this even a legal statement?

To answer the last question: yes, this is a legal statement. You can assign an object of a subclass to a variable declared of a super class. The declared type of a variable is called its *static type*, and the actual assigned type is called its *dynamic type*. The static type is fixed and cannot be changed once it's declared. However, the dynamic type can change during runtime when the program is running. You actually already saw an example of this in the GrossPay method in Figure 14-9, where the loop declares a variable called e of the type Employee (static type), but because the employees collection contains both HourlyEmployee and SalariedEmployee objects, the type of the object actually assigned to each will either be HourlyEmployee or SalariedEmployee (dynamic type).

It's important to understand that the compiler will only check that the static types are applied correctly. This means that if you try to call a method on an object, the static type check must satisfy that a method with that signature is available to be called on an object of that type. Consider again the GrossPay method. This method is declared as abstract in the Employee class, which is good enough for the compiler to be satisfied with the static type check. The compiler knows that an actual object assigned to a variable of type Employee will have a GrossPay method to call, because a subclass of Employee would have to either implement GrossPay or itself be declared as abstract, in which case no object could be created from this class.

The dynamic type of an object only comes into play during runtime where the big question is to determine which version of a method is called when a method is declared in both a super class and subclass. This is determined by a process called dynamic dispatch, which you will explore further in the following tutorial.

Review Questions

14.7 Given class Video (a subclass of class Media), which of the following are legal?
- `Video v1 = new Video ();`
- `Video v2 = new Media();`
- `Media m1 = new Media();`
- `Media m2 = new Video();`

14.8 For each of the declarations in the previous question, indicate the static and dynamic type of the variable.

Tutorial 3: Dynamic Dispatch and the ToString Method

In this tutorial, you will improve on the way the employee detail information is displayed. Currently it is limited to only display information and call methods that are available in the Employee class. So, the Name is included, since that is a property of Employee, along with the Gross Pay, since that is an abstract method in Employee. See lines 43–50 in Figure 14-9 for the code and Figure 14-11 for the output from calling this method on each of the employees.

While the method is elegant in that it allows a uniform display for all employees, it is also limiting in that it doesn't allow for including any information specific to hourly and salaried employees. To fix this, you will be exploring the ToString method. ToString is defined in the Object class, which you will recall is the super class of any class that doesn't have an explicit super class declared. It is thus the root of the entire inheritance hierarchy in C#. This means that you can call ToString on any object and know that the ToString method is always available. You can then override it to provide the functionality you want.

Step 3-1: Override ToString in Employee.

Open the Employee class and add this method:

```csharp
public override string ToString()
{
    return "Name: " + Name;
}
```

The override key word means that this method is overriding an inherited method. It is strictly not required to include the override key word, but you will get a warning if you don't.

Step 3-2: Implement ToString in HourlyEmployee.

Switch to HourlyEmployee and add this method:

```csharp
public override string ToString()
{
    string result = base.ToString();//Call super class version of method
    result += string.Format(". Hourly pay rate: {0:c}", PayRate);
    return result;
}
```

Since HourlyEmployee inherits from Employee, this method overrides the version in Employee. However, in the first line of the method, the call base.ToString() allows for calling the overridden version of the method. Note that the key word base can be used to call any method in the super class.

Step 3-3: Change output method.

In EmployeeInheritanceForm.cs, replace the line of code in the employeeDetails method as shown below:

```csharp
private string employeeDetails()
{
    string result = "";
    foreach (Employee e in employees)
    {
        // result += string.Format("Name: {0}, Gross Pay: {1:c}\n", e.Name, e.GrossPay());
        result += e.ToString() + "\n";
    }
    return result;
}
```

This will call ToString on every Employee object (whether they're hourly or salaried) and output to the form.

If you run the program now, you should see output similar to the following:

Tutorial 3: Dynamic Dispatch and the ToString Method

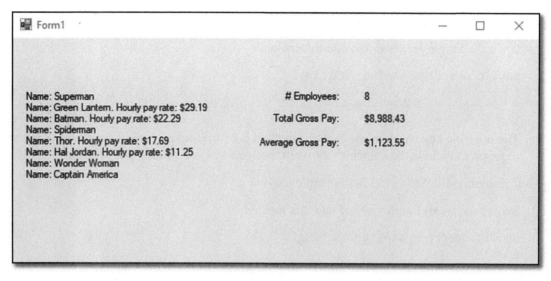

In this case, two of the employees are hourly and their pay rates are listed. However, for the salaried employees, only the name is listed, since there is no ToString method in SalariedEmployee. You can insert a breakpoint in the employeeDetails method following the execution to see how the ToString method is called when different objects are being displayed.

It's time to practice! Do the following step:

Step 3-4: Implement ToString in SalariedEmployee.

Implement the ToString method in SalariedEmployee to get the following output:

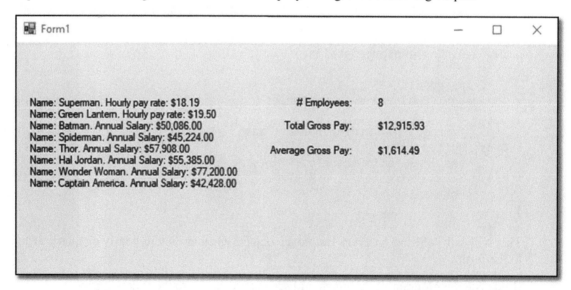

Step 3-5: Explore ToString in the Object class.

We mentioned earlier that the ToString method is implemented in the Object class. However, the implementation in the Object class is usually not what we want to use, but we want to explore it so you know how it works. First, comment out the entire ToString method in the Employee class and run the program again. You should see output similar to the following:

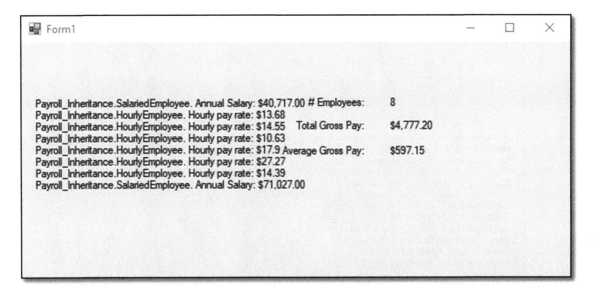

Instead of the name of the employee, the call to base.ToString now produces something entirely different. If you look closely, you will see that what is being produced is the name of the dynamic type of the object (including the namespace). The implementation of ToString in the Object class is simply to return this class name, and in most cases, this is not at all what you want. However, many beginning programmers will make the mistake and try to get the string version of an object without having overridden ToString, and will be confused by this kind of output.

Reinstate the ToString method in Employee.

Step 3-6: Explore automatic calls to the ToString method.

In many situations, the ToString method is automatically called by classes in the .Net library. You will see an example of this here. Start by adding a ListBox to the form, and name it lstEmployees.

Then add the following line of code to the constructor for the form:

`lstEmployees.Items.AddRange(employees.ToArray());`

This adds the entire list of employees to the Items collection for the ListBox. Notice that there is no call to the ToString method. Run the program and you should see the following output:

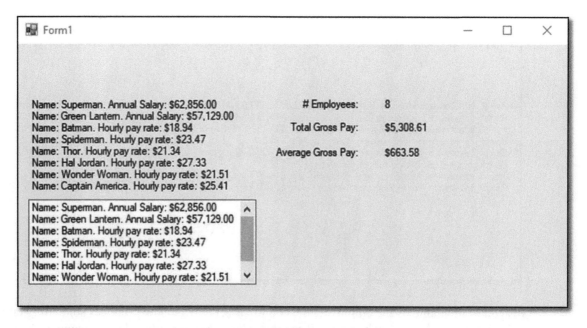

Notice that the output in the ListBox is the same as the original output, yet no calls to ToString were needed. That's because when an item is added to the ListBox, it will automatically call the ToString method on that item. Because the ToString method is implemented in the Object class, there will always be at least that generic version to call.

14.6 Abstract Classes and Interfaces

If we wanted to expand on the Employee system from above, we might consider adding a new type of employee and consider what taxes to withhold on the salary payments made to employees. We might add a ContractEmployee class for contractors where tax shouldn't be calculated or withheld. However, we would want to specify that for salaried and hourly employees, taxes should be withheld. In order to implement this, we have several options:

1. We could implement the tax withholding methods in the Employee class and simply have the method not do anything in the contract employee class. This would make the tax calculations the same in all the classes, which might not be appropriate—and would mean that taxes would be calculated on ContractEmployee.
2. We could instead make the tax methods abstract in Employee and have the implementation in ContractEmployee be empty. While this would work, it isn't a particularly elegant approach, and it infers that the contract employees are in fact taxable. And if at some point in the future, a programmer creates a subclass of ContractEmployee, the functionality could be overridden.
3. Another option might be to simply implement the tax methods in the HourlyEmployee and SalariedEmployee classes where they make sense. While this would prevent taxes from being withheld for ContractEmployee objects, it would also not allow us to have any kind of polymorphism benefits from calculating taxes across a group of employees.

What we need is really the ability for some classes to inherit from another parent; we might call it Taxable. This class could then hold the tax methods, which would be implemented in the proper classes as needed. Then we could have polymorphic collections, method calls and so forth on objects of the Taxable class. However, there is just one problem with this approach: While some languages do allow for this kind of multiple inheritance, C# is not one of them. In C#, each class can only inherit from a single parent.

However, this problem can be solved by using an Interface. Interfaces are similar to abstract classes in that they can specify a collection of methods and properties that must be implemented in classes that implement the interface. However, where a class can only inherit from one class, it can implement multiple interfaces. Interfaces are also types, so we can declare variables and collections to hold objects that implement a particular interface. The following table shows the similarities and differences between abstract classes and interfaces.

	Abstract Class	Interface
Similar	<ul><li>Specify abstract methods</li><li>Are data types for polymorphism</li><li>Can inherit from super classes</li></ul>	
Different	<ul><li>Subclass can only inherit from one abstract class</li><li>Can have non-abstract methods</li><li>Can have public, private and protected members</li><li>Can specify instance variables and properties</li></ul>	<ul><li>Classes can implement multiple interfaces</li><li>All methods are abstract</li><li>All members are public</li><li>Can only specify methods</li></ul>

Review Questions

14.9 Which of the following are NOT valid in C# if Drivable is an interface and Van is a class that implements it?
- `Drivable d;`
- `Drivable d = new Drivable();`
- `Drivable d = new Van();`
- `Drivable d = null;`

14.10 Describe what problems could occur if C# allowed multiple inheritance.

Tutorial 4: Interfaces

In this tutorial, you will expand the Employee system with the ITaxable interface to see how collections can use an interface to call specific methods on a collection of objects. Figure 14-12 shows the class diagram that you will be implementing. You will be adding the new interface, ITaxable, as well as the class, ContractEmployee. You will also be making slight changes to several existing classes. In UML, interfaces are drawn similarly to classes, but only have two sections—one for the interface name, and one for methods, since methods can't specify instance variable. We use a dotted line with an empty head triangle to denote a class that implements an interface. In C#, interface names are by convention preceded by a capital I, so the interface name is ITaxable.

Figure 14-12: Class diagram with interface

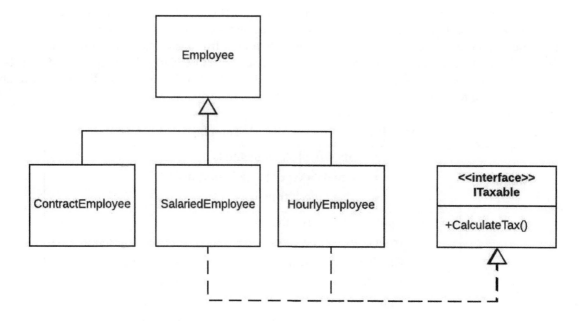

Step 4-1: Add ITaxable interface.

Right click the project in the Solution Explorer and select Add > New Item…

Under Visual C# Items > Code, select Interface and name it ITaxable.

The code changes are very small as shown in Figure 14-13. You just need to make the interface public (line 9) and add the method CalcualteTax in line 11.

Notice that the method isn't declared as public or abstract, even though it is, in fact, both. This is because all method declarations in interfaces can only be public and abstract, so those were deemed unnecessary to have to write.

Figure 14-13: ITaxable interface

```
1   using System;
2   using System.Collections.Generic;
3   using System.Linq;
4   using System.Text;
5   using System.Threading.Tasks;
6
7   namespace Payroll_Inheritance
8   {
9       public interface ITaxable
10      {
11          decimal CalculateTax();
12      }
13  }
```

Step 4-2: Implement ITaxable.

The two subclasses of Employee are both supposed to implement the interface according to Figure 14-12. Start with HourlyEmployee, and add ITaxable in line 9 as shown in Figure 14-14.

Once you do this, you will get an error message that HourlyEmployee doesn't implement ITaxable. To fix this, you need to implement the CalcualteTax method. This can be done by typing lines 42–50 in Figure 14-14, or you can start by choosing the Quick Action to implement the method and use that as a starting point.

Figure 14-14: HourlyEmployee implementing ITaxable

```csharp
 7  namespace Payroll_Inheritance
 8  {
 9      class HourlyEmployee : Employee, ITaxable
10      {
11          public decimal PayRate { get; set; }
12          private int minutesWorked;
13
14          public HourlyEmployee(string name, decimal payRate) : base(name)
15          {
16              this.PayRate = payRate;
17              this.minutesWorked = 0;
18          }
19
20          public void AddTime(int minutes)
21          {
22              minutesWorked += minutes;
23          }
24
25          public decimal HoursWorked()
26          {
27              return minutesWorked / 60.0m;
28          }
29
30          public override decimal GrossPay()
31          {
32              return HoursWorked() * PayRate;
33          }
34
35          public override string ToString()
36          {
37              string result = base.ToString(); //Call super class version of method
38              result += string.Format(". Hourly pay rate: {0:c}", PayRate);
39              return result;
40          }
41
42          /// <summary>
43          /// Tax calculation: For pay rates at or above $20 per hour, taxes are 40% of total pay.
44          /// For pay rates below $20, taxes are 30% of total pay.
45          /// </summary>
46          /// <returns></returns>
47          public decimal CalculateTax()
48          {
49              return (PayRate < 20) ? GrossPay() * 0.3m : GrossPay() * 0.4m;
50          }
51      }
52  }
```

It's time to practice! Do the following step:

Step 4-3: Implement ITaxable in SalariedEmployee.

The tax calculation for salaried employees is based on their annual salary. People who earn more than $65,000 will pay 45%, and those who make less will only pay 35% of their gross pay.

Because of this difference in how taxes are calculated, it isn't possible to provide a single implementation for both types of employees, so it has to be done in each one separately.

Step 4-4: Add ContractEmployee class.

Next, add the ContractEmployee class, which will *not* implement ITaxable. This class is very simple as shown in Figure 14-15. It inherits from the Employee class (line 9) and thus needs to implement the GrossPay method (lines 17–19). This method simply returns the value of the ContractAmount property. The constructor (lines 12–15) takes a name, which is passed to the super class, and a contractAmount, which is passed to the corresponding property. The ToString method is similar to ToString in the other subclasses of Employee, but uses ContractAmount instead of salary or hourly pay rate (line 25).

Figure 14-15: Code for ContractEmployee

```csharp
class ContractEmployee : Employee
{
    public decimal ContractAmount { get; set; }
    public ContractEmployee(string name, decimal contractAmount) : base(name)
    {
        this.ContractAmount = contractAmount;
    }

    public override decimal GrossPay()
    {
        return ContractAmount;
    }

    public override string ToString()
    {
        string result = base.ToString(); //Call super class version of method
        result += string.Format(". Contract amount: {0:c}", ContractAmount);
        return result;
    }
}
```

Step 4-5: Update the user interface.

Add two labels to show the total taxes for all taxable employees as shown on the right side of Figure 14-16.

Figure 14-16: User interface for taxable employees

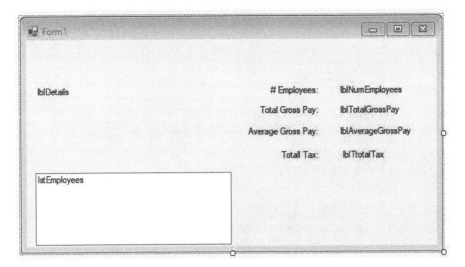

Step 4-6: Update the form code.

Switch to the code file for the form and add the private instance variable to hold a List of ITaxable objects as shown in Line 16 in Figure 14-17. Here you see how ITaxable can be used as a regular class in declaring the type of a variable. The taxEmployees List will be able to hold any object that implements ITaxable.

Add the TotalTax method shown in lines 48–56. This method runs a foreach loop over the objects in the taxEmployees List and calls CalculateTax on each one. Even though the list will actually contain objects of type HourlyEmployee and SalariedEmployee, the only method available is TotalTax, because the List is declared to hold ITaxable objects, and CalculateTax is the only method specified in that interface.

Finally, make changes to the constructor as shown in lines 18, 21, and 28 in Figure 14-17. These are fairly standard changes that allow for a List of taxable employees to be passed to the constructor, assigned to the instance variable, and have the label be updated with the total tax for all taxable employees.

Figure 14-17: Code for EmployeeInheritanceForm

```
11  namespace Payroll_Inheritance
12  {
13      public partial class EmployeeInheritanceForm : Form
14      {
15          private List<Employee> employees;
16          private List<ITaxable> taxEmployees;
17
18          public EmployeeInheritanceForm(List<Employee> employees, List<ITaxable> taxEmployees)
19          {
20              this.employees = employees;
21              this.taxEmployees = taxEmployees;
22              InitializeComponent();
23
24              lblDetails.Text = employeeDetails();
25              lblNumEmployees.Text = employees.Count().ToString();
26              lblTotalGrossPay.Text = string.Format("{0:c}", totalGrossPay());
27              lblAverageGrossPay.Text = string.Format("{0:c}", averageGrossPay());
28              lblTtotalTax.Text = string.Format("{0:c}", totalTax());
29
30              lstEmployees.Items.AddRange(employees.ToArray());
31          }
32
33          private decimal totalGrossPay()...
42
43          private decimal averageGrossPay()...
47
48          private decimal totalTax()
49          {
50              decimal result = 0;
51              foreach(ITaxable e in taxEmployees)
52              {
53                  result += e.CalculateTax();
54              }
55              return result;
56          }
57
58          private string employeeDetails()...
68      }
69  }
```

Step 4-7: Populate the data.

The final step in this tutorial is to populate the list of taxable employees. This is done in Program.cs and shown in Figure 14.18. What you will do is assign each of the employees to one of the now three types of Employee subclasses, and for salaried and hourly employees, also add these objects to a new list to hold all taxable employees. The taxable employees will thus be added to both the list of all employees as well as the list of taxable employees. You will need to make several changes as shown in the following.

Add a List of ITaxable objects (line 19).

Change the random number generator in line 25 to generate numbers between 0 and 2, since we have one more type of employee.

Modify the else in line 34 to an else if, and add the else section in line 40 to deal with contract employees.

For salaried and hourly employees, you will need to add them to the taxEmployee list, as shown in lines 32 and 38. Because the emp variable is declared to be an Employee, you will need to cast this to ITaxable in order to be able to add to the taxEmployees list.

In line 52, add taxEmployees as the second parameter in the call to the constructor for EmployeeInheritanceForm.

You can now run the program and see that the taxes are calculated and added to the user interface. You can set a breakpoint in the beginning of the Main method to follow the execution and see how the execution flows and which values are assigned where.

Figure 14-18: Code for Program.cs to populate taxEmployees list

```csharp
    static class Program
    {
        /// <summary>
        /// The main entry point for the application.
        /// </summary>
        [STAThread]
        static void Main()
        {
            Random r = new Random(); //Random number generator - to be used several times below
            List<Employee> employees = new List<Employee>();
            List<ITaxable> taxEmployees = new List<ITaxable>();
            string[] employeeNames = {"Superman", "Green Lantern", "Batman", "Spiderman",
                            "Thor", "Hal Jordan", "Wonder Woman", "Captain America"};
            foreach (string name in employeeNames)
            {
                Employee emp = null;
                int salOrHrl = r.Next(0, 3); //Random number between 0 and 2
                if (salOrHrl == 0) //Create hourly employee
                {
                    decimal payRate = (decimal)r.NextDouble() * 20 + 10; //Random number between 10 and 30
                    HourlyEmployee hourly = new HourlyEmployee(name, payRate);
                    hourly.AddTime(r.Next(30, 480)); //Add some random time for this employee
                    emp = hourly; //Assign hourly to the Employee variable that is added to the list later
                    taxEmployees.Add((ITaxable)emp); //cast to ITaxable, which will work because HourlyEmployee is ITaxable
                }
                else if(salOrHrl==1) //create salaried employee
                {
                    decimal salary = r.Next(40000, 80000); //Random number between 40,000 and 80,000
                    emp = new SalariedEmployee(name, salary);
                    taxEmployees.Add((ITaxable)emp);
                }
                else if (salOrHrl == 2)
                {
                    decimal contractAmount = (decimal)r.NextDouble() * 300 + 100; //Between 100 and 400
                    emp = new ContractEmployee(name, contractAmount);

                }
                employees.Add(emp); //Add the employee to the list

            }

            Application.EnableVisualStyles();
            Application.SetCompatibleTextRenderingDefault(false);
            Application.Run(new EmployeeInheritanceForm(employees, taxEmployees));
        }
    }
```

14.7 Using Interfaces to Sort Collections

Along with inheritance, interfaces play a major role in the .Net framework classes. As an example of this, we will take a look at a few ways to sort collections using implementation of interfaces.

At its most basic level, in order to sort something, you will need to know how to rank two elements in the collection relative to each other. Which is the larger of the two? The .Net framework bases its sorting of collections on the ability to be able to compare two objects of the same type to each other. This is facilitated by the interface called IComparable. This interface has one method called CompareTo, which takes the object of the type and compares it to the object that the method is called on. The return value from CompareTo then determines which object is ranked higher.

Once a class implements the IComparable interface, its elements can be sorted if they are contained in a collection like List, by simply calling the Sort method on the list. So, if you have a collection of strings, you can simply call the Sort method on that collection since String already implements IComparable:

```
List<String> colors = new List<String> { "red", "blue", "green" };
colors.Sort();
```

While we don't know how the Sort method works, it somehow uses the CompareTo method to establish the rank order between two of the objects in the list—and then does that repeatedly until the list is sorted. Thus, the CompareTo method is called like this:

```
obj1.CompareTo(obj2);
```

The output from the CompareTo method is an integer that indicates which of the objects are 'bigger'. Table 14-1 shows the output from the CompareTo method.

Table 14-1: Result of the CompareTo method

Comparison	Output
obj1 is bigger than obj2	Positive (1)
obj1 and obj2 are equal	Zero (0)
obj1 is smaller than obj2	Negative (-1)

You can make your own classes sortable by implementing the IComparable interface the CompareTo method. All you have to do is ensure that the method returns a positive, negative or zero integer based on the comparison. Imagine you had created a Message class that contained two properties, Text and Date, and objects of this class would be sorted by the Date. You could write code like that shown in Figure 14-19, creating a list of three messages. Then sorting the list would bring the list in order by date.

Figure 14-19: List of Messages being sorted

```
List<Message> messages = new List<Message>()
{
    new Message{Text="First message", Date = new DateTime(2018, 7, 10)},
    new Message{Text="Second Message", Date = new DateTime(2018, 8, 9)},
    new Message{Text="Third Message", Date = new DateTime(2019, 5, 3)},
    new Message{Text="Fourth Message", Date = new DateTime(2019, 5, 3)},
    new Message{Text="Fifth Message", Date = new DateTime(2019, 5, 3)}
};
messages.Sort();
```

The Message class itself would then look like Figure 14-20. As you can see, the CompareTo method returns an integer and takes in another Message as its argument. Then it compares the value of the Date properties in the two objects and returns 0, 1 or -1 depending on the outcome of the comparison.

Figure 14-20: Message class

```csharp
class Message : IComparable<Message>
{
    public string Text { get; set; }
    public DateTime Date { get; set; }

    public int CompareTo(Message other)
    {
        if (this.Date == other.Date)
            return 0;
        else if (this.Date > other.Date)
            return 1;
        return -1;
    }
}
```

While the code in Figure 14-20 isn't very complicated, it could be simplified by using the fact that the DateTime class itself implements IComparable, and thus implements the CompareTo method. This would allow for simplifying the Message class as shown in Figure 14-21. For most simple sort criteria, this is the preferred format, as the risk of making mistakes is far less.

Figure 14-21: Using CompareTo in the DateTime class

```csharp
class Message : IComparable<Message>
{
    public string Text { get; set; }
    public DateTime Date { get; set; }

    public int CompareTo(Message other)
    {
        return this.Date.CompareTo(other.Date);
    }
}
```

What if you wanted to sort by multiple criteria? For instance, say we wanted to sort messages with the same date, according to the text in the message. This can be done easily by expanding the if statement where the first search criteria are the same (the one that returns 0 in Figure 14-20), to compare the secondary sort criteria. An example of this is shown in Figure 14-22.

Figure 14-22: Multiple sort criteria

```
public int CompareTo(Message other)
{
    if (this.Date == other.Date)
        return this.Text.CompareTo(other.Text);
    else if (this.Date > other.Date)
        return 1;
    return -1;
}
```

Using the IComparable interface and the CompareTo method works well with objects that have a natural sort order, such as strings, which can be sorted alphabetically. However, with more complex objects that can be sorted in different ways, we need to use a different approach. Consider, for example, a Book class that contains the name of an author, a title and the copyright year of the book. Objects of such a class could be sorted according to any one of the three criteria (or perhaps by one criteria inside another criterion—such as sorting by author name within year). With such a class, instead of implementing IComparable, we can create a new class that implements the IComparer interface. This interface contains a single method called Compare, which works very similarly to CompareTo, except it takes two objects of the same class as shown in Figure 14-23.

Figure 14-23: IComparer class

```
class MessageDateComparer : IComparer<Message>
{
    public int Compare(Message m1, Message m2)
    {
        return m1.Date.CompareTo(m2.Date);
    }
}
```

Using this form requires instantiating an object of the class that implements IComparer and then passing that object to the collection's Sort method, like this:

```
messages.Sort(new MessageDateComparer());
```

Sorting by different criteria then involves creating a new class for each possible sort criteria and passing an object from that class to the Sort method.

Review Questions

14.11   Why is it enough for the Sort method to know how to rank order two objects?

14.12   What are the benefits and disadvantages of using the Sort method compared to writing your own sort algorithm?

Tutorial 5: Sorting Employees

In order to illustrate how to use interfaces to sort objects, you will add sorting capabilities to the payroll system you've been building. In this tutorial, you will add controls to allow for sorting the employees in the system. However, we first have to determine what it means to sort an employee. The abstract Employee class has one property, Name, and a method for Gross Pay along with a method, ToString, which returns a string representation of the object. Either of those three could be used to sort by, but we will just consider Name and Gross Pay for this tutorial.

Step 5-1: Make Employees comparable.

With two different items to sort on, we would have to determine which one to use—Name or Gross Pay. Let's arbitrarily choose Gross Pay. The first thing we would want to do is to make Employee objects comparable by Gross Pay. This is easily done by implementing the IComparable interface in the Employee class as shown in Figure 14-24.

Figure 14-24: Employee class implementing IComparable

```csharp
public abstract class Employee: IComparable<Employee>
{
    public string Name { get; set; }

    public Employee(string name)
    {
        this.Name = name;
    }

    public abstract decimal GrossPay();

    public override string ToString()
    {
        return "Name: " + Name;
    }

    public int CompareTo(Employee other)
    {
        //throw new NotImplementedException();
        return this.GrossPay().CompareTo(other.GrossPay());
    }
}
```

Start by specifying that the class should implement IComparable as shown in line 9. This will give you an error message that the class doesn't actually implement the interface. You can fix this by clicking the lightbulb and choosing Implement Interface. This will create the CompareTo method (lines 25–29).

Comment out the throw statement and add line 28, which takes advantage of the fact that the Decimal class also implements IComparable, and thus makes CompareTo available.

Finally, to ensure that we can see the Gross Pay in the output, change line 22 to also output Gross Pay.

Tutorial 5: Sorting Employees

Step 5-2: Set up Form.

Switch to EmployeeInheritanceForm and remove lblDetails. Make the whole form and the listbox wider. Select all the labels in the top right corner and move them to the right. Move the list box up add a button next to the Listbox as shown in Figure 14-25. Name the button btnSortGrosspay.

Switch to the code behind file and comment out the line in the constructor that populates lblDetails.

Figure 14-25: Form for sorting

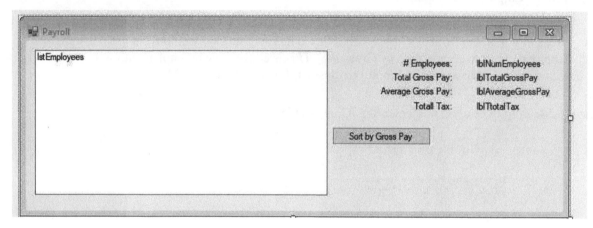

Step 5-3: Implement sorting.

Double click the button and implement the action method as shown in Figure 14-26. This code is straightforward, starting with calling the Sort method on the list of employees. The Sort method will invoke the CompareTo method in the Employee class as it works to rearrange the objects in the list.

Figure 14-26: Code to sort the employees list and add to ListBox

```
private void btnSortGrosspay_Click(object sender, EventArgs e)
{
    employees.Sort(); //Sort the list according to natural sort order
    lstEmployees.Items.Clear(); //Clear list box
    lstEmployees.Items.AddRange(employees.ToArray()); //Repopulate list box
}
```

Step 5-4: Run the application.

When you run the program, and click the Sort by Gross Pay button, you will see the output as shown in Figure 14-27. Since the salary numbers are randomly generated, your specific numbers will be different, but they should be sorted in ascending order.

Figure 14-27: Form for sorting

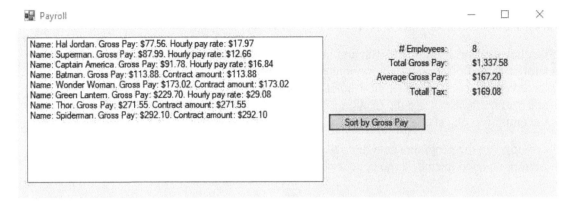

It's time to practice! Do the following step:

Step 5-5: Sort by Name.

> Change the implementation of ComnpareTo in the Employee class to sort by Name instead of GrossPay. Then run the program to observe the changes when clicking the button. Be sure to change the implementation of CompareTo back before continuing.

Step 5-6: Sort by multiple criteria.

> The current implementation of sorting has a significant problem; we can only sort by a single criterion at a time, and switching between criteria requires changing the code. That's obviously not a good situation, so we will look at how to allow for multiple sort criteria without having to update the code.

> To do this, you start by adding a new class to the project, named NameComparer. The code for the class is in Figure 14-28. Start by specifying that the class should implement the IComparer interface as shown in line 9.

> You can then use the lightbulb to implement the interface, which will generate the Compare method in lines 11–14. Notice how this method is similar to the CompareTo method, but instead of taking a single Employee as a parameter, it takes two.

> The implementation of the Compare method in line 13 is also similar to what you've seen before. It relies on the CompareTo method, that is already implemented in the String class to compare the Name of Employee x to the Name of Employee y.

Figure 14-28: NameComparer class.

```
9       class NameComparer : IComparer<Employee>
10      {
            1 reference
11          public int Compare(Employee x, Employee y)
12          {
13              return x.Name.CompareTo(y.Name);
14          }
15      }
16  }
```

Step 5-7: Set up the user interface.

Add a button to the user interface below the Sort by Gross Pay button that will invoke sorting by name. Add the text Sort by Name and give it the ID btnSortName.

Step 5-8: Use the NameComparer class to sort.

Double click the Sort by Name button and implement the code as shown in Figure 14-29. This is very similar to the code you wrote previously to sort by gross pay. The only difference is in line 78, where you create an object of the NameComparer class and pass to the Sort method. This way, the Sort method will use the Compare method in NameComparer when determining how to rank the employee objects.

You can now run the program again and see that you can sort by different criteria by clicking each button; no need to change the code to re-sort.

Figure 14-29: NameComparer class.

```
76      private void btnSortName_Click(object sender, EventArgs e)
77      {
78          employees.Sort(new NameComparer()); //Pass NameComparer object to the Sort method
79          lstEmployees.Items.Clear();
80          lstEmployees.Items.AddRange(employees.ToArray());
81      }
```

It's time to practice! Do the following step:

Step 5-9: Add another IComparer class.

Change the sorting by gross pay to also use an IComparer object.

Step 5-10: Reverse Sort.

Now that you have the ability to sort the list, it is really simple to reverse the sort. To illustrate how to do this, add another button to the interface, named btnReverse with the text Reverse Sort. Add the code as shown in Figure 14-30.

Figure 14-30: NameComparer class.

```
83      private void btnReverse_Click(object sender, EventArgs e)
84      {
85          employees.Reverse(); //Reverse the order of elements in the list
86          lstEmployees.Items.Clear();
87          lstEmployees.Items.AddRange(employees.ToArray());
88      }
```

Running the program now, you will be able to reverse whatever sorting has been applied.

14.8 Access Modifiers in C#

One of the benefits of object-oriented programming is that it allows for encapsulating data and only exposing a public interface of available operations on any given object. You have seen that methods and properties can be marked private or public to distinguish between what can only be accessed from within the containing unit or also from the outside. Thus, anything declared as public makes up the public interface for the class, whereas things that are declared as private are hidden (or encapsulated). This allows for protecting data from being inadvertently changed in unanticipated ways. For example, a class might contain a counter that indicates how many times something important happened. If the counter value was publicly available, it could be set to some value that didn't match the actual number of times the event happened. By marking it private, it is easier to protect the value.

Over time, as object-oriented languages have evolved, language designers discovered that only being able to specify an attribute or method as public or private didn't always provide sufficiently granular levels of access control. For example, there are situations where subclasses could be granted closer access than other classes outside the inheritance hierarchy. Similarly, classes within the same project can often also be granted wider access than classes in other parts of the system.

So, in C#, we have a few more access modifiers available to us, including *internal* and *protected*. The internal modifier is used to allow access to members within the same Visual Studio project (sometimes also called a component). This could be useful in large systems made up of multiple closely related classes. For example, in a system controlling an autonomous vehicle, you might have several components such as engine control, steering, navigation and so on. Within each component, there are likely operations that are used by other classes in the component that shouldn't be used directly from other components. Those operations can be marked as internal and only be available to other classes within the component.

The *protected* access modifier is used to allow for access to a super class from a subclass. The default in C# if you do not provide an access modifier is protected.

Figure 14-30 shows all the access modifiers in the context of the simplified payroll system we have been building. All the access modifiers are relative to a property in the Employee class. If something is declared as private in Employee, it will only be available inside that class. However, if something is declared as protected, it is also available to methods in the HourlyEmployee and SalariedEmployee classes since they are subclasses of Employee. Anything declared as internal in Employee will also be available to other classes in the Payroll system, whereas classes in other systems, such as the BillingSystem, only have access to items declared as public in Employee.

Figure 14-31: Access Modifiers in C#

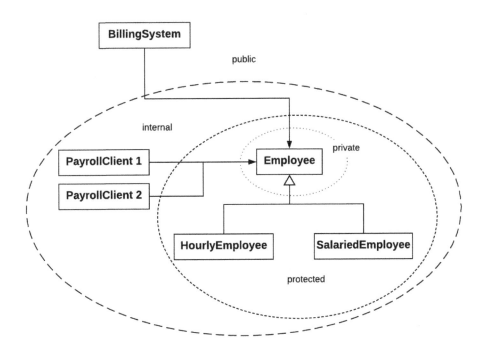

While inheritance and polymorphism are powerful concepts, in most business systems these concepts play a fairly small role. Most business problems do not lend themselves particularly well to this kind of modeling. In fact, if you find yourself modeling your problem domain with multiple levels of inheritance and interfaces, you should consider whether there are more straightforward approaches that rely less on inheritance. However, both inheritance and interfaces play a big role in the .Net framework classes, so it is vitally important to understand how they work and to be able to apply the principles.

Exercises

Exercise 14-1

Draw a class diagram for the following classes to be used in an animal simulation program: Bird, Dog, Parakeet, Hippopotamus, Animal, Snake, Cow, Canine, Lion, Mammal, Vulture.

Exercise 14-2

Come up with another five animal classes you can add at different levels to the diagram you produced in Exercise 14.1.

Exercise 14-3

Come up with a few pieces of data to store in each class from Exercise 14.1.

Exercise 14-4

Come up with a few methods that can model appropriate behavior in each class from Exercise 14.1.

Exercise 14-5

Consider the diagram you created in Exercise 14.1. How can you add the concept of a Pet to this diagram?

Exercise 14-6

Consider the Auto Repair system from Exercises 13.3–13.5. Add two classes as subclasses of Repair: ListRepair and TimeAndMaterialRepair. A ListRepair is a predefined repair that has a fixed price, so it is similar to Repair in that it simply has a description and an amount. The TimeAndMaterialRepair instead has an amount that is calculated based on the time that a mechanic works on a car and any materials used in the repair. Add forms to allow for working with both ListRepair and TimeAndMaterialRepairs, and associate them with specific cars. By keeping an Amount property in the Repair class, you should not need to modify any of the code calculating statistics related to the amount for each repair. Each of the two classes should also include overridden ToString methods that provide some specifics of the detail of each object.

Exercise 14-7

What is the output from running the Main method in the Program class?

```
public class A
{
    public int X { get; set; }

    public void m()
    {
        Console.WriteLine("m in A");
        X++;
    }
    public void p()
    {
        Console.WriteLine("p in A");
        X++;
    }
    public void q()
    {
        Console.WriteLine("q in A");
        X++;
    }
}

public class B : A
{
    public int Y { get; set; }

    public void m()
    {
        Console.WriteLine("m in B");
        Y++;
    }
}

public class C : B
{
    public int Z { get; set; }
    public void p()
    {
        Console.WriteLine("p in C");
        Z++;
    }
}
```

```
public class Program
{
    public static void Main(String[] args)
    {
        C c = new C();
        A a = new B();

        c.p();
        c.m();
        c.q();
        a.m();
        a.q();
        a.p();
        Console.WriteLine(c.X +" "+ c.Y +" "+ c.Z);
    }
}
```

Exercise 14-8

Create a Person class with two attributes: Name and Age. Create several objects of this class and add to a collection. Implement sorting by each of the two attributes to sort the collection in both ascending and descending order. Add the ability to sort the objects by Name within Age (if multiple people have the same age, sort them by name).

Chapter 15

Introduction to Web Applications Development

In previous chapters you used C# to build Windows form applications, which are installed on individual client computers and run on Windows. This chapter applies the knowledge you gained earlier to develop web applications that are installed on a web server and run on browsers on client computers. You will build web applications using the ASP.Net platform, which is part of the .Net Framework, in the familiar Visual Studio environment.

Learning Objectives

After studying this chapter, you should be able to:

- Describe the terms *web server, HTTP request and response, Web Forms* and *MVC approaches*, and *code-behind file*.
- Describe the differences between web application and Windows form application, static and dynamic web pages, client-side and server-side scripts, stateless and stateful modes, File System and Local IIS (HTTP) websites, HTML controls and web server controls.
- Create a web page with web server controls.
- Use C# code-behind files to respond to events on a web form.
- Validate data using Validation controls.

Topics

15.1	Introduction to Web Applications	15.4	Working with Web Server Controls
15.2	Creating a Single-Page ASP.NET Website	15.5	Validating Data using Validation Controls
15.3	Introduction to HTML		

15.1 Introduction to Web Applications

Windows form applications like those you developed in previous chapters are installed and run on client computers with the Windows operating system. Hence, to access a Windows form application from a computer, the application must be installed on that computer.

Web applications, on the other hand, are installed and run on a server(s) and are accessed through a browser from clients like laptops, smartphones and tablets over the Internet. Figure 15-1 shows the components of a simple web application that consists of clients, the Internet and a **web server** that stores, maintains and delivers web pages to the clients. To carry out these functions, the server is configured with *web server software* like the Microsoft Internet Information Server (IIS), Apache or Nginx. The term *web server* is often used to refer to the server hardware and/or the software.

Because web applications are accessed through a browser, the clients don't have to be Windows machines.

Figure 15-1: Components of a web application

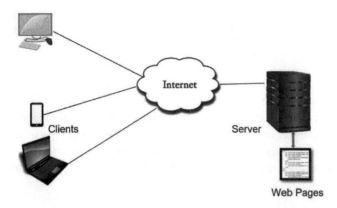

How Does a Client Display a Web Page from the Server?

From a user's perspective, displaying a web page on a client is simple. You type the URL of the page into a browser, select a bookmark, or click a link. For instance, the following URL would display the *ordering* page on the **domain** *prospectpressvt* within the top level domain *.com*:

    http://www.prospectpressvt.com/ordering

The first two components of the above URL are:
- http: The **web server protocol** used by the client and the server to communicate with each other. Other types of protocols include SMTP (e-mail server protocol) and FTP (File Transfer Protocol).
- www The default **host name**. If you don't specify a host name, it is assumed to be www. Thus, *yahoo.com* is the same as www.yahoo.com. However, *finance.yahoo.com* finds the domain *yahoo.com* on the host *finance*.

Now let's look at what happens behind the screen when you display a web page. We will start with the simpler case of **static** pages, and then look at the more general **dynamic** pages.

Static and dynamic web pages

A **static** web page remains the same each time it is displayed, unless it is manually altered. Thus a web page that displays product prices would be a static page if the prices are "hard-coded" into the contents of the page. However, it would be a dynamic page if the prices are read from a file/database and incorporated into the web page each time it is displayed. A dynamic web page typically is created on-the-fly when the page is requested. Dynamic pages may also accept user input and process it.

In addition to static and dynamic web pages, a website might contain additional static files like .css files that contain formatting information, .js (JavaScript) files that contain client-side scripts and .cs files containing server-side C# code.

Displaying static web pages

To understand how static web pages are displayed, let's look at how a static page is stored on the server. Though the browser displays a web page with formatted text and graphics, a static web page is stored on the server as HTML markup that specifies the contents and format of the web page. Note that HTML is referred to as a markup language rather than a programming language, because it is used primarily to control the presentation of information. The files that contain the HTML markup generally have .htm or

.html extensions. The HTML code for a static page typically includes the text displayed on the page. As an example, here is a segment of the text displayed on the Prospect Press website:

> **ETextbooks:**
>
> DIRECT-TO-STUDENTS: Prospect Press sells eTextbooks directly to students via Redshelf and Vital Source. To order a title, go to the Prospect Press TITLE page and click on the version you want.
>
> Or visit Red Shelf at http://redshelf.com or Vital Source at http://bookshelf.vitalsource.com/, search for the title, and order there. See below for a comparison of Vital Source and Red Shelf:

The corresponding HTML markup looks like this:

```
<p><strong><u>ETextbooks</u></strong><strong>:</strong></p>
<p>DIRECT-TO-STUDENTS:  Prospect Press sells eTextbooks directly to students via Redshelf and Vital
Source. To order a title, go to the Prospect Press TITLE page and click on the version you want.</p>
<p>Or visit Red Shelf at <a href="http://redshelf.com"
target="_blank">http://redshelf.com</a> or Vital Source at <a
href="http://bookshelf.vitalsource.com/">http://bookshelf.vitalsource.com/</a>, search for the title,
and order there.  See below for a comparison of Vital Source and Red Shelf:</p>
```

We will discuss the structure of HTML later in this chapter.

What happens when you click on a link or type in a URL and press the Enter key to display a static web page? The browser on the client opens the connection to the server and sends a **request** (**HTTP request**, to be more precise) to the web server, which essentially tells the web server, "*Hey, get me the file specified in this URL.*" The server retrieves the file that contains the HTML for the page and any related files, and sends a **response** (**HTTP response**) that includes the files back to the browser. Thus a static page is delivered to the browser exactly as stored on the server.

When the browser receives the response, it closes the connection, creates the web page as specified in the HTML code and displays it. The user won't be able to modify or interact with the static page displayed in the browser. The combination of request and response is called a *transaction*. Figure 15-2 shows the process.

Figure 15-2: HTTP request and response

Displaying dynamic web pages

The process of displaying dynamic and static web pages involves an HTTP request by the client and a response from the server. However, displaying dynamic pages involves additional processing on the server because they are created on demand. That is, segments of the page are generated dynamically using **scripts** embedded in HTML or stored in a separate file(s). The scripts may do tasks like looking up information from a database, displaying dynamic data like inventory levels and account balances and customizing the web page. So, a dynamic page may look different each time it is displayed, depending on various factors like who the user is, the date/time, changes in the data read by the script and the parameter values passed through the script.

What is a script? A script is essentially a type of code that is written in interpretive scripting languages like JavaScript, PHP and CGI, to control the behavior of applications. For instance, you may validate input data on a web form using a script written in JavaScript and executed by the browser. A program, on the other hand, is code written in programming languages like Java and C# that are compiled. Scripts are easier to develop, but slower to run. In addition to code written in scripting and programming languages, web pages also use code written in markup languages like HTML and CSS (Cascading Style Sheet), which are described later.

It should be realized that the difference between scripting and programming languages is becoming blurred. Though, strictly speaking, the term *script* means code written in interpretive languages, script is generally used to represent any code that works with a web page, including code written in a programming language like C#.

There are two approaches to running scripts. If the script is run on the server, it is called *server-side scripting* or *server-side coding*. If the script is run on the browser, it is called *client-side scripting* or *client-side coding*. Typical web applications today use both server-side and client-side coding.

Server-side scripting

In the ASP.NET environment, a single server plays the role of a web server (that handles HTTP requests), as well as an application server that provides an environment to execute script and create dynamic web pages. In addition, it becomes a database server by running SQL Server DBMS on it. However, generally each of these three functions could be performed on a separate physical server to improve security and performance.

When a web server receives a request for a web page that includes a server-side script, the server executes the script, generates the HTML code for the web page and sends it to the browser. Because the server-side script is run on the server and only the result is sent to the browser, server-side scripting has the advantage that the users won't be able to view the script in the browser. However, it puts additional burden on the server.

Server-side scripts typically include ASP.NET code and code written in languages like ASP.Net, PHP, CGI and JSP, which are not compiled, and programs written in programming languages like C#. It should be noted that ASP.NET is a framework for developing web applications, and not a programming language. In this chapter, you will use IIS (Internet Information Services) Express as the web server, ASP.NET as the application server to build ASP.NET applications and SQL Server Express as the DBMS for the database server.

Client-side scripting

A client-side script could be embedded in the HTML code for a page or in a separate file. But, it is run on the client (in the browser) when the browser receives the web page from the web server. That would mean you can also view the script by displaying the source code of a page in the browser. JavaScript is currently the most popular language for client-side scripting.

Client-side scripts are typically used to make web forms more interesting, user-friendly and interactive. Applications of client-side scripts include data validation, running multimedia and creating gadgets like calculators. Running client-side script is simpler because it doesn't involve the application server.

Stateless and stateful modes

HTTP is a stateless protocol. That means as soon as a client receives a response for its request, the connection between the client and server is terminated, and the server keeps no information about the client. So, if the same client makes another request, the server has no way to know whether it is the same client or another client. The stateless mode of communication between the server and the client is fine for applications involving only simple static pages. However, it does not meet the requirements of today's typical dynamic applications.

ASP.Net provides several features to maintain the state, including view state, session state, application state and server-side caching. You will learn more about view state and session state later. In the stateful mode, the server keeps track of information such as who the user is and what he does from one page to the next until the user logs out.

ASP.NET Environment

ASP.NET is an application server platform to build web applications. It is based on the .NET Framework that you have used in previous chapters; so, all .NET Framework features like type safety, inheritance and compatibility with .NET languages are available for ASP.NET development. In addition to Visual Studio, you may use the free standalone product, **Visual Studio Community**, to develop ASP.NET applications.

As described earlier, ASP.NET applications are typically developed in a network environment using an Internet or Intranet. However, you may use a standalone development environment where a single computer serves as the client and server. In the standalone environment, you may use the *IIS Express web server* that comes with Visual Studio as the web server and **SQL Server Express** as the database server.

Review Questions

15.1   What is the domain name in the URL: https://www.createspace.com? What is the host name?

15.2   How does a dynamic web page differ from a static web page?

15.3   What is an HTTP request?

15.4   What are two differences between server-side and client-side scripts?

15.5   What is the drawback of the stateless mode of communication between the server and the client?

15.2 Creating a Single-Page ASP.NET Website

In Visual Studio, a website is referred to as a web project. You can build two types of web projects: *Web Application Project*, which is the older ASP.NET project type, and *Web Site Project*, which was introduced later with ASP.NET 2.0. The process of developing web pages and code is essentially the same in both types of projects. So, everything you learn in this chapter can be applied to both types. However, there are some differences between the two models in how the project is compiled and deployed.

In a Web Application Project, all class files, including code-behind files, are pre-compiled and deployed as a single assembly on the server. In a Web Site Project, typically the source code is stored on the server and compiled dynamically at run time, which takes extra time and makes the code less secure. However, Web Site Projects make it easier to update and test parts of the website.

In this book, we develop websites using the *Web Application Project* model, which is the model recommended by Microsoft for new development. The terms **web application, web project** and **website** are used interchangeably to refer to websites.

Typically, an ASP.NET web project is developed in a network environment involving a server that is configured with IIS as the web server software. However, it is more convenient to develop a website in a standalone environment where a single computer serves as both the client and the server. The process of developing websites is essentially the same in both environments. For the sake of convenience, you will use the **standalone environment**.

Tutorial 1: Developing a Simple Web Page: Ice Cream Cost

This tutorial develops a web page corresponding to the Ice Cream Cost Windows form, shown in Figure 15-3, which you created in Tutorials 1 and 2 in Chapter 1. The user interface for the web page is similar, except that there is no Exit button:

Figure 15-3: The Ice Cream Cost Windows form

You will use the familiar Visual Studio environment to create a website and then add a web page to it. Before you create the website, you need to set the Visual Studio settings to *Web Development* to optimize the environment for web development.

> **Step 1-1:** Set Visual Studio settings to *Web Development*:
> Open Visual studio and select **Tools** > **Import and export settings**.
> On the *Welcome to the... Wizard* page, click **Reset all settings** > **Next**.
> On the *Save Current Settings* page, click **Yes, save my current settings** > **Next**.
> On the *Choose a Default Collection of Settings* page, select **Web Development** > **Finish**.

If you need to restore the saved settings, use the **Import selected environment settings** menu item.

Step 1-2: Create the website:
   To create a web application project that contains a default web page or no web page, select
      File > New > Project
   Or, select **New Project** link on the *Start Page*.
   (**Do not** select New > Web Site, which would create a Web Site project, described earlier.)
   You will see the New Project window, similar to the one shown in Figure 15-4.

Figure 15-4: New Project window

Select **Installed > Templates > Visual C# > ASP.NET Web Applications**.
Enter **IceCream** for the *Name* field.
Create a folder named **ASPwebsites** on your drive.
The default location of projects is MyDocuments\Visual Studio 20xx\Projects.
Click *Browse* and select ASPwebsites folder for *Location*.

You also may change the default location as follows:
   Select **Tools > Options**.
   Expand **Projects and Solutions**, and select **General**.
      Change **Project Location** to **ASPwebsites**.
Make sure the CheckBox for **Create directory for solution** is checked.
Click *OK*.
You will see the New ASP.NET Project window, shown in Figure 15-5, which lets you select a template for your project.

Templates for web projects

Figure 15-5 shows the different templates available to create a web project. The **Empty** template and the **Web Forms** template essentially use the same approach called *Web Forms* approach to develop web pages. Note that a web page developed using the Web Forms approach is also called a **web form** or **Web Form page**. Web Forms approach to development, similar to Windows Forms development, uses a design surface and drag-and-drop controls (rather than writing HTML code) to rapidly build rich user interfaces. A web form may contain HTML markup, CSS, server-side scripts and client-side scripts. Ease of development is the key advantage of the Web Forms approach. However, ease of maintenance, reuse of code and shared development of a page by multiple developers may suffer due to the tight coupling between the code and the use interface.

Figure 15-5: Templates for projects

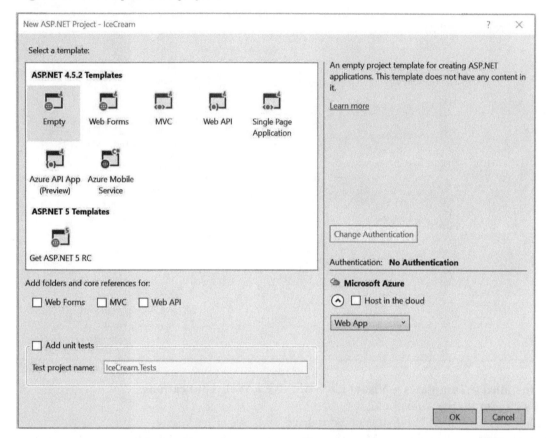

You will use the **Empty** template in this book because of its simplicity. The Empty template, as the name indicates, creates a web project with no web pages, whereas a project created using the Web Form template includes a default page and folders/files needed to provide additional functionality.

You may select the Empty template and still add a subset of the folders and references that are included in the Web Forms template by checking the *Web Forms* CheckBox in the bottom section entitled *Add folders and core references for*. You will use this option for the projects in the next chapter. The rest of the templates are not used in the projects you create in this book. A brief description of the commonly used templates follows.

The **MVC** template supports the ASP.NET MVC (Model-View-Controller) software design architecture that divides an application into three different components:
      Model: the business entity, including the data, represented in the application
      View: data as presented in the user interface
      Controller: the component that maintains the model and the view

This approach makes it easier to do *separation of concerns* (i.e., separation of different aspects of software functionality like business logic, user interface and data access) that enhances maintenance and reuse of code and shared development where different developers can work on the model, view and controller, simultaneously. So, it is particularly suitable for large projects involving multiple developers. Further, this approach provides some performance advantages and gives full control over generated HTML markup.

The **Web API** template lets you create and consume Web API (Application Program Interface) Services. The **Azure API App** extends the capabilities of Web API using Microsoft's Azure cloud platform, and the **Azure Mobile Service** helps mobile developers to take advantage of the Azure platform.

The **Single Page Applications** (SPAs) helps create highly interactive applications that include a single page. Rather than sending new pages to the client, the server sends data that dynamically updates the single page based on user interactions.

Step 1-3: Select a template for the project:
Select **Empty** template. Click OK.

If you open ASPwebsites folder, you will see a new folder named IceCream that holds the Solution file (IceCream.sln) and the project folder, also named IceCream, which holds configuration files and files that store information on web pages.

Adding a Web Form to a Website

A web form typically includes two types of code: the C# (or other compatible language) code that deals with the application logic and the HTML and CSS that deals with the user interface. The C# code is generally stored in a separate file called a **code-behind file**, which has the extension .aspx.cs (or .aspx.vb for Visual Basic, and so on). C# code also may be embedded in the HTML as **in-line code**.

The two-file system, which uses a separate code-behind file, is recommended because it separates the application logic from the user interface, making it easier to understand and maintain the code.

Step 1-4: Add a web form to the website:

Right click the project name IceCream (not the Solution, IceCream) in the Solutions Explorer.
Select **Add > Add New Item > Web Form**.

The Add New Item window appears. Change the name of the web form to **Default.aspx**, as shown in Figure 15-6. The aspx file stores the ASP.NET and HTML codes that specify the form. In addition to the Default.aspx file, there are two other files: the code-behind file, Default.aspx.cs, which contains the C# code, and Default.aspx.designer.cs, which stores certain code generated by Visual Studio.

An alternative is to select **Add > Add Webform**, which doesn't give you the option to add items other than web forms.

Figure 15-6: Add New Item window

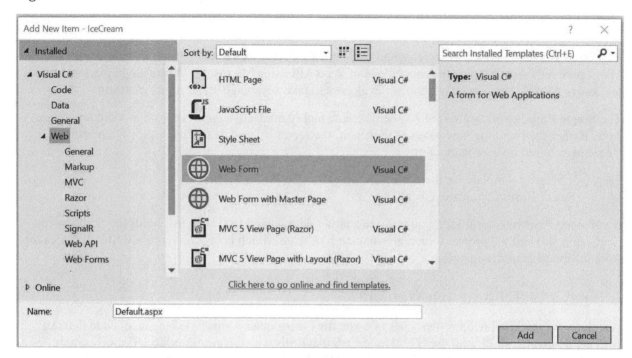

When you add a new web form, Visual Studio displays the **Web Forms Designer** window in **Source** view that shows the source code for the page, which is generated by Visual Studio and stored in the **Default.aspx** file. Figure 15-7 displays the code. For convenience, we will refer to the source code within the Default.aspx file as "**aspx code**."

You may change the view to **Design** or **Split** by clicking the corresponding buttons at the bottom of the window. You may display the C# code window by clicking the Default.aspx.cs tab at the top of the window.

Note that the ASPwebsite\IceCream\IceCream folder now includes the additional files Default.aspx, Default.aspx.cs and Default.aspx.designer.cs.

Figure 15-7: The Designer window in Source view

To help familiarize you with the code in the aspx file, we will take a look at certain basic aspects of HTML in the next section.

Closing and Opening a Website

To close a website, select **File > Close Solution**.

To open a website from Visual Studio,
  Select **File > Open > Project/Solution**, or, **File > Open Project** in VS Community.
  (**Do not** select Open > Web Site.)
  Select the Solution file, IceCream.sln.
  You also may select the Solution file from the list displayed by selecting
  File > Recent Projects/Solutions.

When Visual Studio is closed, you may double click the Solution file IceCream.sln from the folder where it is stored (ASPwebsites/ IceCream).

Review Questions

15.6   What is the difference between the *Web Forms* approach and *MVC* approach to developing web pages?

15.7   Describe the differences between Web Application project and Web Site project.

15.8   How does code-behind differ from in-line code?

15.3 Introduction to HTML

Essentially, HTML (Hypertext Markup Language) is used to tell the browser how to structure and present content in a web page. A closely related language is **XHTML** (EXtensible HTML), which is a version of HTML with stricter rules on the structure to conform to **XML** (EXtensible Markup Language). XHTML is almost identical to HTML, and it is supported by all major browsers.

HTML is fairly easy to understand. It is made up of **elements**, which start with an **opening tag (start tag)** and end with a **closing tag (end tag)**, with contents, if any, in between. Here is an example of the **title** element of an HTML document:

    <title>Ice Cream Cost</title>

Here, <title> is the opening tag, "Ice Cream Cost" is the content and </title> is the ending tag.

Opening tag: The opening tag consists of the name of the element and optional attributes, enclosed in a pair of angle brackets. In this example, *title* is the name of the element.

Closing tag: The closing tag is similar to the opening tag, except that there is a forward slash before the name of the element.

Content: The content, if any, is the content of the element.

HTML is **not** case sensitive. So the element name *TITLE* is the same as *title*. However, XHTML is case sensitive and requires tag and attribute names to be in small letters.

It should be noted that the term tag is often used in place of element, which could be confusing. We will use the term tag to mean opening or closing tag, not element.

There is a large variety of elements to specify various aspects of the contents, like the appearance, emphasis and importance. The paragraph (<p>) and bold (<b>) elements are two examples:

Paragraph (<p>)

Markup: <p> Figure 15-7: The Designer window in Source view</p>
Output: Figure 15-7: The Designer window in Source view

Bold (<b>)

Markup: <p>Figure 15-7 displays the initial <b>aspx</b> code </p>
Output: Figure 15-7 displays the initial **aspx** code

Elements may be nested as illustrated in the previous example, where the <b> element is nested inside the <p> element.

Note that HTML5, which is the current version of HTML, also has the **<strong>** element to indicate strong importance. Though <strong> and <b> have the same effect on the content, web crawlers recognize the importance of the content of <strong> element. Similarly, there is an *italics* (<i>) element and an *emphasis* (<em>) element, both of which have the same effect, but <em> indicates emphasis. The general trend in HTML is to specify meaning rather than look of the content.

Note that some elements may not have an end tag.

Attributes of an element

An opening tag of an element may contain one or more attributes that provide information about the element. The syntax of an attribute can be seen in the following example of a hyperlink element (<a>) that is used to create a link to another web page.

<a href="http://www.prospectpressvt.com/">Prospect Press</a>

The attribute name (href) is followed by an equal sign and the value of the attribute (http://www.prospectpressvt.com) inside quotes.

Appendix D lists an extensive (but incomplete) set of commonly used HTML tags to give you an understanding of the type of elements.

Aspx and HTML

Figure 15-7 presented the initial source code, referred to as aspx code, which was generated by ASP.NET when you added the Web Form to the website. The code is copied here.

```
Default.aspx.cs    Default.aspx*
    <%@ Page Language="C#" AutoEventWireup="true" Code

<!DOCTYPE html>

<html xmlns="http://www.w3.org/1999/xhtml">
<head runat="server">
    <title></title>
</head>
<body>
    <form id="form1" runat="server">
    <div>

    </div>
    </form>
</body>
</html>
```

The aspx code essentially specifies the web page. All files with the .aspx extension, like the Default.aspx file, are processed on the server when the page is loaded.

The aspx code contains two types of code:
1. Code that is processed on the server, called ASP.NET code, which the browser doesn't understand. The first line of code, which starts with "<%@ Page," is an example of such code, as described shortly. ASP.NET code is typically translated (rendered) by the server to produce HTML markup that the browser understands.
2. Code that is **not** processed on the server. This code consists of HTML markup, CSS and any client-side script, which the browser understands. So, the browser can process it directly without having the server translate it.

@Page directive

The first element in the aspx code is an **ASP.NET directive** called the @Page directive, which is a special type of ASP.NET code that provides information for the server:

    <%@ Page Language="C#" AutoEventWireup="true" CodeFile="Default.aspx.cs" Inherits="_Default" %>

The *Language* attribute says C# is the language, *AutoEventWireup* enables the event-driven feature, *CodeFile* says Default.aspx.cs is the **code-behind file** (a separate file that contains the code in C# or other languages), and *Inherits* specifies the _Default class. Because directives are information for the server, no corresponding markup is sent to the browser.

HTML document

Other than the @Page directive in the first line, the rest of the aspx file is an **HTML document**.

General structure of an HTML document

The general structure of an HTML document has the pattern shown in Figure 15-8.

Figure 15-8: Structure of an HTML document

```
<!DOCTYPE html>
<html>
    <head>
        <title>Title of the page</title>
        <meta attributes>
    </head>
    <body>
        (Contents of the page)
    </body>
</html>
```

DOCTYPE declaration

The HTML document starts with the **DOCTYPE** declaration, which tells the browser that the version of HTML used is html, which means HTML5.

    `<!DOCTYPE html>`

html element

The entire HTML code is wrapped in the **html element** as follows:

```
<html xmlns="http://www.w3.org/1999/xhtml">
    (contents)
</html>
```

The start tag includes an attribute, **xmlns,** which is optional in HTML5, but required in XHTML to specify the XML namespace. So, you may delete this parameter without having an effect on the page.

Head element

The head element contains information about the page that is not displayed in the page for the user. The head must contain the **title element**, which specifies the title that is displayed on the browser tab and in search engine results.

The head may contain formatting styles and additional **meta elements** to specify meta data, which is information about the document, like the name of the author, key words and description that appears in search engine results. The meta element includes the *name* and *content* attributes.

The following head element shows the runat="server" attribute, the title element and examples of some common meta elements.

```
<head runat="server">
    <title>Ice Cream Cost</title>
    <meta name="description" content="Compute Ice Cream Cost">
    <meta name="keywords" content="web form, Compute Cost, Ice Cream Cost">
    <meta charset="utf-8"
</head>
```

The runat="server" attribute

The **runat** attribute specifies that the code is run on the server. Specifying the runat attribute for the head makes the head accessible by code on the server so that changes can be made to the head using code. The runat attribute is required for web server controls, as you will see in the body element.

Step 1-5: Insert the title, "Ice Cream Cost" in the title element, so that it looks like this:

```
<title>Ice Cream Cost</title>
```

Body element

The body contains everything you see on the web page. It contains the form element that includes two parameters, id and runat, as illustrated here.

```
<body>
    <form id="form1" runat="server">
        <div>

        </div>
    </form>
</body>
```

The parameter **id** specifies the name of the form, and **runat** specifies that the form is processed on the server so you can refer to the form and the controls on the page using server-side code. The form element for the empty page doesn't have any content other than the div element.

A form element is required if a page contains web server controls. The server controls must be placed inside the form element. Though other HTML markup may be placed outside the form element, it is recommended that all content displayed on the page be placed inside.

The div element specifies a generic section to group elements. HTML5 also provides more meaningful new elements like header, footer, nav, article, figure and section. You may delete the div element because it doesn't serve any purpose in this application.

Step 1-6: Delete the div element in the Source view so that the body element appears as follows:

```
<body>
    <form id="form1" runat="server">

    </form>
</body>
```

Review Questions

15.9   In general, what are the components of an HTML element?

15.10   What is the difference between ASP.NET code and standard HTML markup that is part of the HTML language?

15.11   What is the function of the head element of an HTML document? Name two contents of the head.

15.12   What is the effect of specifying the runat="server" attribute for the form element?

15.4 Working with Web Server Controls

Working directly with the aspx code in Source view to add controls and to format the page gives you more control over the design of the page. Since the focus here is ASP.NET and C# programming, for the most part, we will take the easier path of using the drag-and-drop method of adding controls in the Design view, and let Visual Studio generate the code for the design you created. We will, however, examine the generated code.

Let's develop the Ice Cream Cost application in Design view, as shown in Figure 15-9.

Figure 15-9: The IceCreamCost form in Design view

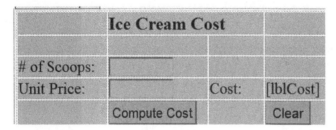

Positioning Text and Controls on a Web Form

Because positioning text and controls using spaces and tabs in HTML may not be reliable across different client environments, a table is often used as a quick and easy solution. An alternative is to use a tool like **Bootstrap** that utilizes **CSS** and JavaScript to provide extensive functionality for layout and formatting, including mobile friendliness. Again, because our focus is on ASP.NET and C# programming, we will use the simpler approach of using a simple table to help with layout.

You will first add a table with the required number of rows and columns. The text, like "Ice Cream Cost" and "# of Scoops," is typed directly into the cells of the table. Controls like TextBoxes and Buttons are dragged and dropped into the cells.

Step 1-7: Add a table with 5 rows and 4 columns:
     Select Design view by clicking the Design button at the bottom of the Editor window.
     On the Menu bar, select **Table > Insert Table**.
     On the Insert Table window, change Rows to 5 and Columns to 4.

     Specify the width of the table to be 500 pixels, rather than as a percent of the width of the browser window, as demonstrated in Figure 15-10. You may change it later by adjusting the width of the table in Design view.

Figure 15-10: The Insert Table window

Provide a solid border for the table:
    Select the table (click inside the table and press Esc key twice).
    Select *Format > Borders and Shading > Box > Solid > OK*.

You can specify various other properties of the table like background color and picture, alignment and cell spacing. For now, you will keep the defaults.

Click OK.

Working with a table

Working with a table in Design view is similar to working with MS Word tables. You select the table by clicking the border. You may adjust the size of the table by dragging it by the handlers.

To adjust the width of a column, you drag the right border of the column. Similarly, to adjust the height of a row, you drag the bottom border of the row.

You select multiple cells by dragging the pointer across the cells. You may change the properties of the selected cell(s) using the selections from the *Format* menu, *Table* menu or the *shortcut* menu displayed when you right click the selection.

Step 1-8: Enter the heading and other text into the table:
    Click the border of the table to select it, and adjust the width of the table to an initial desired size, if necessary.
    Set the Font for the entire table:
        Select the table and select **Format > Font**.
        Set the **Font, Font style** and **Font size** to default values. The default Font size is *small*.
    Merge the second and third cells of the first row:
        Select the two cells and right click. Select **Modify > Merge Cells**.

If you type in text into any cell, you will find that the width of the column increases and the widths of the remaining columns are automatically adjusted to be equal. This could be a nuisance. To minimize this problem, adjust the widths of the columns by dragging the column edges, starting with the first column, followed by the second and third columns, **before you start typing in the texts**. The table in Figure 15-9 has column widths of about 120, 120, 75 and 75 for columns 1, 2, 3 and 4, respectively

After adjusting the column widths, type in the heading "Ice Cream Cost" into the merged cell.

Change the font for the heading to *large* and *bold* style.

Type in the text "# of Scoops:", "Unit Price:" and "Cost:" into the cells in columns 1 and 3, as shown in Figure 15-9. You may have to readjust the column widths after you type in the text into the cells. You may format individual items by selecting *Format* from the menu.

Adding Web Server Controls to a Web Form

ASP.NET provides a variety of web server controls (also called **server controls** and **Web Forms controls**), including TextBox, Label, Button, RadioButtonList, DropDownList, ListView and GridView, to help you create the user interface without having to write HTML code.

Web server controls have functionality and properties similar to those of the familiar Windows Forms controls. But they differ from Windows Forms controls because of their special features to work with web pages. For instance, TextBoxes of both types are used for user input, and mostly have similar properties, though server controls have *ID* in place of the *name* property of Windows controls. Server controls have additional properties, including **AutoPostBack** and **Runat**, which are discussed later.

Server controls generally have a limited set of events compared with Windows Forms controls. For example, both types of TextBoxes have the TextChanged event. However, the TextBox server control doesn't have several events like Click, KeyPress, Enter, Leave and Validating, which are available for the Windows Forms Text Box.

You add server controls to a Web Form by dragging the control from the Toolbox and dropping it on the form. Because you are using a table to position the controls, you will drop the controls to appropriate cells in the table. After you drop it in the table, you can select a control and set its properties just like in a Windows form.

Step 1-9: Add controls to the form:
First, drag and drop a TextBox from the Toolbox's **Standard** tab into the cell at the intersection of column 2 and row 3.
Set its Text property to blank, and ID to txtScoops. Adjust the width as necessary.

Note that the *TextMode* property of the TextBox is set to *SingleLine*. Other options for TextMode, which determines the type of TextBox, include *Email, Url, Date* and *Password*.

Add the remaining controls to appropriate cells, and set their *Text* property as specified in Figure 15-9. Set the IDs of the controls as specified in Table 15-1. Adjust the width of the controls and the cells as necessary.

Table 15-1: The controls on the Web Form and their properties

Purpose	Type of control	Property	Setting
Enter number of scoops	TextBox	ID	txtScoops
Unit price	TextBox	ID	txtUnitPrice
Display cost	Label	ID	lblCost
Compute cost	Button	ID	btnComputeCost
Clear controls	Button	ID	btnClear

Switch to Source view. The Source view should display the ASP.NET code for the controls you added. Web server controls, like txtScoops, are signified by the "asp:" prefix, as in this code:

```
<asp:TextBox ID="txtScoops" runat="server" Width="94px"></asp:TextBox>
```

In addition to changing the properties of controls in Design view, you also may change properties like *Width*, in the ASP.NET code.

Categories of web server controls

You will find other commonly used server controls on the **standard** tab of the Toobox. In addition, there are several other Toolbox tabs that contain different groups of controls, including:
    Login: Controls that enable you to create a login page and build secure web applications
    Data: Controls that help access data from databases and display the data
    Validation: Controls that validate input data
    Navigation: Controls that enable you to create navigation aids including menus
    WebParts: Controls to create web pages that let users personalize web pages from the browser
    Ajax Extensions: Controls that help manage client-side scripts

You will use some controls from the Validation group later in section 15.4.

Adding C# Code to a Web Form

C# code is used in a Web Form just like in a Windows form to specify the actions to be taken in response to events. The events could be those associated with controls (e.g., Click, Validating, SelectedIndexChanged) or with the page (e.g., Load, Init). In a Web Form, controls also have additional properties that determine the timing of execution of event handlers, as you will see later.

Next, you will insert C# code into the click event handler of the buttons to compute the cost and to clear controls. The C# code you type in will be stored in the code-behind file, Default.aspx.cs.

Step 1-10: Double click the button, btnComputeCost.
Visual Studio opens the Default.aspx.cs file and displays the click event handler of the button. Insert the following code (previously presented in Chapter 1) into the click event handler to compute the cost:

```csharp
protected void btnComputeCost_Click(object sender, EventArgs e)
{
    int scoops;
    double unitPrice, cost;

    // Get input data from Text Boxes:
    scoops = int.Parse(txtScoops.Text);
    unitPrice = double.Parse(txtUnitPrice.Text);

    cost = unitPrice * scoops;              // Compute result
    lblCost.Text = cost.ToString("C");      // Display Result
}
```

Step 1-11: Insert this code into the click event handlers of btnClear:

```csharp
private void btnClear_Click(object sender, EventArgs e)
{
    txtScoops.Text = null;
    txtUnitPrice.Text = null;
    lblCost.Text = null;

}
```

Running a Web Form

To run the website for testing in the default browser, you click the browser name in the toolbar. You may change the browser by selecting a browser from the dropdown list. If you are asked whether you want to modify the web.config file to enable debugging, click OK.

Closing the Web Form: The C# code that you enter in the event handlers runs on the server. So, you cannot use C# methods like Close() to close the form. You can use Javascript that runs on the client, or click the Stop Debugging button to stop the application in Visual Studio. To close the browser, click the Close button of the browser.

Step 1-12: Run the form, and click the *Compute Cost* button to execute the C# code that computes the cost. Fix errors, if any. Close the form, and stop the application.

AutoPostBack property of server controls

When you click the *Compute Cost* button to execute the C# code, the click event of the button is raised on the client. But, the C# code, which is server-side code, is executed on the server. How does raising an event on the client result in executing the code on the server?

When you click a button in the browser, the page is automatically posted back to the server. That is, the client sends a message to the server, which includes the input data like the number of scoops and the unit price that you entered into TextBoxes. The server executes the C# code and sends the HTML markup for the page that includes the computed cost.

The automatic post back is a default feature of Buttons. This feature is also available for other server controls that accept user input, like TextBox and DropDownList. To enable this feature, you set the AutoPostBack property of the control to *True*. If the AutoPostBack property of a TextBox is set to True, the web page will be posted back whenever the TextBox loses focus, like when you press the Tab or Return key. Similarly, changing the value of a DropDownList, RadioButtonList or CheckBox also results in postback of the page, as you will see in Tutorial 2.

The aspx Code for IceCreamCost Form

Let's take a peek at the aspx code generated by Visual Studio to see how auto-styles are used to specify the formatting of the table, and understand how the ASP.NET code represents different server controls. First, we will look at the head of the HTML.

Head element and auto-styles

Figure 15-11 displays a part of the head element of the HTML that defines several **auto-styles** named auto-style1, auto-style2 and so on, which are used in the HTML body to specify the format for each cell in the table. (Your code may not look as clean as the formatted code shown here.) Where did these styles come from? Visual Studio generates an auto-style for each unique set of styles (properties like font size, width and height) you choose for the cells. Thus each style specifies a set of properties, and each set is given a specific **auto-style** number. Auto-styles in your form may be different, depending on the properties you specified.

Figure 15-11: Head element that uses auto-styles

```
<html xmlns="http://www.w3.org/1999/xhtml">
<head runat="server">
    <title>Ice Cream Cost</title>
    <style type="text/css">
            .auto-style1 {width: 23%;}
            .auto-style2 {width: 51px;}
            .auto-style5 {font-size: large; font-weight: bold;}
            ...
    </style>
</head>
```

An advantage of defining each style in the head element is that if multiple cells have the same style, the same style can be specified in each cell, instead of repeating the set of properties. You will see the use of these styles in the body element. For instance, the first cell of each row uses auto-style9.

Turning off auto-styles

If you do not want the auto-styles to be generated, you can turn off this feature, as specified here. If you do that, the element for each cell and the table will have the actual properties specified instead of using an auto-style, possibly repeating the same set of properties in multiple elements.

> Select **Tools > Options > Web Forms Designer**.
> First, select **CSS Styling > Manual Style Application**.
> Next, select **CSS** and set all six properties on the right half of the window to **CSS (inline styles)**. Click OK.

The tutorials in this chapter assume that the auto-styles are **on**, and, for the sake of simplicity, you will use the auto-generated styles that are embedded in the Head element. Auto-styles can also be renamed to make it easier to interpret the code. However, it is preferred to use an external style sheet that keeps the styles separate from the aspx code. You can specify your own styles and create such an external style sheet using CSS, which is outside the scope of this book.

Body element

When you add a table, Visual Studio generates the <table> element that specifies the properties of the table, a <tr> element for each row and a <td> element for each cell. Figure 15-12 displays the elements for rows 1, 3 and 5, each with a different type of content.

Figure 15-12: Body of HTML for IceCreamCost form

HTML elements **body, form and table**:
```
<body style="height: 299px">
    <form id="form1" runat="server">
        <table class="auto-style1">
```

HTML element for **row 1** that includes the text "Ice Cream Cost" in the second cell:
```
            <tr>
                <td class="auto-style9"> </td>
                <td class="auto-style5" colspan="2">Ice Cream Cost</td>
                <td class="auto-style2"> </td>
            </tr>
```

HTML element for **row 3** that includes the TextBox txtScoops:

```
<tr>
    <td class="auto-style9"># of Scoops:</td>
    <td class="auto-style8">
        <asp:TextBox ID="txtScoops" runat="server" Width="94px" </asp:TextBox>
    </td>
    <td class="auto-style2"> </td>
    <td class="auto-style2"> </td>
```

HTML element for **row 5** that includes the buttons btnComputeCost and btnClear:

```
        <tr>
            <td class="auto-style9"> </td>
            <td class="auto-style8">
                <asp:Button ID="btnComputeCost" runat="server" Text="Compute Cost"
                            Width="137px" OnClick="btnComputeCost_Click" />
            </td>
            <td class="auto-style7">
                 </td>
            <td class="auto-style2">
                <asp:Button ID="btnClear" runat="server" Text="Clear" OnClick="btnClear_Click" />
            </td>
        </tr>
    </table>
</form>
</body>
```

As discussed earlier, one type of aspx code is static HTML markup that the browser understands, and is not processed on the server. The following *table*, *tr* and *td* elements are examples of static markups:

```
<table class="auto-style1">
    <tr>
        <td class="auto-style9"> </td>
        <td class="auto-style5" colspan="2">Ice Cream Cost</td>
        <td class="auto-style2"> </td>
    </tr>
```

The second *td* element displays the static text "Ice Cream Cost."

Note that the parameters for the *table* and *td* elements include an auto-style, as in <table class="auto-style1">. These are formatting styles defined in the head element. The coding " " specifies a nonbreaking space.

The second type of aspx code is ASP.NET code that the browser doesn't understand, and therefore is processed on the server to translate (render) the code to HTML markup.

The following elements for the form, TextBox and Button, which have the **runat="server"** attribute are ASP.NET code, which are processed on the server.

```
<form id="form1" runat="server">
<asp:TextBox ID="txtScoops" runat="server" Width="94px" </asp:TextBox>
<asp:Button ID="btnComputeCost" runat="server" Text="Compute Cost"
            Width="137px" OnClick="btnComputeCost_Click" />
```

Among the three controls, the TextBox and Button elements have an **"asp:"** prefix, indicating that they are web server controls.

Processing ASP.NET Code

When the user requests a web page, the ASP.NET code for the page is rendered (converted) to HTML markup by the server so that the browser can understand it. So, the page that is sent to the client contains only HTML markup, and client-side scripts, if any. To view the HTML that the browser receives, run the website, right click the page and select **View page source**. In general, an understanding of the rendered HTML is not necessary for web development. The following example is intended to show the relationship between web server controls and the rendered HTML, and to familiarize you with two important HTML elements: the *form* and *input*.

The HTML elements rendered from the ASP.NET code for the form, TextBox and Button that were discussed in the previous section look like this:

```
<form method="post" action="./Default.aspx" id="form1">
<input name="txtScoops" type="text" id="txtScoops" style="width:94px;" />
<input type="submit" name="btnComputeCost" value="Compute Cost"
       onclick="javascript:WebForm_DoPostBackWithOptions(…)"
       id="btnComputeCost" … style="width:137px;" />
```

Note that the HTML form element has two new attributes: **method** and **action**. The action attribute specifies where to send the form data when the form is submitted. The method attribute specifies how the form data is sent to the page specified in the action attribute:

    method="get": Form data is sent by appending the data to the URL, which limit the size of the data, and make the data visible in the URL.

    Method="post": Form data is appended inside the HTTP request and is not visible in the URL.

In the previous example, each server control, like TextBox and Button, is represented by a single HTML <Input> element (discussed next). However, because web server controls are not part of the HTML language, some web server controls use multiple HTML elements to expand the functionality of individual HTML controls. For example, the **DropDownList** control that you will use in the next section is represented by two HTML elements, <select> and <option>, as follows:

```
<select name="ddlPlays" … class="auto-style68" style="width:184px;" >
    <option value="3"> Les Miserables </option>
    <option value="1"> Mamma Mia </option>
    <option value="2"> Rock of Ages </option>
</select>
```

HTML controls

In addition to the web server controls, the HTML tab of the Toolbox provides several HTML controls that you can use to create the user interface:

These controls are part of the HTML language. So each control is represented by a single HTML element. When you add an HTML control to a page by dragging and dropping it in Design view, Visual Studio adds the corresponding HTML element to the aspx code. You may also add an HTML control by directly typing in the HTML element in the aspx file in Source view.

As an example, let's look at the **Input(Text)** control, which is the HTML control corresponding to TextBox. Adding this control to the form adds an <Input> element to the aspx code that looks like this:
```
<input id="Text1" type="text"/>
```

By contrast, the web server control, TextBox, is represented by the ASP.NET code
```
<asp:TextBox ID="TextBox1" runat="server"></asp:TextBox>
```
which is translated by the server to the HTML element
```
<input id="Text1" type="text"/>
```

The HTML controls are **client-side** controls, which means they are not accessible by programs running on the server. So, you cannot use the data entered into an HTML **Input(Text)** control (also called **Text Field**) to do calculations using server-side code like the Click event handler of btnComputeCost. Further, the data you enter into a Text Field is lost when the page is refreshed as a result of a page postback.

A benefit of HTML controls is that because they are part of the HTML language, there is no need to render them to HTML, as it is necessary for web server controls. If there is no need to access the contents on the server, HTML controls are recommended because they have less overhead and they improve the performance of the web page compared with web server controls. So, an HTML control is preferred in special cases where you need only client-side functionality for the control. For example, to do animations with images using client-side JavaScript, you may use the HTML control, Image.

However, keep in mind that the HTML controls may not meet the requirements in most situations.

HTML server controls

You can easily convert an HTML controls to an HTML server control by adding the **runat** attribute to the HTML element as in this code:

    <input id="Text1" type="text" **runat="server"**/>

HTML server controls, like web server controls, are processed on the server and the data you enter into them can be accessed by server-side code; in addition, the data is not lost when the page is refreshed.

However, HTML server controls have less functionality than Web server controls. You also cannot double click an HTML button or submit control to generate the click event handler header in the code file, aspx.cs. You typically will have to add a function to the aspx file within a script element.

Web server controls have several other advantages over HTML server controls:
- Ability to detect the capabilities of the browser and render HTML markup accordingly. Thus a control may look different in a mobile browser than in a laptop.
- A richer object model
- Ability to define custom layout for the controls using templates
- Ability to define a consistent look for controls using themes

If the content of a control needs to be accessed by code on the server, **the web server control is preferable** because of its advantages over HTML server controls. However, understanding of HTML server controls would be beneficial because they were used in older applications. Developers may use HTML server control when they want to write their-own HTML markup and/or when they want every browser to see the same set of HTML markup.

<Input> element

Several **HTML controls** available in the Toolbox, like Input(Submit), Input(Text) and Input(CheckBox), are represented by the <Input> element. You can add these controls also by typing in the <Input> elements into the aspx code. For example, you can create a Text Field named textScoops, instead of a TextBox, to enter data using the following markup:

    <input name="txtScoops" type="text" id="txtScoops" style="width:94px;" />

You may convert these HTML controls to HTML server controls by adding the runat="server" attribute.

The value of the **type** attribute determines the type of input control created. The default type "text" creates a Text Field. Table 15-2 provides a sample of the types of input controls.

Table 15-2: Type attribute of Input element

Type attribute of Input element	Type of control created
checkbox	check box*
date	control to enter date
file	control that lets the user select a file
number	control to enter a number
submit	button that submits the form
text	text field*
* These controls are similar to their Web Server and Windows Forms counterparts, but they are limited in functionality.	

To test the behavior of HTML controls, let's temporarily replace a TextBox with an HTML Input(Text) control.

Step 1-13: Verify that HTML controls are not processed on the server:
    Delete the TextBox, txtUnitPrice from the table, and add the HTML control Input (Text) from the HTML tag of the Toolbox.
    Set the ID property of the control to txtUnitPrice.
    Open the code-behind file, Default.aspx.cs.
    You will see a red line under txtUnitPrice.Text, indicating that there is an error.
    Move the cursor over the red line to view the error message,
        "The name txtUnitPrice does not exist in the current context."

What causes the error though txtUnitPrice is a valid name? The C# code is server-side code, which is processed on the server; but the HTML control, txtUnitPrice, is a client-side control, which cannot be accessed on the server.

Step 1-14: Verify that the value you enter into an HTML Input control is lost when the page is refreshed as a result of a page postback:
    Comment out the code in the click event handler of btnComputeCost, and run the page.
    Enter values for number of scoops and unit price.
    Click the Compute Cost button. Since the code is commented out, no cost is displayed.
    Note that the value of unit cost disappears. Why?

Clicking a Button causes a postback of the page. A web server control retains its value when a page is refreshed as a result of a page postback. But an HTML control like txtUnitPrice doesn't retain its value.

Step 1-15: Verify that an HTML **server** control can be accessed on the server and that it retains its value when the page is posted back:
    Change the HTML control, txtUnitPrice, to an HTML server control by adding the runat="server" attribute to the Input element. The Input element looks like this:
```
<input id="txtUnitPrice" class="auto-style10" runat="server" type="text" />
```
    Uncomment the code in the click event handler of txtComputeCost.
    There is no more red line under the name, txtUnitCost. But there is a red line under the Text property.
    Move the cursor on the red line to view the error message,
        "htmlInputText does not contain a definition for 'Text' ..."
    As the message says, the Input(Text) control doesn't have a Text property. Instead it has a Value property.
    Change the Text property of txtUnitrPrice to Value, like this:
```
unitPrice = double.Parse(txtUnitPrice.Value);
```

Run the page, enter values for number of scoops and unit price. Click the Compute Cost button. The program works correctly, and the HTML server control, txtUnitPrice, doesn't lose its value when the page is posted back.
Thus, for this application, the HTML server control works well. However, because of the advantages of web server control discussed earlier, we will use web server controls for user input.

Step 1-16: Delete the HTML server control, txtUnitPrice and add a TextBox in its place.
    Set the ID property of the TextBox to txtUnitPrice.
    Open the code-behind file and change the Value property to Text, like this:
        unitPrice = double.Parse(txtUnitPrice.Text);
    Now the page is back to where it was before the experimentation with HTML control.

In summary, stay away from the HTML controls in the Toolbox while developing web forms in this chapter.

It's time to practice! Do Exercise 15-1.

Review Questions

15.13   What is the effect of setting the AutoPostBack property of a web server control to True?

15.14   Consider the ASP.NET code:
        `<asp:TextBox ID="txtScoops" runat="server" Width="94px" </asp:TextBox>`

        The corresponding HTML markup is:
        `<input name="txtScoops" type="text" id="txtScoops" style="width:94px;" />`

        Explain why the runat="server" attribute is missing in the HTML markup.

15.15   How do you convert an HTML control like Input(Text) control to an HTML **server** control?

15.16   List two ways HTML controls differ from web server controls.

15.17   List three advantages of web server controls over HTML server controls.

15.5 Validating Data Using Validation Controls

ASP.NET provides several easy to use and powerful controls to validate input data using client-side validation and display error messages, if any. Table 15-2 describes the ASP.NET Validation controls.

Table 15-2: ASP.NET Validation controls

Validation control	Description
RequiredFieldValidator	Makes sure that the user does not skip an entry
RangeValidator	Checks whether an entry is between specified limits
CompareValidator	Compares an entry to a specified value, including the value of another control. You may also compare the datatype of an entry to a specified data type.
RegularExpressionValidator	Checks whether an entry matches a specified pattern like a phone number, zip code or e-mail address
CustomValidator	Validates entries using custom code in the ServerValidate event handler of the control

It is important to note that if there are no data in a control, none of the validators, except the RequiredFieldValidator, are evaluated.

The available validation controls can be found in the **Validation** tab in the Toolbox. This group also includes the **ValidationSummary** control, which doesn't do any validation, but is used to display the messages from all validation controls on the page together.

First, let's use the RequiredFieldValidator and CompareValidator to make sure that the user enters the number of scoops into txtScoops and that it is a valid integer.

15.5 Validating Data Using Validation Controls

Step 1-17: Add the RequiredFieldValidator and CompareValidator to the form:
To place the RequiredFieldValidator, add a new row to the table and merge the four columns of the row together.
Drag the RequiredFieldValidator control from the Validation tab of the Toolbox into the newly added row. Set its ID to **RequiredFieldScoops**.

Add another row to the table and merge the columns of the row together.
Drag the CompareValidator into the new row. Set its ID to **CompareScoops**.

To validate a control, set the properties of the Validation control as shown in Table 15-3:

Table 15-3: Properties of Validation controls

Property	Description
ControlToValidate	The ID of the control to be validated; for example, txtScoops.
Display	Specifies how the error message is displayed. There are three settings: **Static:** Screen space for error message is taken even when there is no error. **Dynamic:** Space for error message is taken only when there is an error. **None:** Display the error in a ValidationSummary control.
ErrorMessage	The message to be displayed in the validation control

Step 1-18: Set the properties of the RequiredFieldScoops validator, like this:

Property	Value
ControlToValidate	txtScoops
Display	Dynamic
ErrorMessage	Please enter number of scoops

Similarly, set the properties of the CompareScoops validator to the same values as shown, except that the error message is "Please enter an integer for number of scoops."

In addition, set the *Operator* property of CompareScoops validator to *DataTypeCheck*.

Note that you can have multiple validators for the same control.

The form should look as shown in Figure 15-13.

Figure 15-13: Ice Cream Cost form with validators

Unobtrusive Validation

In addition to setting properties of validation controls, you need to choose from one of two different methods of client-side validation. As stated earlier, validation controls use client-side validation.

One option is to choose the traditional approach that was used prior to release of .Net Framework 4.5. In this approach, the validation control generates a lot of JavaScript that is rendered to HTML and sent to the browser for client-side validation. The generated inline JavaScript significantly increases the size of the code.

The second option is to use a relatively new method called **unobtrusive validation**, which is the default method in .NET Framework 4.5 and later releases. This method uses the JavaScript from the JQuery library without having to add significant amount of JavaScript to the page.

To use unobtrusive validation, first you need to install jQuery library because websites created from the *Empty Web Site* template do not have jQuery installed. An easy way to install and configure jQuery is to use the Visual Studio extension called, which can be used to install and configure several packages that improve the functionality of web projects. It is fairly easy to install jQuery, as detailed in Step 1-24.

Both methods require that the browser supports JavaScript. First you will use the traditional method. To use the traditional method, you need to disable unobtrusive validation for the web page using the following code in the Load event handler of the page:

```
UnobtrusiveValidationMode = UnobtrusiveValidationMode.None;
```

Step 1-19: Disable Unobtrusive validation by inserting code into the Load event handler of the page as follows:
```
protected void Page_Load(object sender, EventArgs e)
{
    UnobtrusiveValidationMode = UnobtrusiveValidationMode.None;
}
```

Before you run the page, set initial focus to txtScoops, and set the tab order.

Step 1-20: Set the initial focus to txtScoops by adding the following statement to the Load event handler of the page:
```
txtScoops.Focus();
```

Set the TabIndex property of the controls txtScoops, txtUnitPrice, btnComputeComputeCost and btnClear to 1, 2, 3 and 4, respectively.

Step 1-21: Run the web page.

Enter no value for number of scoops, but enter a valid number for unit price.
Click the Compute Cost button. You should see the message "Please enter # of scoops."

Enter a noninteger value for number of scoops and click the Compute Cost button. You should see the message "Please enter an integer for scoops."

Timing of Validation

In general, validation of data takes place both on the client and the server. Client-side validation takes place in an input control that has a validator associated with it, when you make a change in the data and move focus away from the control. Thus, if you enter a noninteger into txtScoops and Tab to the next field, the Compare validation will take place, resulting in displaying the error message. An advantage of client-side validation is that it helps give immediate feedback to the user without a postback of the page that causes a roundtrip to the server.

However, client-side validation won't identify missing data, unless a value was entered and then deleted, because validation will take place only if you change the data and move focus away from it. So, if you move focus away from txtScoops without entering any data, no client-side validation will take place because you haven't made any change to the data.

Server-side validation rechecks the data using the same validation criteria as the client-side validation, but it is done on the server. This ensures that browser incompatibilities with the client-side validation scripts don't cause the submission of invalid data. Server-side validation take place in every control that has a validator associated with it, whenever you click a button whose *CauseValidation* property is set to *True*. Hence, if you click the Compute Cost or Clear button without entering the number of scoops, the RequiredField validation will take place, and you will get the message to enter the number of scoops.

Server-side validation takes place when the page is posted back by clicking the button. The validations take place before processing the code in the event handlers. If you do not want validation to take place when you click a button like the Clear button, you can set the *CauseValidation* property of the button to *False*.

Setting the **EnableClientScript** property of the validator to *False* would turn off client-side validation.

Let's check it out.

Step 1-22: Test the behavior of validators:
Click the Stop button if the web page is running, and then run it.
Make sure that txtScoops has the focus.

Press Tab to move focus away from txtScoops without entering any data.
No message is displayed, indicating that the data in txtScoops was not validated because you haven't yet changed the value of txtScoops.

Click inside txtScoops so that it has the focus. Enter a decimal number like 3.5, and press Tab. You will see the error message "Please enter an integer for scoops."

Next, delete the invalid number, and press Tab. You will see the error message "Please enter number of scoops." The RequiredFieldValidator was evaluated this time because the data in txtScoops was changed.

Next, you use Unobtrusive validation.

Step 1-23: Comment out the statement in the Load event handler, which disables Unobtrusive validation.

Step 1-24: Install and configure jQuery library:

Right click the project IceCream in the Solution Explorer.
Select **Manage NuGet Packages.** You will see the NuGet IceCream window.
Select **Browse**.

Chapter 15: Introduction to Web Applications Development

Type "**aspnet.scriptmanager.jquery**" in the search box. The name of the package will appear in the bottom left pane.
Select the package and click **Install**. Click OK.

Step 1-25: Run the web page, and repeat the test as before with no value for number of scoops, and also with noninteger values. You should see the same results as before.

Here is the complete C# code for the Ice Cream Cost form.

```csharp
1   using System;
2   using System.Collections.Generic;
3   using System.Linq;
4   using System.Web;
5   using System.Web.UI;
6   using System.Web.UI.WebControls;
7
8   namespace IceCream
9   {
10      public partial class Default : System.Web.UI.Page
11      {
12          protected void Page_Load(object sender, EventArgs e)
13          {
14              txtScoops.Focus();
15              //UnobtrusiveValidationMode = UnobtrusiveValidationMode.None;
16          }
17          protected void btnComputeCost_Click(object sender, EventArgs e)
18          {
19              // Declare variables
20              int scoops;
21              double unitPrice, cost;
22              // Get input data from Text Boxes:
23              scoops = int.Parse(txtScoops.Text);
24              unitPrice = double.Parse(txtUnitPrice.Text);
25              // Compute and display cost
26              cost = unitPrice * scoops;
27              lblCost.Text = cost.ToString("C");
28          }
29          protected void btnClear_Click(object sender, EventArgs e)
30          {
31              txtScoops.Text = null;
32              txtUnitPrice.Text = null;
33              lblCost.Text = null;
34          }
35      }
36   }
```

Additional information on validators and examples of their use can be found at
https://msdn.microsoft.com/en-us/library/ms972961.aspx

In the next chapter, you will learn how to develop multipage web applications that access databases.

It's time to practice! Do Exercise 15-2.

Review Questions

15.18   When does client-side validation takes place? When does server-side validation take place?

15.19   What are the advantages and disadvantages of client-side validation compared with server-side validation?

15.20   How do you prevent validation when you click a button?

Exercises

Exercise 15-1: Compute job cost

The following windows form lets the user enter the labor and material costs for a job into two TextBoxes and display the subtotal, sales tax and total.
Subtotal = Labor cost + Material cost
Sales tax rate is 6%.
Total cost is the sum subtotal and sales tax.

You are to create a Web Form page that has the same functionality as the above Windows form with a similar look, allowing for differences in display in a browser. Create a website named Exercises_Ch15, within the folder AspWebSites. (See Tutorial 1, Step 1-3.) Add a Web Form named JobCost. (See Tutorial 1, Step 1-4. Make sure you change *Itemname* to JobCost.) Add the necessary controls and C# code.

Exercise 15-2: Validating unit price

This exercise provides additional validation of input data in the Ice Cream Cost form presented in Figure 15-9. Use validation controls to make sure that the user enters a value for unit price and that it is between 0 and 10. Use the Minimum and Maximum properties of the RangeValidator to specify the range for unit price. Provide appropriate error messages.

Chapter 16

Web Development: Multipage and Database Applications

Chapter 15 provided you with the basic skills to build a single-page web project. In this chapter, you will learn to access SQL Server databases and develop multipage web projects.

Learning Objectives

After studying this chapter, you should be able to:

- Look up information from an SQL Server database table.
- Explain how a dynamic web page is created and displayed.
- Use view state and session state to preserve data.
- Describe the use of query strings to transfer data.
- Create multipage websites.
- Move to another page using Redirect, Transfer and Cross-page PostBack methods.

Topics

16.1	Accessing Databases from Web Forms	16.3	Multipage Websites and Session State
16.2	Preserving Data During Postbacks Using View State	16.4	Setting a Page to Its Previous State

16.1 Accessing Databases from Web Forms

Dynamic web pages typically access databases and other sources to look up information that is used to change the contents of a web page. To help you learn how to access databases using ASP.NET, you will work with a modified version of the Theater Tickets application from Chapter 6. The application will access ticket prices from an SQL Server database.

You will also learn how to use additional web server controls like DropDownList and RadioButtonList to enhance the user interface of a web page, and learn the use of the AutoPostBack feature of server controls.

To start with, you will develop the **Reservation** page displayed in Figure 16-1, which lets the user select the play, zone and membership status; enter the number of tickets; and compute the cost. In the next three sections, you will add a confirmation page and modify the Reservation page to add a button that takes the user to a confirmation page. The confirmation page displays the selected options and lets the user confirm the selections or change them by re-visiting the page that would display the selections.

The entire project is divided into four phases corresponding to the four sections (16.1, 16.2, 16.3 and 16.4) of the chapter. Thus, this section includes tutorials to develop **Phase 1** of the project.

Figure 16-1: The Reservation Web Form with data

A major difference between the Windows Form application from Chapter 6 and the current Web Form application is that this application accesses a database to find and display the base price for the play selected by the user from the DropDownList. To find and display the base price, you use C# code that is attached to the SelectedIndexChanged event of the DropDownList. To execute the server-side C# code when the user selects a play, you set the AutoPostBack property of the DropDownList to True, which would cause a postback of the page.

The base price doesn't include an additional charge (zone price) that depends on the selected zone. The zone price is computed and displayed whenever the user selects a zone. The discount that is based on the member status is displayed when the user checks/unchecks the CheckBox. The zone price and member discount are hard-coded. To execute the C# code to compute the zone price and the discount, you set the AutoPostBack properties of the RadioButtonList and the CheckBox to True.

Additional details on setting up the DropDownList and RadioButtonList are provided in Tutorial 1.

Tutorial 1: Accessing a Database: Theater Tickets Application

First you will create a website named Theater and add a Web Form named Reservation. The process is similar to the one you used to create the IceCream application in Tutorial 1 (steps 1-2 to 1-4), so the steps are more condensed in this tutorial.

Create the project and add a form.

Step 1-1: Develop a project named Theater using essentially the same steps as in the IceCream project, **except for the project template**, as follows:

> From Visual Studio menu, select: **File > New > Project**.
> Select **Installed > Templates > Visual C# > ASP.NET Web Applications**.
> Enter **Theater** for *Name*.
> Select **ASPwebsites** folder for *Location*.
> Make sure the CheckBox for **Create directory for solution** is checked. Click OK.
> On the New ASP.NET Project window, select **Empty** template.

Chapter 16: Web Development: Multipage and Database Applications

In addition, check the WebForms CheckBox in the section entitled *Add folders and core references for*, which adds the *App_Data* and *Models* folders and the Global.asax and Global.asax.cs files to the Project:

The *App_Data* folder is used to store database files, and the *Model* folder stores any classes you create. The *Global* files are used to respond to application-level and session-level events like the start/end of an application or session. You also may add your-own folders to the Solution Explorer.

Step 1-2: Add a Web Form named **Reservation**.

Right click the project name Theater in the Solution Explorer window.
Select **Add > Add Web Form**.
Change **Item name** to **Reservation.** Click **Add**.

Step 1-3: Change to Source view, and insert the title, "Theater Ticket Reservation," in the title element so that it looks like this:
`<title>` Theater Ticket Reservation `</title>`

Delete the div element in the Source view so that the body element looks as follows:
```
<body>
        <form id="form1" runat="server">

        </form>
</body>
```

Next, you will add a table and add the text and controls into the cells of the table.

Add controls to the form.

Step 1-4: Change to Design view, and add a table with 7 rows and 5 columns and 700 pixels wide. Note that the first column is very narrow and it is used for margin.
Select the table (click inside the table and press the Esc key twice).

Select *Format > Borders and Shading > Box > double > OK*.
Add the text and the server controls as in Figure 16-2, **except** the DropDownList and RadioButtonList. Figure 16-3 presents the purpose and the properties that you need to set for each control

Figure 16-2: The Reservation page in Design view

```
              Theater Ticket Reservation
  Play:  [Databound    ▼]      Base Price  [lblBasePrice]
  Zone:   ○Zone1  ○Zone2  ○Zone3   Zone price:  [lblZonePrice]
          □Member?               Discount:    [lblDiscount]
  # of  [        ]
  Tickets:                       Cost/Ticket: [lblUnitPrice]
          [ Continue ]
  SqlDataSource - SqlDataSource_Theater
```

You will add the RadioButtonList, DropDownList and the SqlDataSource and set their properties in the steps to follow.

Figure 16-3: Controls on the Reservation page

Purpose	Type of control	Property	Setting
Select play	DropDownList	ID	ddlPlays
		AutoPostBack	True
Select zone	RadioButtonList	ID	rblZones
		AutoPostBack	True
		RepeatDirection	Horizontal
Specify member status	CheckBox	ID	chkMember
		Text	Member?
		AutoPostBack	True
Enter number of tickets	TextBox	ID	txtNoOfTickets
Redirect to Confirmation page	Button	ID	btnContinue
Display base price	Label	ID	lblBasePrice
Display zone price	Label	ID	lblZonePrice
Display discount	Label	ID	lblDiscount
Display unit cost	Label	ID	lblUnitPrice

Setting up the RadioButtonList

Chapter 6 discussed the use of a group of RadioButtons to let the user select a single item from a group. In such a group, to find which RadioButton was selected by the user, you need to examine the Checked property of each button in the group. ASP.NET provides the RadioButtonList control that makes it easier to identify the selection made by the user. Similar to a ComboBox, you can use the **SelectedValue** property of the RadioButtonList to get the value associated with the button selected by the user.

You use the **Items** collection of the RadioButtonList to create a group of RadioButtons. The **RepeatDirection** property of the control determines whether the RadioButtons are arranged horizontally or vertically.

Step 1-5: Add a RadioButtonList and add three buttons:
Drag and drop the RadioButtonList from the ToolBox into the cell at row 3 and column 3 of the table.
Set the ID property of the control to rblZones.
Select RepeatDirection property, and set it to Horizontal.
Select the **Items** property of rblZones, and click on the ellipses (…) on the right side.
On the *ListItem Collection Editor*, click the **Add** button to add a RadioButton. Type in "Zone1" for Text and Value properties, as shown in Figure 16-4.

Figure 16-4: ListItem Collection

Repeat the previous step to add RadioButtons for Zone2 and Zone3. Click OK.

Similar to RadioButtonList, ASP.NET also provides a **CheckBoxList** that lets you add a group of CheckBoxes to the control so the user can select multiple items from a list.

Working with DropDownLists in Web Forms

The DropDownList is similar to the ComboBox with the DropDownStyle property set to DropDownList. That is, the list of items is displayed by clicking the down arrow and the text portion is not editable. Just like in a ComboBox, you use the **SelectedIndexChanged** event handler to execute code when you select an item from a DropDownList. The ComboBox control that you have used in Windows forms is not available in ASP.NET.

Step 1-6: Add a DropDownList into the cell at row 2 and column 3 of the table.
Set its ID to ddlPlays.
Set the AutoPostBack property to *True*.

You will use the DropDownList to display the play names from the **Plays** table (**PlayId, PlayName, BasePrice**) in the SQL Server database **Theater.mdf** by binding the control to the field. When the user selects the name of a play, the corresponding base price is displayed in the Label, lblBasePrice.

As discussed in Chapter 12, in order to access data from a database, you create a data source, configure it, and bind the DropDownList to the data source. Next, we will take a brief look at the steps, which were also presented in Chapter 12.

Creating and Configuring a Data Source

You will create the data source from the existing database **Theater.mdf**. If you would like to create an SQL Server database, you may follow the steps given in Appendix B. Because the database is accessed by ASP.NET running on a server, you cannot store the database on the client. You need to create an **ASP.NET folder** named **App_Data** that becomes a part of the application so that it can be accessed by the server.

Create the data source

Step 1-7: If App_Data folder was not created in Step 1-1, create it. Then copy Theater.mdf into the folder:
Right click the project name Theater in the Solution Explorer.
Select **Add > Add ASP.NET Folder > App_Data**.

Caution: Creating a standard folder and naming it to App_Data won't work.

Copy Theater.mdf from Project_starts/DataFiles to the App_Data folder.

Step 1-8: Add the **SqlDataSource** control to the Web Form:
Merge together the first three cells of the last row.
Drag the **SqlDataSource** control from the **Data** tab of the ToolBox and drop it in the merged cell.
Change the ID of the data sourc to **SqlDataSource_Theater**.

Configure the data source

Step 1-9: Specify the database, table, fields and selection criteria, if any:
Select **Configure Data Source** from the smart tag menu of the data source.
Click **New Connection** button.
On the *Add Connection* window, make sure that the DataSource is SQL Server Database File (SqlClient).
Click Continue.
Click Browse, and select **Theater.mdf** from the **App_Data folder**.

Caution: It is important that you select the database file from the App_Data folder; otherwise your application won't work if moved to a different location.

You may click **Test Connection** button to test the connection, and click **OK**.
Click **OK** on the *Add Connection* window.
On the Configure Data Source window, click **Next**.
Leave the **Yes** box checked to save the connection to the application configuration file.
Click **Next**.
Leave the default selections as indicated in Figure 16-5 to specify that all columns of the Plays table are selected. If only certain columns of the table are to be selected, you would check those individual columns.

Figure 16-5: Configure Data Source: selecting the fields

To sort the plays by play name, select **ORDER BY** button, and select **PlayName** from the first dropdown list. Click **OK**. Note that the SQL now includes the ORDER BY clause.

Click **Next**, and click **Test Query** button. You should see the records from the Plays table. Click **Finish**.

Open the Web.Config file from the Solution Explorer to view the connection string:

"Data Source=(LocalDB)\MSSQLLocalDB;AttachDbFilename=|DataDirectory|\Theater.mdf;Integrated Security=True;Connect Timeout=30" providerName="System.Data.SqlClient"

Note that the directory for the database file name is |**DataDirectory**|, not the local folder. Again, if you connect to a database in a local folder, the application won't work correctly if moved to a different directory or computer.

Binding a DropDownList to a Data Source

Binding a DropDownList to the data source involves setting three properties of the DropDownList as specified here:

    DataSourceId: SqlDataSource_Theater
    DataTextField: PlayName
    DataValueField: PlayId

DataSourceId is the name of the data source.

DataTextField specifies the name of the field that is **displayed** in the DropDownList. This field is set to PlayName so that the dropdown list displays the play names.

DataValueField specifes the name of the field whose value is returned by the SelectedValue property. We set the DataValueField to PlayId so that we can use the PlayId to find the price of the selected play, though we let the user select the play name.

DataValueField could be set to the same field as the DataTextField. Thus both properties could be set to PlayName, if you want to use the play name rather than the Id.

These three properties could be set in the Properties window of the DropDownList, or using the wizard or code. You will use the wizard and then view the property settings in the Property window.

Step 1-10: Bind the DropDownList control to SqlDataSource_Theater:
Select **ChooseDataSource** from the smart tag menu of the DropDownList.
Select *PlayName* for the field to display, and *PlayId* for the value of the DropDownList, like this:

```
Select a data source:
SqlDataSource_Theater

Select a data field to display in the DropDownList:
PlayName

Select a data field for the value of the DropDownList:
PlayId
```

Click OK.

Note that the Properties window displays these settings:
DataSourceId: SqlDataSource_Theater
DataTextField: PlayName
DataValueField: PlayId

Step 1-11: Run the Web Form and verify that the DropDownList lists the names of the plays, and that when you select a play, the name of the play is displayed.

Next, you will use the PlayId of the selected play to find the price for the play.

Looking Up a Value from a SqlServer Table

Chapter 12 discussed different ways to select records and look up information from a table:

1. **Use a data table in combination with the table adapter and the binding source.**
   The table adapter uses SQL to select data from the database table and load the data into the data table (maintained in the client's memory).
   The binding source allows the user to navigate the data table and filter records.
   Thus records can be found by specifying the criteria in the SQL (parameter query), or by specifying the filter in the binding source.
   The use of data tables provides a convenient way to navigate through the records, add, update and delete records and look at different groups of data.

2. **Use the SqlCommand object in combination with a DataReader.**
   The SQL associated with the SqlCommand object is used to select records, and the DataReader is used to sequentially access the data retrieved by the SQL.

There is a third method to find records. You will use this method to find the price for a play.

Retrieving records using the Select method of the SqlDataSource

The third method to look up information, like the price for a play, uses the Select method of the SqlDataSource, which involves four steps:
1. Use the Select method to retrieve the records specified in the data source, and store the data in a **DataView** object.
2. Apply a filter to select the row for the play selected by the user.
3. Store the selected row of the DataView in a **DataRowView** object.
4. Get the base price from the DataRowView and store it in a variable.

The Select method has the following pattern:
```
DataSourceName.Select(ParameterValues)
```

Here is an example that uses the Select method of SqlDataSource_Theater to retrieve the records using the SQL specified in the data source, SELECT * FROM [Plays] ORDER BY [PlayName]:
```
SqlDataSource_Theater.Select(DataSourceSelectArguments.Empty)
```

When the SQL statement doesn't include any parameters the Empty argument DataSourceSelectArguments.Empty is specified. The above statement would execute the SQL statement and return the result set.

To find the selected play, the result set returned by the above Select method is cast to a DataView and then stored into a variable of DataView type, named dvPlaysTable. The statement looks like this:
```
DataView dvPlaysTable = (DataView)
        SqlDataSource_Theater.Select(DataSourceSelectArguments.Empty);
```

To refresh your memory, a DataView consists of rows, where each row is of type DataRowView.

After you store the data in a DataView, you can specify the criterion to select a row (or multiple rows) using the RowFilter property of the DataView. The following code uses the PlayId of the play selected from the DropDownList to filter the rows:
```
dvPlaysTable.RowFilter = "PlayId = " + ddlPlays.SelectedValue.ToString();
```

To get the base price from the selected row, first you store selected row in a DataRowView named drvPlayRow:
```
DataRowView drvPlaysRow = (DataRowView)dvPlaysTable[0];
```

The index 0 is used to select the first (and only) row from the DataView.
Next, you get the base price from the BasePrice column of the DataRowView and store it in the variable basePrice:
```
double basePrice = (double) drvPlaysRow["BasePrice"];
```

Figure 16-6 combines the above statements together.

Figure 16-6: Code to find the base price for the selected play

```
DataView dvPlaysTable    =
        (DataView)SqlDataSource_Theater.Select(DataSourceSelectArguments.Empty);
dvPlaysTable.RowFilter   = "PlayId = " + ddlPlays.SelectedValue.ToString();
DataRowView drvPlaysRow  = (DataRowView)dvPlaysTable[0];
double basePrice         = (double)drvPlaysRow["BasePrice"];
```

Now let's look at how to put this code to work.

Event handler to execute the code

Which event handler would you use to execute the above code in Figure 16-6 to find the price for the play selected in the DropDownList? If this were a Windows form, you would place the code in the SelectedIndexChanged event handler (which may have some unintended consequences under certain conditions) or in the SelectionChangeCommitted event handler of the control.

The Web Form has the SelectedIndexChanged event, but not the SelectionChangeCommitted event. So, you will use the SelectedIndexChanged event. But there is a minor problem that needs to be addressed. The SelectedIndexChanged event is not fired the first time the page is loaded. So, if you place the code in this event handler, the page won't automatically display the price for the play initially displayed in the DropDownList (the first play in the list). That is, the DropDownList will display a play, but its price won't be displayed in the form.

To address this issue, you add an empty item as the first item in the list of plays, the first time the page is loaded. This requires adding code to the Load event handler of the page to do the following:
1. Use the IsPostBack property of ASP.NET to check whether the form is loaded for the first time. Unlike in a Windows Form, the Load event of a Web Form is fired each time the user selects an item from a control like the DropDownList whose AutoPostBack property is set to *True*, which causes a page postback. So, it is necessary to check whether the form is loaded for the first time or whether it is a postback.
2. The first time the form is loaded (that is, when it isn't a postback), bind the DropDownList to the data source and add an empty item to the list.

The code in lines 15–18 in Figure 16-7 implements the above steps. Line 12 declares variables that are used to compute the unit price later.

Figure 16-7: Load event handler of Reservation form

```
10      public partial class Reservation : System.Web.UI.Page
11      {
12          decimal basePrice, zonePrice, discount, unitPrice;
13          protected void Page_Load(object sender, EventArgs e)
14          {
15              if (!IsPostBack)      // If it is the first time the page is loaded,
16              {
17                  ddlPlays.DataBind();    // bind the DropDownList
18                  ddlPlays.Items.Insert(0, new ListItem("", "")); // add an empty item
19              }
20          }
```

IsPostBack Property of ASP.NET

The code in line 15 uses the IsPostBack property of ASP.NET page to check whether the page is being loaded for the first time. A value of True indicates that the page is being loaded (rendered) in response to a postback. A value of false means that the page is being loaded the first time.

DataBind() method of DropDownList

Line 17 uses the DataBind() method of the DropDownList to bind the control to the data source, which is specified by the value of the DataSourceId of the control. This is necessary because when the Load event is fired, the DropDownList is not yet bound to the data source.

Line 18 inserts an empty item into the DropDownList. To insert an item into the list at a specified position, you use the Insert method of the Items class:

```
ddlPlays.Items.Insert(0, new ListItem("", ""));
```

The first parameter of ListItem represents the string returned by the SelectedItem property of Items collection, and the second parameter is the value returned by the SelectedValue property.

Step 1-12: Add code to the Load event handler to bind the DropDownList to the data source and to add an empty item:

In Design view, click the form outside the table and click the View Code button in the Solution Explorer. Add the code from Figure 16-7 to the Load event handler of the Reservation form. In addition, add the code from line 12 to declare variables for later use.

The next step is to add the code to find the base price when the user selects a play from the DropDownList.

Step 1-13: Add the code to the SelectedIndexChanged event handler of ddlPlays to display the base price:

Double click the DropDownList to create the SelectedIndexChanged event handler and insert the code from lines 24–31 in Figure 16-8 into the event handler. You will add the code in lines 32–34 to compute the unit price later.
Add the following *directive* that specifies the name space for the *DataView* class:
```
using System.Data;
```

Figure 16-8: SelectedIndexChanged event handler of the DropDownList

```
21
22    protected void ddlPlays_SelectedIndexChanged(object sender, EventArgs e)
23    {
24        // Find the base price for the selected play
25        DataView dvPlaysTable =
26            (DataView)SqlDataSource_Theater.Select(DataSourceSelectArguments.Empty);
27        dvPlaysTable.RowFilter = "PlayId = " + ddlPlays.SelectedValue.ToString();
28        DataRowView drvPlaysRow = (DataRowView)dvPlaysTable[0];
29        basePrice = (decimal)drvPlaysRow["BasePrice"];
30        lblBasePrice.Text = basePrice.ToString();    // Display base price
31        // Compute and display unit price:
32
33
34
35    }
```

Step 1-14: Run the website, select different plays and make sure the right price is displayed as follows.

Theater Ticket Reservation

Play: Rock of Ages   Base Price 80

Working with RadioButtonList and CheckBox

When the user selects a zone by clicking a RadioButton, the additional price (over and above the base price) for the zone is displayed in lblZonePrice.

Theater Ticket Reservation

Play: Rock of Ages   Base Price 80

Zone: ○ Zone1   ● Zone2   ○ Zone3   Zone price: 10

Each zone has a fixed price that doesn't depend on the selected play. Figure 16-9 presents the SelectedIndexChanged event handler that computes and displays the zone price whenever the user selects a zone.

Figure 16-9: Code to compute zone price

```
protected void rblZones_SelectedIndexChanged(object sender, EventArgs e)
{
    // Compute zone price
    switch (rblZones.SelectedValue.ToString())
    {
        case "Zone1":
            zonePrice = 0;
            break;
        case "Zone2":
            zonePrice = 10;
            break;
        case "Zone3":
            zonePrice = 20;
            break;
    }
    lblZonePrice.Text = zonePrice.ToString(); // Display zone price
    // Compute and display unit price


}
```

Figure 16-10 presents the CheckedChanged event handler of the CheckBox, which displays the discount whenever the user checks or unchecks the CheckBox.

Step 1-15: Create the SelectedIndexChanged event handler of rblZones. Add the code from Figure 16-9.

Step 1-16: Create the CheckedChanged event handler for chkMember. Add the code from Figure 16-10.

Step 1-17: Run the website and test the code, by selecting different plays, zones and member status.

Figure 16-10: Code to compute discount

```
58
59      protected void chkMember_CheckedChanged(object sender, EventArgs e)
60      {
61          // Compute and display discount for members
62          if (chkMember.Checked)
63              discount = 10;
64          lblDiscount.Text = discount.ToString(); // Display discount
65          // Compute and display unit price:
66
67
68
69      }
```

Note that each time the user selects a play, zone or member status, the page is posted back to the server because, by default, the AutoPostBack property of the DropDownList, RadioButtonList and CheckBox is set to True. ASP.NET preserves the user selections and the values displayed in Labels when the server renders the page and sends it back to the browser. However, the values of variables are not automatically preserved. Next we will look at the ASP.NET features that allow you to preserve values and share data within a page, within a session and within an application.

Preserving Data between PostBacks

The previous steps developed the code to display the prices/discount when the user makes a selection, as shown in Figure 16-11. But you haven't computed the cost/ticket yet.

Figure 16-11: Partially completed Reservation form

In this section you will compute and display the cost/ticket (unit price) every time the user makes a selection. Unit price is given by the formula: Unit price = Base price + Zone price − Discount.

Recall that every time the user selects a play, zone or the member status, the page is posted back to the server so that ASP.NET can execute the code in the event handler that was fired. But, you cannot use the class-level variables basePrice, zonePrice and discount to compute unit price because the values of variables are not preserved when the page is posted back.

Computing the cost/ticket using the prices and discount displayed in the Labels may look like an obvious choice. In general, programs that are tightly linked to the user interface are difficult to reuse and maintain. For example, a program that uses the data from the Labels won't work if the user interface is changed in such a way that a ListBox is used in place of the Labels to display data, or the input data items are not displayed at all.

Problem with using class (field) level variables to preserve data in a Web Form

Let's see what happens if we use the data stored in class level variables, also referred to as field variables, as we do in Windows Forms.

The class level variables for the prices and discount are already declared at the class level in line 12, above the Load event handler in Figure 16-7, as shown here:

```
12      decimal basePrice, zonePrice, discount, unitPrice;
```

We will use these variables in the event handlers to compute the unit price when the user changes an input data item.

Step 1-18: Insert the code from lines 33 and 34 of Figure 16-12 at the end of the SelectedIndexChanged event handler of the DropDownList. Put a break in line 35.

Figure 16-12: SelectedIndexChanged event handler of the DropDownList.

```
22      protected void ddlPlays_SelectedIndexChanged(object sender, EventArgs e)
23      {
24          // Find the base price for the selected play
25          DataView dvPlaysTable =
26              (DataView)SqlDataSource_Theater.Select(DataSourceSelectArguments.Empty);
27          dvPlaysTable.RowFilter = "PlayId = " + ddlPlays.SelectedValue.ToString();
28          DataRowView drvPlaysRow = (DataRowView)dvPlaysTable[0];
29          basePrice = (decimal)drvPlaysRow["BasePrice"];
30          lblBasePrice.Text = basePrice.ToString();    // Display base price
31          // Compute and display unit price:
32
33          unitPrice = basePrice + zonePrice - discount;   // uses class level variables
34          lblUnitPrice.Text = unitPrice.ToString();       // May give incorrect unit price
35      }
```

Step 1-19: Insert the same two lines of code (from lines 33 and 34 of Figure 16-12) into lines 55 and 56 in the SelectedIndexChanged event handler of the RadioButtonList, and put a break at the closing brace of the event handler:

```
53          // Compute and display unit price
54
55          unitPrice = basePrice + zonePrice - discount;
56          lblUnitPrice.Text = unitPrice.ToString();
57      }
```

Insert the same code into lines 67 and 68 in the CheckedChange event handler of the CheckBox, and put a break on line 69.

Step 1-20: Run the website.
Select Rock of Ages for play and display the corresponding base price of 80 in the Label.
The program breaks in line 35. Hover the pointer over unitCost, baseCost, zoneCost and discount.

Chapter 16: Web Development: Multipage and Database Applications

The unitPrice (= basePrice + zonePrice − discount) is correctly computed as 80 because basePrice has the value 80, and the variables zonePrice and discount still has their default value of zero. So, everything is cool.

Remove the break from line 35. Click the Continue button to display the values as in the following illustration.

Theater Ticket Reservation

Play:	Rock of Ages ▼			Base Price 80
Zone:	○ Zone1	○ Zone2	○ Zone3	Zone price:
	☐ Member?			Discount:
# of Tickets:				Cost/Ticket: 80

The code in the SelectedIndexChanged event handler of the DropDownList was run on the server to find the value of base price, assign it to the variable basePrice, use it to compute the value of unit price and display it in the Label. So there was no need to persist data during a postback.

Next, select Zone2 from the RadioButtonList. The program breaks at line 57 in the SelectedIndexChanged event handler of the RadioButtonList. Hover the pointer over the four variables in line 55.

The zonePrice has the value 10 as expected. But notice that the class level variable **basePrice shows a value of 0,** resulting in the wrong unitPrice value of 10.

Remove the break from line 55. Click Continue to display the values as shown.

Theater Ticket Reservation

Play:	Rock of Ages ▼			Base Price 80
Zone:	○ Zone1	⦿ Zone2	○ Zone3	Zone price: 10
	☐ Member?			Discount:
# of Tickets:				Cost/Ticket: 10

Why did the value of basePrice change from 80 to 0 when you selected a zone? When you select a zone, the page is posted back to the server so that ASP.NET can run the code in the SelectedIndexChanged event handler of the RadioButtonList. After running the code, the server generates the HTML markup for the page that displays the unit cost and preserves the properties of the server controls. However, the values of variables declared in the code are not preserved.

Click Continue on the Toolbar, and check the CheckBox for member status. The program breaks in line 69. The values of variables basePrice and zonePrice are zero, indicating that the values of these class level variables are not preserved when the page is posted back. Click Continue to display the values. Remove the break from line 69.

So, how do you preserve the prices and discount between postbacks of the page? You saw that using class level variables won't work.

Next you will develop phase 2 of the application that shows how to use view state to preserve the values of variables when the page is posted back.

It's time to practice! Do Exercise 16-1, Part A.

Review Questions

16.1   What is the difference between DataTextField and DataValueField of a DropDownList?

16.2   What is the relationship between a DataView and DataRowView?

16.3   What does a value of True for the IsPostBack property of a web page indicate?

16.4   Why are class level variables not used to share data between event handlers in a web page?

16.2 Preserving Data during Postbacks Using View State

Because HTTP is a stateless protocol, each HTTP request for a page is treated as an independent request. The server does not retain information from previous requests from a browser by a user.

ASP.NET features that help address this limitation include view state, session state and application state. Storing data in view state allows you to access the data from a single page during a session, whereas storing data in session state lets you access the data from multiple pages during a session. Application state preserves data across all sessions of an application, for all users. The tutorials in this section develop **Phase 2** of the project, which uses view state to preserve the values of variables when the page is posted back. The next section (16.3) contains the tutorials for Phase 3 that use session state to share data between pages.

View state

View state stores data within a page (in hidden fields) to be accessed only from the same page. Storing data within the page makes view state less secure than session state, which stores data on the server. Because view state is stored within the page, it increases the page size, which may affect the performance negatively. It should be noted that the MVC approach doesn't support view state.

As stated earlier, in the web forms approach, data entered by the user into TextBoxes and other server controls are automatically preserved between page posts by ASP.NET. So, if you enter data into TextBoxes and click a button to do a computation, the page is automatically posted back to the server with the data you entered, and the page that the server sends back to the browser will preserve the data. ASP.NET also preserves the values of page properties and server controls, but ASP.NET uses view state to do it. So, you don't have to do any coding to persist the values of page properties or server controls between page posts.

However, the values of variables are not preserved when a page is posted back. You can preserve a data item like base price by storing it in a view state item, which is also referred to as a **view state variable**.

You store data in view state items by adding an item to the *view state object*. You access the view state object using the ViewState property of the page. The view state object stores a collection of items of the key/value type, using the pattern:

```
ViewState.Add (name, value);
```
or
```
ViewState["name"] = value;
```

For instance, the following statement adds the value of basePrice to the collection using the item name (view state variable) BasePrice.

```
ViewState.Add("BasePrice", basePrice);
```
or
```
ViewState["BasePrice"] = basePrice;
```

An item could be of any .Net Framework type, including value types, arrays and certain types of collections and objects.

Note that you don't have to declare the view state variable, BasePrice. However, when you retrieve its value, you need to cast it to the appropriate type, determined by the type of the value assigned to it.

To retrieve the value from view state object, use the name of the item, like this:
```
double basePrice = (double) ViewState["BasePrice"];
```

Note that the item should be cast to the appropriate type.

Because view state is stored within the page, data stored in view state variables is transferred between the browser and the server every time the page is posted back. So, storing large amounts of data in view state may affect the performance of the page. Table 16-1 describes the properties/methods of ViewState.

Table 16-1: Properties/methods of ViewState objects

Properties/Methods	Description
Count	Number of items in the collection e.g., int count = ViewState.Count;
Add (name, value)	Adds an item to the collection, or replaces if it exists e.g., ViewState.Add("BasePrice", basePrice);
Clear()	Removes all items from the collection e.g., ViewState.Clear();
Remove (name)	Removes the specified item from the collection e.g., ViewState.Remove("BasePrice");

Using view state to preserve data in Reservation form

Because the prices and discount data are needed only in the Reservation page, you will use view state in this initial version of the website to store the prices and discount. When the website is expanded (in phase 3) to include a confirmation page that displays the data from the Reservation page, you will use session state instead of view state.

Currently the following statement computes the unit price in each of the three event handlers (lines 33, 55 and 67):
```
unitPrice = basePrice + zonePrice - discount;
```

The value of each variable on the right side of the equal sign is computed in a separate event handler. But they need to be made available in all three event handlers to compute the unit price. As discussed earlier, these class level variables do not retain their values when the page is posted back to the server and resent to the browser. So, the values of these variables need to be stored in ViewState when they are computed within the event handlers so that the view state variables can be used to compute the unit price. For the event handler of the DropDownList, the above statement that computes the unit price will be replaced by the following two statements:

```
ViewState["BasePrice"] = basePrice;
ViewState["UnitPrice"] = (decimal)ViewState["BasePrice"] +
        (decimal)ViewState["ZonePrice"] - (decimal)ViewState["Discount"];
```

The first statement adds an item named BasePrice to the view state, and stores the value of basePrice in the item.

The second statement computes the unit price using the zone price and discount that are stored in view states in the event handlers of the RadioButtonList and CheckBox. The computed unit price is stored in the view state item named UnitPrice.

To display the value of the view state in lblUnitPrice, the value needs to be converted to a string:
```
lblUnitPrice.Text = ViewState["UnitPrice"].ToString();
```

The complete event handler for the DropDownList, which includes the three statements, is presented here.

```
protected void ddlPlays_SelectedIndexChanged(object sender, EventArgs e)
{
    // Find the base price for the selected play
    DataView dvPlaysTable =
        (DataView)SqlDataSource_Theater.Select(DataSourceSelectArguments.Empty);
    dvPlaysTable.RowFilter = "PlayId = " + ddlPlays.SelectedValue.ToString();
    DataRowView drvPlaysRow = (DataRowView)dvPlaysTable[0];
    basePrice = (decimal) drvPlaysRow["BasePrice"];
    lblBasePrice.Text = basePrice.ToString();    // Display base price
    // Compute and display unit price:
    ViewState["BasePrice"] = basePrice;
    ViewState["UnitPrice"] = (decimal)ViewState["BasePrice"] +
            (decimal)ViewState["ZonePrice"] - (decimal)ViewState["Discount"];
    lblUnitPrice.Text = ViewState["UnitPrice"].ToString();
}
```

You need to make similar changes to store zone price and discount in view states in the event handlers for the RadioButtonList and the CheckBox, respectively. The following three statements store the zone price in view state and use it to compute and display unit price in the event handler of the RadioButtonList:

```
ViewState["ZonePrice"] = zonePrice;
ViewState["UnitPrice"] = (decimal)ViewState["BasePrice"] +
        (decimal)ViewState["ZonePrice"] - (decimal)ViewState["Discount"];
lblUnitPrice.Text = ViewState["UnitPrice"].ToString();
```

For the CheckBox, the corresponding code appears as follows:

```
ViewState["Discount"] = discount;
ViewState["UnitPrice"] = (decimal)ViewState["BasePrice"] +
        (decimal)ViewState["ZonePrice"] - (decimal)ViewState["Discount"];
lblUnitPrice.Text = ViewState["UnitPrice"].ToString();
```

Note that when the user selects a play, the user might not have selected the zone or the member status. So the ZonePrice and Discount items may not exist in the view state when the event handler for the DropDownList is executed. Because the user may make selections in any order, or may re-select an item, it is necessary to create view state items for base price, zone price and balance in the Load event handler of the page, when the page is loaded the first time in a session. The following code creates the view state variables:

```
ViewState["BasePrice"] = (decimal)0;
ViewState["ZonePrice"] = (decimal)0;
ViewState["Discount"]  = (decimal)0;
```

Figure 16-13 presents the complete code.

Figure 16-13: The code for Reservation page using view state

```csharp
decimal basePrice, zonePrice, discount;
protected void Page_Load(object sender, EventArgs e)
{
    if (!IsPostBack)     // If it is the first time the page is loaded,
    {
        ddlPlays.DataBind();     // bind the DropDownList
        ddlPlays.Items.Insert(0, new ListItem("", "")); // add an empty item
        // Initialize view state
        ViewState["BasePrice"] = (decimal)0;
        ViewState["ZonePrice"] = (decimal)0;
        ViewState["Discount"] = (decimal)0;
    }
}

protected void ddlPlays_SelectedIndexChanged(object sender, EventArgs e)
{
    // Find the base price for the selected play
    DataView dvPlaysTable =
        (DataView)SqlDataSource_Theater.Select(DataSourceSelectArguments.Empty);
    dvPlaysTable.RowFilter = "PlayId = " + ddlPlays.SelectedValue.ToString();
    DataRowView drvPlaysRow = (DataRowView)dvPlaysTable[0];
    basePrice = (decimal)drvPlaysRow["BasePrice"];
    lblBasePrice.Text = basePrice.ToString();   // Display base price
    // Compute and display unit price:
    ViewState["BasePrice"] = basePrice;
    ViewState["UnitPrice"] = (decimal)ViewState["BasePrice"] +
            (decimal)ViewState["ZonePrice"] - (decimal)ViewState["Discount"];
    lblUnitPrice.Text = ViewState["UnitPrice"].ToString();
}
```

```csharp
   42     protected void rblZones_SelectedIndexChanged(object sender, EventArgs e)
   43     {
   44         // Compute zone price
   45         switch (rblZones.SelectedValue.ToString())
   46         {
   47             case "Zone1":
   48                 zonePrice = 0;
   49                 break;
   50             case "Zone2":
   51                 zonePrice = 10;
   52                 break;
   53             case "Zone3":
   54                 zonePrice = 20;
   55                 break;
   56         }
   57         lblZonePrice.Text = zonePrice.ToString(); // Display zone price
   58         // Compute and display unit price
   59         ViewState["ZonePrice"] = zonePrice;
   60         ViewState["UnitPrice"] = (decimal)ViewState["BasePrice"] +
   61             (decimal)ViewState["ZonePrice"] - (decimal)ViewState["Discount"];
   62         lblUnitPrice.Text = ViewState["UnitPrice"].ToString();
   63     }
   64
   65     protected void chkMember_CheckedChanged(object sender, EventArgs e)
   66     {
   67         // Compute and display discount for members
   68         if (chkMember.Checked)
   69             discount = 10;
   70         lblDiscount.Text = discount.ToString(); // Display discount
   71                                                 // Compute and display unit price:
   72         ViewState["Discount"] = discount;
   73         ViewState["UnitPrice"] = (decimal)ViewState["BasePrice"] +
   74             (decimal)ViewState["ZonePrice"] - (decimal)ViewState["Discount"];
   75         lblUnitPrice.Text = ViewState["UnitPrice"].ToString();
   76     }
```

Step 1-21: Make changes in the code to preserve the data in view state, as shown in Figure 16-13:

Add the statements shown in lines 17 to 22 to the Load event handler of the page.
Make the changes to the SelectedIndexChanged event handler of ddlPlays, as shown in lines 36 to 39 in Figure 16-13, to use view state variables in place of class level variables.
Similarly, make the changes to the event handler of rblZones, as shown in lines 59 to 62, and make changes to the event handler chkMember, as shown in lines 72 to 75.

Step 1-22: Run the website and test the code. Make sure the unit price (= Base price + Zone price − Discount) is computed correctly.

It's time to practice! Do Exercise 16-1 Part B.

Review Questions

16.5 Describe the characteristics of view state.

16.6 What is the effect of executing this statement?
```
ViewState.Add("BasePrice", basePrice);
```

16.3: Multipage Websites and Session State

Developing websites involving multiple pages introduces additional features of ASP.NET. In order to understand how to open a web page from another page and share data between pages using session state, we will expand the current Theater website to include a second page named Confirmation that is shown in Figure 16-14. This page displays the selections made by the user on the Reservation page, computes and displays the total cost (= Number of tickets × Unit price), and gives the user the option to change the reservation or confirm it. The tutorials in this section that implement these changes constitute phase 3 of the project.

Figure 16-14: Confirmation form with data

Step 1-23: Add a new Web Form named Confirmation:
Right click the project name, Theater.
Select **Add > Web Form**.
Change Item Name to **Confirmation**. Click **OK**.

Step 1-24: Add controls to the Confirmation page as shown in Figure 16-15:

Figure 16-15: Confirmation page in Design view

Add a table with 6 rows and 2 columns.
Leave the first column blank, and insert text/controls into the second column as specified here:

Row 1: Enter "Confirmation."
    Select **Heading 2** from the *Block Format* dropdown list on the Toolbar.
    Select Format > Justify > Center.
Row 2: Enter "Your selection:"
    Highlight the text "Your selection:" and select Format > Font > Bold.
Row 3: Add a Label named lblReservation.
Row 4: (Leave blank.)
Row 5: Split the second column into two, and add two buttons named btnChange and btnConfirm into the two columns, with captions as shown in Figure 16-15.
Row 6: Add a Label named lblConfirm.

Select the table (click inside the table and press *Esc* key twice).
Select *Format > Borders and Shading*.
Select *Box* from *Border* tag, and *solid* for style.

Using Session State to Share Data between Pages

Session state stores information that needs to be shared between multiple pages during a specific session of a user. For example, the session state may be used to store a user's shopping cart as the user navigates the web pages. Unlike the view state, session state stores the data in the server memory, and the data is available till the user closes the browser or the session is timed out.

To identify a session, ASP.NET uses the value of the *sessionID* property of the Session object. If the request from the browser doesn't include a sessionID (e.g., in the first request for the page), ASP.NET starts a new session and sends the sessionID to the browser with the response. The browser returns the sessionID value with each request. If the time between requests exceeds a *time-out* value, a new session is started.

By default, the session ID value is stored in a *cookie* (text consisting of name value pairs that is stored on a user's machine to be retrieved by the website later). Alternatively, the application can be configured for a cookieless session that stores the sessionID in the URL of the page. This can be done by using the following setting in the web.config file:

```
<sessionState cookieless="true" />
```

Storing and retrieving data using session variables

Just like adding items to the ViewState object, you can add items to the collection contained in the Session object, which is accessed using the Session property of the Page class. You use the Session property to add an item using a syntax similar to the one used for view state:

```
Session.Add("name", value);
```
or
```
Session["name"] = value
```

Here is an example of each syntax:
```
Session.Add("BasePrice", basePrice);
```
or
```
Session["BasePrice"] = basePrice;
```

Both statements use an item named BasePrice to store the value of the variable basePrice.

An item that is added to the session state is often referred to as a **session variable**. In the previous example, BasePrice is the session variable.

Again, you don't have to declare the session variable. When you retrieve it, you cast it to the appropriate type, determined by the type of the value assigned to it. To retrieve the value of the session variable, BasePrice, you may use its name or its index in the collection:

```
double price = (double)Session["BasePrice"];
```
or
```
double price = (double)Session[0]; // assumes that the index for BasePrice is zero
```

You may access the Session object from a class other than the Page class (or a class that inherits the Page class), using the Session property of the HttpContext object:

```
HttpContext.Current.Session["BasePrice"] = basePrice;
```

Table 16-2 describes the properties/methods of Session object.

Table 16-2: Properties/methods of Session objects

Properties/Methods	Description
Count	Number of items in the collection e.g., int count = Session.Count;
Add (name, value)	Adds an item to the collection, or replaces if it exists e.g., Session.Add("BasePrice", basePrice);
Clear()	Removes all items from the collection e.g., Session.Clear();
Remove (name)	Removes the specified item from the collection e.g., Session.Remove("BasePrice");

Next you will learn how to move from one page to another, and use session state to share data between pages.

Moving from One Web Page to Another

There are several methods to display a page (target page) from another page (source page). We will look at four commonly used methods.

1. **Redirect:** This method uses the Redirect method of the HttpResponse class. You invoke the method using this syntax:
    `Response.Redirect (URL)`

   To redirect to the Confirmation page (target page) from the Reservation page (source page), you would code the Redirect method in the Reservation page like this:
    `Response.Redirect("Confirmation.aspx");`

   How does the Redirect method work? When this method is executed on the server, it sends a *redirect message* to the browser. The browser sends a *request* (HTTP GET) for the target page, just like when you click a hyperlink. You may ask, "Why not have the server send the Confirmation page the first time?" If the server sends a page without an HTTP request, the browser wouldn't know that it is a new page and hence would display the original URL (URL of the Reservation page) when it displays the Confirmation page. This would be misleading.

   The Redirect method also lets you pass variables to the page using a **query string** using the following pattern:
    `Response.Redirect (URL?variableName=value&variableName=value& …)`

   The "?" indicates the start of the query string and variableName and value indicate the name and value of the variable to be passed to the page. You may pass multiple variables using "&" between variables, but the size of the query string is limited.

   The following statement would pass the selected play and selected zone using query string:
    `Response.Redirect("Confirmation.aspx?play=" + ddlPlays.SelectedItem.ToString() +`
         `"&zone=" + rblZones.SelectedItem.ToString());`

   So, this method allows you to programmatically redirect to another page using code, by dynamically specifying the target URL and the query string.

   On the target page (Confirmation), you may get the value of the variable, play, from the QueryString collecton using the syntax
    `Request.QueryString["play"]`

   Query string is an alternative to using session variables to share data. When you pass variables using query string, the variables and their values are visible on the target page. So, this method is not appropriate for passing sensitive information. Further, if the same data is needed in multiple pages, you need to pass the data from page to page, whereas session variables can be stored once, and then accessed from any page. However, unlike session variables, query strings do not take up room in the server memory. The links can also be shared by users.

2. **RedirectPermanent:** When a page is permanently moved to a different location or is renamed, you use a different method of the HttpResponse class:
    `Response.RedirectPermanent (URL)`

   The RedirectPermanent works similar to Redirect, except that the server sends a different HTTP status code to the browser.

3. **Transfer:** A third option is to use the Transfer method of the HttpServerUtility class using this syntax:
   ```
   Server.Transfer (URL)
   ```

   To transfer from the Reservation page to the Confirmation page, you would code the Transfer method in the Reservation page, like this:
   ```
   Server.Transfer ("Confirmation.aspx");
   ```

   Similar to the Redirect method, the Transfer method allows you to programmatically redirect to another page, giving you the ability to dynamically specify the target URL and to include query string.

   This method, however, works differently from the Redirect method. When the Transfer method is executed on the server, it loads the Confirmation page and sends it to the browser. Since the page was sent by the server without an HTTP request, the browser is unaware of the transfer, and it would continue to display the previous URL—that is, the target page would display the source page's URL. Further, the target page must be in the same website as the source page.

   The Transfer method also allows the target page to reference the source page using the **PreviousPage** property exposed by the Page class. Thus the target page can access the controls and public properties on the source page. (Exposing controls as public properties would make it easy to access the controls on the previous page.) You will use the PreviousPage property later in this chapter.

4. **Cross-page posting:** In this method, you set the PostBackUrl property of a button to the URL of the page to be viewed, as shown here:

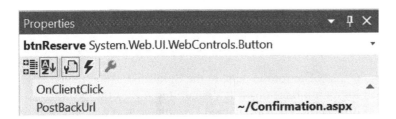

The tilde (~) operator is automatically inserted when you enter the URL. You may also specify a query string with the URL. Note that the link is hard-coded.

Essentially, cross-page posting posts the form data from the source page to the target page. When you click the button, the client posts the page to the server (by sending an HTTP POST command), which sends the values of controls on the source page to the target page, and sends the target page to the browser. The browser displays the target page and its URL.

The cross-page posting, similar to the Transfer method, allows the target page to reference the source page using the **PreviousPage** property of the Page class. Thus, if you use cross-page posting or Transfer method to display Confirmation page, you may access the Reservation page control txtNoOfTickets from the Confirmation page and display the number of tickets in the Label, lblReservation, using the code
```
TextBox txtNumTickets =
        (TextBox) PreviousPage.FindControl("txtNoOfTickets");
lblReservation.Text = txtNumTickets.Text;
```

The first statement uses the PreviousPage property to get the control, txtNoOfTickets; cast it to a TextBox; and assign it to the variable, txtNumTickets. The second statement uses the Text property of the TextBox to display the number of tickets.

Redirect, Transfer and cross-page posting can access information stored in session state and also use query string to pass data between two pages. But only the Transfer method and cross-page posting can use the PreviousPage property to access the source page.

Moving from Reservation Page to Confirmation Page

Which of the four methods discussed in the previous section should you use to display the Confirmation page? The RedirectPermanent method doesn't apply to the current example. The Transfer method has the drawback that it doesn't update the URL after a transfer. Both the Redirect method and cross-page posting would work.

To help you familiarize with both methods, you will use the Redirect method to display the Confirmation page, and cross-page posting to go back to the Reservation page. Session variables will be used to make reservation information available to Confirmation page and also to the Reservation page when the user revisits the page. Note that cross-page posting makes data from only the previous page available to the target page.

Step 1-25: Use the Redirect method in the Reservation page to display the Confirmation page:

Insert the following code into the Click event handler of btnContinue on the Reservation page:
```
protected void btnContinue_Click(object sender, EventArgs e)
{
        Response.Redirect("Confirmation.aspx");
}
```

Specifying the start page

To specify the start page, you right click the .aspx file for the page in the Solution Explorer and select *Set as Start Page*. If no start page is specified for a multipage project, Visual Studio runs the currently selected web form, if any. If no web form is selected, the form named Default.aspx, if any, is displayed. Running a project with no start page specified, no form selected and no Default.apsx file results in an error when you run the project. To view the different start options, right click the project name Theater and select *Start Options*.

Step 1-26: Specify the start page and run the website:

To specify Reservation page as the start page, right click Reservation.aspx and select *Set as Start Page*.
Run the website and click the *Continue* button to display the Confirmation page. The Confirmation page should open with no data in the Label, lblReservation.

Next, you will learn how to share data between the Reservation page and the Confirmation page. Recall that the Redirect method doesn't allow you to access the previous page using the PreviousPage property of the Page class. So you will use session variables.

Chapter 16: Web Development: Multipage and Database Applications

Storing reservation information in session variables

We will use session variables to store all reservation information that is needed for display and computation on the Confirmation page. This includes:

Selected play, selected zone and **member status** that are displayed on Confirmation page.
Unit price and **number of tickets** to compute the total cost for display on Confirmation page.

In addition, if the user comes back to the Reservation page from the Confirmation page, it is necessary to re-display the current values of controls on the Reservation page. Though these values persist during a postback, they are not automatically preserved when you leave the page and re-load the page. So, the following additional data need to be stored in session variables:

Base price, zone price, and **discount** to be re-displayed in Labels on the Reservation page
Index of selected play and **index of selected zone**, which are to be used to re-display the selections made by the user from the DropDownList and RadioButtonList

Currently the prices and discount are stored in view state variables. These four view state variables can be easily changed to session state variables by doing a search and replace of all occurrences of "ViewState" by "Session" in the entire C# code. The remaining data needs to be stored in session variables within the click event handler of btnContinue, as shown in Figure 16-16:

Figure 16-16: Code for the Click event handler of the Continue button

```
77
78      protected void btnContinue_Click(object sender, EventArgs e)
79      {
80          // Store the selected play, zone, member status and number of tickets
81          // in session variables
82          Session["SelectedPlay"] = ddlPlays.SelectedItem.ToString();
83          Session["SelectedZone"] = rblZones.SelectedItem.ToString();
84          Session["IsMember"] = chkMember.Checked;
85          Session["NoOfTickets"] = int.Parse(txtNoOfTickets.Text);
86          // Store indices of selected play and selected zone in session variables
87          Session["PlayIndex"] = ddlPlays.SelectedIndex;
88          Session["ZoneIndex"] = rblZones.SelectedIndex;
89
90          // Move to confirmation page
91          Response.Redirect("Confirmation.aspx");
92      }
```

Lines 82 to 85 store selected play, selected zone, member status and the number of tickets in session variables. These data items are needed on the Confirmation page.

Lines 87 and 88 store the indices of the selected play and zone in session variables to help redisplay the selections if the user revisits the Reservation page.

Line 91 uses the Redirect method to display the Confirmation page.

Step 1-27: Replace all occurrences of "ViewState" by "Session" in the C# code in Reservation.aspx.cs. Figure 16-17 shows the updated code with Session variables.

Step 1-28: Insert the code from Figure 16-16 to the Click event handler of btnContinue on the Reservation page.

Display Reservation data on the Confirmation page

The Confirmation page gets the Reservation data that are stored in session variables and displays them in the Label, lblReservation. Figure 16-17 presents the code.

Figure 16-17: Code in the Load event of the Confirmation page

```
12      protected void Page_Load(object sender, EventArgs e)
13      {
14          // Get selected play, zone, member status, # of tickets & unit price
15          //      from session variables and store them in simple variables
16          string selectedPlay = (string)Session["SelectedPlay"];
17          string selectedZone = (string)Session["SelectedZone"];
18          bool isMember = (bool)Session["IsMember"];
19          int noOfTickets = (int)Session["NoOfTickets"];
20          decimal unitPrice = (decimal)Session["UnitPrice"];
21          decimal totalCost = noOfTickets * unitPrice;
22
23          // Concatenate the reservation data into a single string with line breaks
24          // using the HTML tag, <br />
25          string reservation =
26              "Play:      " + selectedPlay.ToString() + "<br />" +
27              "Zone:      " + selectedZone.ToString() + "<br />";
28          if (isMember)
29              reservation = reservation + "Member?    Yes <br />";
30          else
31              reservation = reservation + "Member?    No <br />";
32          reservation = reservation + "# of tickets: " + noOfTickets.ToString()
33              + "<br />" + "Total Cost:" + totalCost.ToString();
34
35          // Display the string in the Label, lblReservation
36          lblReservation.Text = reservation;
37      }
```

Line 16 casts the value of the session state item (session variable), SelectedPlay, to string and stores it in the string variable selectedPlay. Lines 17 to 20 gets the values of the remaining session variables.

Lines 25 to 33 concatenate the reservation information into a single string with line feeds inserted using the HTML tag <br />. Line 36 displays the reservation data.

Note that the Label and the table cell where the Label is placed expands automatically depending on the number of lines displayed. So, unlike viewing information displayed in a ListBox, the user doesn't have to scroll when the size of the display exceeds the size of the control.

Confirming selection

Clicking the Confirm button on the Confirmation page displays a message in the Label, lblConfirm. Figure 16-18 shows the code in the Click event handler of btnConfirm.

Figure 16-18: Code for the Click event handle of btnConfirm

```
39  protected void btnConfirm_Click(object sender, EventArgs e)
40  {
41      lblConfirm.Text = "Your reservation is confirmed!";
42  }
```

Step 1-29: Insert the code from Figure 16-17 into the Load event handler of Confirmation page.

Step 1-30: Insert the code from Figure 16-18 into the Click event handler of btnConfirm on the Confirmation page.

Run the website. Make selections on the Reservation page and click the Continue button. Make sure the Confirmation page displays the data, as shown earlier in Figure 16-14.
Click the Confirm button to display the confirmation message.

Note that if you access the Confirmation page directly, or from a page other than the Reservation page, you will get an error because the session state items that are used in the Confirmation page are created in the Reservation page. Here we assume that the Confirmation page is accessed only from the Reservation page. In the next section you will learn how to check what the source page is.

It's time to practice! Do Exercise 16-2 Part A.

Review Questions

16.7   What would be a drawback of using the Transfer method from the user's perspective?

16.8   What are two differences between the Redirect and Transfer methods in functionality?

16.9   A website has three pages. The user enters an order using the first two pages in sequence, and the third page displays the data entered on the first two pages for confirmation. What is the drawback of sharing data between the pages using cross-page posting (without using session variables or query string) compared to the Redirect method with session variables?

16.10 A website has three pages. The user enters an order using the first two pages in sequence, and the third page displays the data entered on the first two pages for confirmation. What is the drawback of sharing data between the pages using the Transfer method with the query string compared to the Redirect method with session variables?

16.11 Describe the session state.

16.4 Setting a Page to Its Previous State

Users may want to go back to a page that was visited earlier and view previously made selections and data entered by the user. To enable this, it is necessary to preserve the data associated with the controls on the page. Further, when the user displays such a page, it is necessary to check whether the page was previously visited. Next you will learn how to re-display the user selections on the Reservation form, if the user goes back to it from the Confirmation page. Phase 4 of the project in the tutorials in this section will allow you to implement these enhancements.

Returning to Reservation Page

The Confirmation page has a button, btnChange, that lets the user go back to the Reservation page. If the user goes back to the Reservation page, the controls on the Reservation page should display the values previously set by the user, so that the user can make changes. However, if the user displays the Reservation page from a page other than the Confirmation page, or displays it directly, there are no previously set values to display. So, the Reservation page should identify the source page, and set the properties of the controls only if the source page is Confirmation.

Both the Transfer method and cross-page posting provides you with the PreviousPage property of the Page class to get access to the previous page from the Confirmation page. But, because the Transfer method doesn't update the URL, we will use **cross-page posting** to go back to the Reservation page. This involves setting the PostBackUrl property of btnChange.

Step 1-31: Set the PostBackUrl property of btnChange on Confirmation page to Reservation.aspx, as illustrated here. You may type in **Reservation.aspx,** or click the ellipses (…) on the right side and select Reservation.aspx. The characters "~/" are automatically inserted.

OnClientClick	
PostBackUrl	~/Reservation.aspx
SkinID	

Next, we will look at the code required in the Reservation page to check whether the previous page, if any, is the Confirmation page, and if so, to set the controls to the selections previously made by the user.

Getting the URL of the previous page

To check whether there is a previous page (source page), you use **PreviousPage** property of the current page. The PreviousPage property returns the previous page if there is one, and returns null if there is none. For instance, you may use the following if statement to take an action based on the existence of a previous page:

```
if (PreviousPage != null)
```

To get the name of the previous page, you use the UrlReferrer property of the HttpRequest class. The UrlReferrer returns the URL of the previous page. To get the previous page name from the URL, use the AbsolutePath property of the UrlReferrer:

```
Request.UrlReferrer.AbsolutePath
```

Note that you use the Request property of the current page to access the HttpRequest class. The AbsolutePath returns the previous page name with a slash (/) preceding the name, as in "/Confirmation.aspx."

The following code would get the name of the previous page, if it exists:

```
string previousPageName;
if (PreviousPage != null)
    previousPageName = Request.UrlReferrer.AbsolutePath;
```

The value of previousPageName could be null, "Confirmation" or some other page name.

If the name of the previous page is Confirmation, you need to display previously selected values in the controls on the Reservation page as explained next.

Setting Control Properties to Previously Selected Values

To help set the properties of controls on the Reservation page to their previously set values, you have already stored the required values in session variables. Now you will include the code in the Load event handler of Reservation page to use those values to set the controls to their previous state. Figure 16-19 presents the code.

The *if* statement in line 15 ensures that the code in the Load event handler is executed only when the Reservation page is loaded the first time, either directly or from another page.

As discussed earlier, the *if* statement in line 21 gets the name of the previous page, if there is one.

The *if* statement starting on line 25 specifies that if it is not a transfer from Confirmation page (i.e., if the Reservation page is displayed directly or from a different page), the session variables are created and set to zero in lines 27–29.

If the Reservation page is displayed from the Confirmation page, the control properties are set to previously selected values using the statements in lines 33-41.

Step 1-32: Make changes in the Load event handler of Reservation page, as shown in Figure 16-19.

Figure 16-19: Load event handler of Reservation page

```
12      decimal basePrice, zonePrice, discount;
13      protected void Page_Load(object sender, EventArgs e)
14      {
15          if (!IsPostBack)     // If it is not a postback of the page,
16          {
17              ddlPlays.DataBind();    // bind the DropDownList
18              ddlPlays.Items.Insert(0, new ListItem("", "")); // add an empty item
19              // if transferring from another page, get the name of that page
20              string previousPageName = "";
21              if (PreviousPage != null)
22                  previousPageName = Request.UrlReferrer.AbsolutePath;
23
24              // if not transferring from Confirmation page, create session variables
25              if (previousPageName != "/Confirmation.aspx")
26              {
27                  Session["BasePrice"] = (decimal)0;
28                  Session["ZonePrice"] = (decimal)0;
29                  Session["Discount"] = (decimal)0;
30              }
31              else            // if transferring from Confirmation page
32              {   // Set control properties to previously selected values
33                  ddlPlays.SelectedIndex = (int)Session["PlayIndex"];
34                  rblZones.SelectedIndex = (int)Session["ZoneIndex"];
35                  chkMember.Checked = (bool)Session["IsMember"];
36                  txtNoOfTickets.Text = Session["NoOfTickets"].ToString();
37                  // Re-display computed values
38                  lblBasePrice.Text = Session["BasePrice"].ToString();
39                  lblZonePrice.Text = Session["ZonePrice"].ToString();
40                  lblDiscount.Text = Session["Discount"].ToString();
41                  lblUnitPrice.Text = Session["UnitPrice"].ToString();
42              }
43          }
44      }
```

Figure 16-20 presents the remaining code for the Reservation page.

Figure 16-20: Updated code with Session variables for all event handlers

```csharp
45
46  protected void ddlPlays_SelectedIndexChanged(object sender, EventArgs e)
47  {
48      // Find the base price for the selected play
49      DataView dvPlaysTable =
50          (DataView)SqlDataSource_Theater.Select(DataSourceSelectArguments.Empty);
51      dvPlaysTable.RowFilter = "PlayId = " + ddlPlays.SelectedValue.ToString();
52      DataRowView drvPlaysRow = (DataRowView)dvPlaysTable[0];
53      basePrice = (decimal)drvPlaysRow["BasePrice"];
54      lblBasePrice.Text = basePrice.ToString();    // Display base price
55                                                    // Compute and display unit price:
56      Session["BasePrice"] = basePrice;
57      Session["UnitPrice"] = (decimal)Session["BasePrice"] +
58              (decimal)Session["ZonePrice"] - (decimal)Session["Discount"];
59      lblUnitPrice.Text = Session["UnitPrice"].ToString();
60  }
61
62  protected void rblZones_SelectedIndexChanged(object sender, EventArgs e)
63  {
64      // Compute zone price
65      switch (rblZones.SelectedValue.ToString())
66      {
67          case "Zone1":
68              zonePrice = 0;
69              break;
70          case "Zone2":
71              zonePrice = 10;
72              break;
73          case "Zone3":
74              zonePrice = 20;
75              break;
76      }
77      lblZonePrice.Text = zonePrice.ToString(); // Display zone price
78      // Compute and display unit price
79      Session["ZonePrice"] = zonePrice;
80      Session["UnitPrice"] = (decimal)Session["BasePrice"] +
81              (decimal)Session["ZonePrice"] - (decimal)Session["Discount"];
82      lblUnitPrice.Text = Session["UnitPrice"].ToString();
83  }
```

```csharp
84
85      protected void chkMember_CheckedChanged(object sender, EventArgs e)
86      {
87          // Compute and display discount for members
88          if (chkMember.Checked)
89              discount = 10;
90          lblDiscount.Text = discount.ToString(); // Display discount
91          // Compute and display unit price:
92          Session["Discount"] = discount;
93          Session["UnitPrice"] = (decimal)Session["BasePrice"] +
94              (decimal)Session["ZonePrice"] - (decimal)Session["Discount"];
95          lblUnitPrice.Text = Session["UnitPrice"].ToString();
96      }
97
98      protected void btnContinue_Click(object sender, EventArgs e)
99      {
100         // Store the selected play, zone, member status and number of tickets
101         // in session variables
102         Session["SelectedPlay"] = ddlPlays.SelectedItem.ToString();
103         Session["SelectedZone"] = rblZones.SelectedItem.ToString();
104         Session["IsMember"] = chkMember.Checked;
105         Session["NoOfTickets"] = int.Parse(txtNoOfTickets.Text);
106         // Store indices of selected play and selected zone in session variables
107         Session["PlayIndex"] = ddlPlays.SelectedIndex;
108         Session["ZoneIndex"] = rblZones.SelectedIndex;
109
110         // Move to confirmation page
111         Response.Redirect("Confirmation.aspx");
112     }
```

Step 1-32: Run the website. Test the code by specifying the reservation information and transferring to the Confirmation page.

Next, click the Change Reservation button to go back to the Reservation page. Make sure the selections you made previously on the Reservation page are displayed on the page.

Change the selections. Make sure the prices/discount are updated when you change the selections.

Session variables of array type

A session variable could be of any .Net Framework type, including value types, arrays, collections, structures and objects. Rather than storing individual data items in separate session variables, you can store a group of related data items in an array and store the array in a single session variable. The following code illustrates the use of a session variable of array type, which stores several reservation data items.

```csharp
string[] reservationInfo = {ddlPlays.SelectedItem.ToString(),
    rblZones.SelectedItem.ToString(), chkMember.Checked.ToString(),
    txtNoOfTickets.Text, ddlPlays.SelectedIndex.ToString(),
    rblZones.SelectedIndex.ToString()};

Session["ReservationInfo"] = reservationInfo;
```

The first statement in the above code stores the six data items (selected play, zone, member status, number of tickets, index of selected play and index of selected zone) from the click event handler of btnContinue in a string array, reservationInfo. The second statement stores the array in the session variable, ReservationInfo. The above two statements could replace the statements in lines 102 to 108 in btnContinue_Click event handler.

To display the data on the Confirmation page, you may retrieve the data from the session variable and store them in a string array:

```
string[] reservationInfo = (string[]) Session["ReservationInfo"];
```

The data stored in the string array may be displayed in the Label lblReservation on the Confirmation page using this code:

```
string selection = "Play:    " + reservationInfo[0] + "<br />" +
                   "Zone:    " + reservationInfo[1] + "<br />";
if (reservationInfo[2] == "True")
        selection = selection + "Member?  " + "Yes";
else
        selection = selection + "Member?  " + "No";
selection = selection + "# of tickets: " + reservationInfo[3];

lblReservation.Text = selection;
```

The above code uses the index of the array element to access an individual data item.

If you are familiar working with structures and/or classes, you may use a structure or class in place of the array so that you can refer to individual data items by a name rather than by the index.

It's time to practice! Do Exercise 16-2, Part B.

Exercises

Exercise 16-1: Auto rental—single page

This exercise develops a simplified web version of the Ace Auto Rentals system specified in Programming Assignment 2. "Ace Auto Rentals" is a small business that rents automobiles. Ace would like you to develop a web page named AutoRentals, within the website Exercises_Ch16, which allows the rental agency to do the following for each car rental:

Part A

1. Select the type of auto (SUV, minivan, luxury, full size, midsize or compact) from a DropDownList that automatically displays the daily rental rate in a Label or ListBox.
   Each auto type and the corresponding daily rental rate (SUV = $ 80, minivan = $70, luxury = $50, full size = $40, midsize = $35 and compact = $25) are already stored in the SQL Server database table RentalRate (AutoType, RentalRate) within the database file, AutoRentals.mdf.
   Your application must look up the daily rental rate from the RentalRate table.
2. Select one of the discount categories (Normal, State Employees, Business or Favorite Customer) from a RadioButtonList, which automatically displays the discount amount in a Label or ListBox. Discount: Normal = 0%, State employees = 15%, Business = 10%, Favorite = 20%
3. Specify whether the customer needs insurance by checking a CheckBox, which automatically displays the corresponding insurance cost in a Label or ListBox. The cost of insurance is $10/day. There is no discount for insurance cost.

Part B: Do Part A and let the user do the following additional tasks:

4. Enter the number of days of rental into a TextBox.
5. Compute and display the total cost by clicking a button.
   Total cost = (Rental rate − Discount + Insurance) × Number of days
   Store the rental rate, discount and insurance in view state variables when the user makes the selections, and use the variables to compute the total cost.

Exercise 16-2: Auto rental—multipage

Part A

This exercise expands Exercise 16-1 to include two pages. Develop the web page AutoRentals as specified in Exercise 16-1, Part B, except that the page doesn't have the button to compute the total cost (step 5).

Add a new page named RentalCost. Provide a button on AutoRentals page to display RentalCost page.

When you display the RentalCost page, it should display the selections from the AutoRentals page and the number of days, in a ListBox, neatly formatted. Provide a button on the RentalCost page to compute and display the total cost:
   Total cost = (Daily rental rate − Discount + Insurance) × Number of days

Part B

Do Part A and provide a second button on RentalCost page to let the user go back to AutoRentals page and make changes, if any, in the rental information. When the user returns to the AutoRentals page, the page should display the selections previously made by the user, and the data displayed in the controls (daily rental rate, discount amount, insurance cost and number of days).

Use one of the following options to share data between pages:

1. Use individual session variables to store each piece of data.
2. Store all data items in an array and use a single session variable to share the array.
3. Store all data items in a class and use a single session variable to share the class.
4. Store all data items in structure and use a single session variable to share the structure.

Appendix A

Binary Files

A.1 Writing to Binary Files

The key differences between text files and binary files are

1. Text files store all data as text, whereas binary files store data in its original form without converting to text using the standard internal representation of numbers, dates, and so on. Therefore, the contents of text files can be viewed and edited in a text editor like Notepad, but binary files are not readable in a text editor.
2. Data typically is written to text files by combining an entire record into a single string and writing it in a separate line using the WriteLine() method of StringWriter. Data is written to binary files individually using the **Write()** method of BinaryWriter. Data is not grouped into rows.
3. The StreamWriter object to write data to text files can be created without explicitly creating a FileStream object. However, creating a BinaryWriter to write data to a binary file requires creating the FileStream object that is used by the BinaryWriter, as follows:

>   FileStream variableName = newFileStream("FileName", FileMode);
>   StreamWriter variableName = newBinaryWriter(FileStream object);

>   **Example**
>   FileStream examsStream = newFileStream("Exams.bin", FileMode.Append);
>   StreamWriter examsWriter = newBinaryWriter(examsStream);

FileMode.Append: Append specifies that new data is added to existing data. If the file doesn't exist, it will be created.

FileMode.OpenOrCreate would overwrite existing data when new data is written. If the file doesn't exist, it will be created.

Tutorial: Working with Binary Files

Step A-1: Open the project, Ch8_FilesAndArrays. Open the WriteToBinaryFile form that has the same user interface as the WriteToTextFile form that you used in Chapter 8. Open the code window. Figure A-1 shows the complete code to write data from the TextBoxes to a specified file.

Step A-2: Add the code in lines 19 and 20, and in lines 30 and 32 from Figure A-1. Run the project. Select a folder and enter Exams.bin for file name.

Step A-3: Enter the name, score1 and score2, for five students as shown below, save them, and exit the form:

Mike Burns	75	92
Cathy Arnold	86	95
Wesley Mathews	82	90
Jerry Luke	70	89
Gina Rodriguez	77	94

Open Exams.bin with Notepad. (Right click the file, select *Open with,* and select *Notepad.*) Note that the two scores are not readable because they are not stored as text.

Figure A-1: Writing to binary file

```
 9  using System.Windows.Forms;
10  using System.IO;     // Add this line
11  namespace Ch8_FilesAndArrays
12  {
13      public partial class WriteToBinaryFile : Form
14      {
15          public WriteToBinaryFile()
16          {
17              InitializeComponent();
18          }
19          FileStream examsStream;      // Declare variable to represent FileStream object
20          BinaryWriter examsWriter;    // Declare variable to represent BinaryWriter object
21          private void WriteToBinaryFile_Load(object sender, EventArgs e)
22          {
23              ExamsSaveDialog.ShowDialog();   // Display Save As Dialog box
24              // Get the file name selected by the user and save in the variable fileName
25              string fileName = ExamsSaveDialog.FileName;
26
27              if (fileName.Length > 0)    // if the user selected a file
28              {
29                  //Create FileStream object and assign to examsStream:
30                  examsStream = new FileStream(fileName, FileMode.Append);
31                  //Create BinaryWriter objects and assign to examsWriter:
32                  examsWriter = new BinaryWriter(examsStream);
33              }
34              else
35              {
36                  MessageBox.Show("No file selected");
37                  btnSave.Enabled = false;
38              }
39          }
```

A.2 Reading Binary Files

Data is read from a Binary file using the BinaryReader object. Similar to the BinaryWriter object, the BinaryReader object uses the FileStream object, created as follows:

> FileStream*variableName* = newFileStream(@"*FileName*", FileMode);
> BinaryReader*variableName* = newBinaryReader(FileStream);

Example:
> FileStream examsStream = newFileStream(@"E:\Exams.bin", FileMode.Open);
> BinaryReader examsReader = newBinaryReader(examsStream);

FileMode.Open: Opens the file if it exists; if not, an exception is thrown.

The code to read Exams.bin file (which was created in the form WriteToBinaryFile), and compute the average score for each student, is shown in Figure A-2. Exams.bin contains the name, score1 and score2, for multiple students.

Figure A-2: Reading from Binary Files

```
21          private void btnComputePercent_Click(object sender, EventArgs e)
22          {
23              string name, fmtStr;
24              float score1, score2;
25              float percentScore;
26
27              FileStream examsStream = new FileStream("Exams.bin", FileMode.Open);
28              BinaryReader examsReader = new BinaryReader(examsStream);
29              fmtStr = "{0, 15}{1, 6}{2, 6}{3, 6:N1}";
30              lstDisplay.Items.Clear();
31              while (examsReader.PeekChar() != -1)
32              {
33                  // Read the next string (name)from Exams.bin
34                  name = examsReader.ReadString();
35                  // Read the next number (float type) from Exams.bin
36                  score1 = examsReader.ReadSingle();
37                  score2 = examsReader.ReadSingle();
38                  percentScore = 100*(score1 + score2) / 200;
39                  //Display the data and average in the List  box
40                  lstDisplay.Items.Add(string.Format(fmtStr, name, score1, score2, percentScore));
41              }
42              examsReader.Close();
43          }
```

Lines 27–28: Create the FileStream and BinaryReader objects.

Line 31: The BinaryReader doesn't have an EndOfStream property. Instead, it has the PeekChar method that returns -1 when the end of file is reached.

Lines 34–37: Each data item is read separately. There is a separate method to read each type of data:

ReadString to read a string
ReadSingle to read a float type
ReadDouble to read a double type

An advantage of using binary files is that there is no need to convert data items like score1 and score2 from text to numeric types, as is necessary when reading from text files.

Step A-4: Open ReadBinaryFile form. Add the missing code from Figure A-3. Test the code to make sure you get the output as shown below:

Appendix B

Creating a Database

SQL Server databases can be created in Visual Studio because the Express edition of SQL Server DBMS is automatically installed on your computer when you install Visual Studio.

Tutorial: Creating HR Database

This tutorial creates a database called HR (for human resources) that includes a table named Employee (EmpId, LastName, FirstName, DateOfBirth, Phone, DeptNo, CountryId, Salary, FullTime).

Step B-1: Create a new project named Ch11_Databases and rename the form to **EmpDirectory**.

Step B-2: Create a new database as follows:

> Select: **Project, Add New Item, Service-based Database.**
>
> Change the name of the database to: **HR.mdf,** andClick: **Add.**
>
> If the **Data Source Configuration Wizard window** opens, asking you information about the data source, click Cancel because currently there are no tables in the database.
> Notice that the Solution Explorer now includes the HR database file. In addition, you will see the **Server Explorer** (or Database Explorer if you are using Visual Studio Express) on the left side of the form will show the name of the database, shown in Figure B-1. If the Server Explorer is not visible, select *View, Server Explorer.*

Figure B-1: Server Explorer

> Open the project directory, Ch12_Databases. You will see the database file, HR.mdf.

Step B-3: To add a table to the database, right click Tables and select **Add New Table**. You will see the Table Designer window, shown in Figure B-2.

Figure B-2: Table Designer window

Step B-4: Specify the field names, data types and other attributes, and change the table name in the T-SQL statement at the bottom (CREATE TABLE [dbo].[Table]) to **Employee**, as shown in Figure B-3.

Figure B-3: Field names and attributes in Table Design window

Name	Data Type	Allow Nulls	Default
EmpId	int	☐	
LastName	nvarchar(50)	☐	
FirstName	nvarchar(50)	☑	
DateOfBirth	date	☑	
Phone	nvarchar(50)	☑	
DeptNo	int	☑	
CountryId	int	☑	
Salary	float	☑	((0))
FullTime	bit	☑	

CREATE TABLE [dbo].[Employee] (

Step B-5: Click the *Update* button at top left. On the *Preview Database Update* window, click *Update Database* button. You will see the message "Update completed successfully."

Step B-6: Click the Refresh button on the Server Explorer window and expand *Tables*.
Right-click Employee and select *Show Table Data* from the menu.
You now may enter records into the table, as shown in Figure B-4.

Figure B-4: Adding records to table

EmpId	LastName	FirstName	DateOfB...	Phone	DeptNo	CountryId	Salary	FullTime
11002	Iwuji	Chris	11/1/1992	920-426-...	1	62	46200	True
11005	Mathews	Ruth	3/24/1985	646-233-...	4	22	44600	True

You may copy HR.mdf file to a folder outside the current project directory for use in another project. But, before you can copy the file, you need to close the current project.

You may create additional tables within the HR database using the same process.

Appendix C

Answers to Review Questions

Chapter 1

1.1   Input: key word(s) that you type in.
        Process: find web pages related to the key word(s)
        Output: information on related websites

1.2   Input: student Id, course number, section number
        Process: check availability of seats, prerequisites and conflicts with other classes; add student to class list.
        Output: student schedule, class list

1.3   Define the purpose and identify the input, process and output of the program.
Design and develop the Graphical User Interface (GUI).
Identify the components and logic of the program.
Design and develop files/databases, if any.
Write and test the program.

1.4   Syntax error

1.5   Logic error

1.6   False. There could be logic errors that produce wrong results

1.7   Machine Language

1.8   A complier is a special software that translates programs from a high-level language to Machine Language or to an intermediate language to create an executable program.

1.9   An interpreter converts each statement in a program to Machine Language and runs it without producing an executable program.

1.10   Microsoft Intermediate Language Assembly is the executable program created by the compiler in Visual Studio.

1.11   The Common Language Runtime translates each statement from Microsoft Intermediate Language to Machine Language and runs it.

1.12   Text property

1.13   The form specified in the statement Application.Run(new …()).

1.14   TextBox

1.15   Text

1.16 A namespace is a container for a group of classes in the .Net Framework Class Library

1.17 A using directive makes a group of classes from the .Net Framework Class Library available within a form.

1.18 A method is a program unit that performs a task.

1.19 An event handler is a method that executes in response to an event.

1.20 float

1.21 decimal

1.22 float and double types take less memory than decimal, and they are faster to work with.

1.23 The first letter of a variable is lowercase, and the first letter of subsequent words, if any, are capitalized.

1.24 A variable name must start with a letter. A variable can have only letters, numbers and underscore.

1.25 double orderQty = 4.5;

1.26 string lastName = "Jones";

1.27 string firstName = "Jack";

1.28 double totalAmount = 145.50;

1.29 double totalCost = 2145.00;

1.30 double unitPrice = 15.50;

1.31 int quantity;
quantity = 15;

1.32 int quantity = int.Parse(txtQuantity.Text);

1.33 int units = int.Prase(txtUnits.Text);

1.34 currentSales: 0
previousSales: 0

1.35 currentSales: 110
previousSales: 110

1.36 17

1.37 4

1.38 4,525.46

1.39 variable

1.40 A constant must be initialized when it is declared. A constant cannot be assigned a value after it is declared.

1.41 A local variable is a variable that is declared inside a method.

1.42 A field is a variable that is declared at the class level and it is available in all methods within a form. A field exists until the form is closed.

Chapter 2

2.1 decimal

2.2 float and double types take less memory than decimal, and they are faster to work with.

2.3 double

2.4 decimal

2.5 no error

2.6 float cost = **(float)** 2.50;   or   float cost = 2.50F;

2.7 no error

2.8 **short** quantity = 200;

2.9 decimal price = 300;
float currentPrice = **(float)** price;

2.10 no error

2.11 int qty1 = 1000000000;
int qty2 = 2000000000;
long total = (long) qty1 + qty2;

2.12 **double** total = 255.50 + 500;   // or, float total = (float)255.50 + 500

2.13 int totalScore = 25;
int count = 4;
float average = (float)total/count;

2.14 decimal price = 100;
float quantity = **(float)**5.5;
decimal netCost = price***(decimal)**quantity - 10

2.15 (fName.Substring(0, 2) + " " + lName).ToUpper()

2.16 id.Substring(studentId.Length-2, 2);

2.17   Both Today and Now includes today's date. Now also includes the current time, but Today includes only the default time 12:00:00 AM.

2.18   TimeSpan tsRental = DateTime.Now - rentalDate;
lstDisplay.Items.Add(tsRental.Days);
// or,
TimeSpan tsRental = DateTime.Today - rentalDate;
lstDisplay.Items.Add(tsRental.TotalDays);

2.19   TimeSpan tsRentalTime = new TimeSpan(rentalTime.Hour, rentalTime.Minute, rentalTime.Second);
rentalDateTime = rentalDate.Add(tsRentalTime);

2.20   DateTime dueDate = invoiceDate.AddDays(30);

Chapter 3

3.1   5

3.2   0

3.3   Admit

3.4   Deny

3.5   Deny   -- ((ACT >= 24) || (GPA >= 3)) yields true. ! true yields false.

3.6   Admit -- ((ACT >= 24) && (GPA >= 3)) yields false. ! false yields true.

3.7   C

3.8   string grade;
int totalScore = 85;
if (totalScore >= 70)
         grade = "Pass";
else
         grade = "Fail";

3.9   10

3.10 See the diagram that follows:

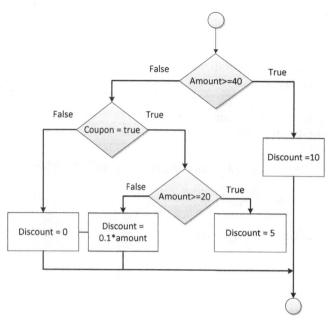

3.11   float amount = 15.0F;
           float discount;
           bool coupon = true;
           if (amount >= 40)
              discount = 10;
           else if (coupon = true)
              if (amount >= 20)
                 discount = 5;
              else
                 discount = 0.1F*amount;
           else
              discount = 0;

3.12   Integer, String and Boolean

3.13   int rating = 2;
           double raise;
           switch (rating)
           {
             case 3:
                raise = 0.25;
                break;
             case 2:
                raise = 0.1;
                break;
             default:
                raise = -1.0;
                break;
           }

3.14   false

3.15   0

3.16   true

3.17   When the user moves focus away from the control

3.18   The CauseValidation property of both controls must be set to true.

Chapter 4

4.1   ```
int yearlyInvestment = 750;
int totalAmt = 0;
int year = 0;
while (year < 4)
{
  year = year + 1;
  totalAmt = totalAmt + yearlyInvestment;
}
```

4.2   750

4.3   2017   76000
      2018   77000

4.4   0

4.5   1

4.6   2016 66000
      2017 67000
      2018 68000

4.7   2018 64000
      2017 63000

4.8   The break statement lets you exit a loop. The continue statement lets you skip statements within the body of the loop but continue the iterations.

## Chapter 5

5.1   A method is a program unit that performs a particular task

5.2   An event handler is a special type of method that is executed when a certain event of a control takes place.

5.3   True
5.4   True

5.5   ComputeCost();
```

5.6 An argument is a data item that is passed to a method by the calling program.

5.7 A parameter is a variable that receives an argument that is passed to a method.

5.8 An argument is associated with the parameter that is in the same position as the argument, unless the name of the parameter is specified along with the argument.

5.9 The scope of a parameter is the method in which it is declared.

5.10 False

5.11 Hi
There

5.12 When an argument is passed by value, only its value is passed to the corresponding parameter.

5.13 0

5.14 Top-down design involves breaking down a larger problem to smaller subtasks, each performed by a separate method.

5.15 Makes software development simpler and easier.
Makes the code more re-usable.
Easier to get a high-level view of the code.

5.16 When an argument is passed by reference, the memory location (address or reference) of the argument variable is passed to the parameter so that the called method can change the value of the argument by changing the value of the corresponding parameter.

5.17 150 100 250

5.18 0 0 25

5.19 An enumeration is a set of named integer constants

5.20 An enumeration provides a convenient way to select from a limited set of numbers. In addition, an enumeration uses meaningful names in place of numbers, which helps minimize errors in making a selection.

Chapter 6

6.1 Maximum property should be set to 10.

6.2 lblUnits.Text=hsbUnits.Value.ToString

6.3 hsbUnits_ValueChanged

6.4 True

6.5 Checked

6.6   rdbDefault.Checked = true

6.7   Two

6.8   cboNames.SelectedItem (and cboNames.SelectedText, but, it is not available in event handlers).

6.9   cboNames.SelectedIndex = -1

6.10 A CheckBox

6.11 A ComboBox

6.12 A set of CheckBoxes

6.13 Basketball

6.14 Cricket

Chapter 7

7.1   A value type variable holds the actual value assigned to that variable, whereas a reference variable holds only the memory location of the object assigned to the variable.

7.2   12

7.3   string[] student = new string[20];

7.4   string[] student;
      student = new string[20];

7.5   string[] weekDay = {"Sun", "Mon", "Tue", "Wed", "Thu", "Fri", "Sat"};

7.6   weekDay[1] = "Monday";

7.7   lstDisplay.Items.Add(weekDay[1]);

7.8   for (int index = 0; index <= weekDay.Length-1; index = index + 1)
          lstDisplay.Items.Add(weekDay[index]);

7.9   The index 7 in testScore[7]) causes a runtime error because 7 is outside the valid range (0-6).

7.10 foreach (string day in weekDay)
          lstDisplay.Items.Add(day);

7.11 string[] workday = new string[5];
      Array.Copy(weekDay, 1, workday,0,5);

7.12 Monday

7.13   Mon

Chapter 8

8.1   System.IO

8.2   The method reads and returns the line at the current read position and then advances the read position to the next line. An exception is thrown if no data are found.

8.3   75648

8.4   extCost = fields[2]*fields[3]

8.5   orderNumbers[index] = fields[1];
       extCosts[index] = extCost;

8.6   foreach (string show in shows)
       cboShows.Items.Add(show);

8.7   int index = Array.IndexOf(months, txtMonthName.Text);
            lstDisplay.Items.Add (unitsSold[index]);

8.8   If ExamWriter is declared in the Load event handler, it won't be available in the btnSave_Click event handler, and in the FormClosing event handler, resulting in error.

8.9   If ExamWriter is declared and created in btnSave_Click event handler, it won't be available in the FormClosing event handler of the form.

8.10   If ExamWriter is declared, created and closed in btnSave_Click event handler, ExamWriter will be created and closed each time the Save button is clicked. This is less efficient than creating it once in the Load event handler, using it multiple times and then closing when the form is closed.

8.11   205

Chapter 9

9.1   You can add items to a collection without specifying the index position. Further, the size of a collection is increased automatically when you insert new items.

9.2   Generic collections are strongly typed; that is, they allow you to specify a single type that can be stored in the collection

9.3   prices.Add(fields);

9.4   string price;
       foreach (string[] row in prices)
       {
            if (row[0] == "1001")
                 price = row[1];
       }

```
            // An alternative is to use the index of the row, as follows:
            string[] row = prices[4];
            string price = row[1];
```

9.5 A Dictionary requires a unique key, which makes it easier to find an item using its key. Further, a Dictionary is implemented as a hash table, whereas a List is implemented as an array.

Chapter 10

10.1 A ListView lets you add multiple data items to a row and lets you directly access each item from a row. A ListBox lets you add only a single string to a row.

10.2 A ListView is a GUI control. Each row of a ListView is a ListViewItem object

10.3 The Lion King

10.4 80

10.5 When you use multiple tab pages on a single form, you can access controls on any one tab page from another page by using the name of the control. By contrast, to access controls on one form from another form, you need to create an instance of the page,

Chapter 11

11.1 DisplayOrders displayOrders = new DisplayOrders();
displayOrders.Show();
float totalCost = float.Parse(displayOrders.lblTotalCost.Text);

11.2 Hiding a form keeps the form and its properties and settings in memory. You don't lose the data entered/displayed in controls at runtime.
Closing a form removes the form from memory. So, you lose all current settings and data.

11.3 The Show method opens the form and C# continues processing the statements that follow the statement. However, with the ShowDialog method, C# opens the form and stops processing the statements that follow the statement until the dialog window is closed.

11.4 Create a Public method within a class and call it from different forms.

11.5 A static method is not associated with a specific instance (object) of the class that contains the method. It is associated only with the class. So, it is invoked using the class name.

To call a nonstatic method, you have to create an object of the class and use the object name in place of the class name.

11.6 A context menu is associated with a control, and it is displayed by right clicking the control.

11.7 Option 3: Selecting an item from a Context menu

Chapter 12

12.1 data tables are temporary tables maintained in memory. Each data table may contain data from one or more database tables that are typically stored on the disk.

12.2 The table adapter connects the data table to the database, and retrieves data from and saves data to the database.

12.3 The binding source links the user interface controls on the form to a data table.

12.4 The Fill method of the table adapter copies data from the database table into the data table.

12.5 The Update method of the table adapter saves the data from the data table into the database table

12.6 A data bound control is a user interface control that is linked to a data source

12.7 The data in a data bound control is linked to the binding source, and the binding source is linked to the data table.

12.8 Set all DataBindings to **None**.
Set DataSource to CustomerBindingSource
Set DisplayMember to CustomerId

12.9 The user will be able to see only employee records with CountryId, 22.
It will have no effect on the records in the data table. The data table still will have all records from Employee table.

12.10 The Employee data table will have only records for employees from DeptNo 11.

12.11 Connection string, SQL

12.12 Creating data access objects at runtime using code makes the code more flexible, efficient and reusable.

12.13 Typed data tables contain fields that are of the same type as the fields in the database table, and they are created typically at design time. Untyped data tables do not contain information on its fields. They are created at runtime.

12.14 Data bound controls are bound to specific fields in the data table through the binding source. So, data on the form is tightly linked to the fields in the data table. The current row is automatically displayed in the controls. Unbound controls are not directly bound to any field.

12.15 Update method of the table adapter.

12.16 Set the DataSource property of the binding source to the data table.

12.17 Help move between rows, get the position/count of rows, and select rows using its Filter property.

12.18 Store the current row in a DataRowView object, and copy the data from the DataRowView to the controls on the form.

12.19   Both codes let the user view IS student records. Code segment #1 will load only IS student records into the data table. Code segment #2 will load all student records into the data table, but will allow the user to view only IS student records.

12.20   The EndEdit method applies the changes in the binding source rows to the data table it is linked to.

12.21   False

12.22   The Update method of the data adapter saves the changes in the data table to the database table.

12.23   A **DataReader** provides an efficient way to sequentially access the data retrieved by an SQL statement from a database

12.24   False

12.25   False

12.26   ExecuteScalar

12.27   ExecuteReader

12.28   ExecuteNonQuery

12.29   Advances the DataReader to the next row

Chapter 13

13.1   Identify several objects in your world from the following classes:
- Book
- Car
- Account
- Student
- Professor

Answers will vary from student to student.

13.2   For each of the classes above, identify a few attributes and behaviors that might be relevant to represent in an information system.

Book: title, author, publication year, number of pages, publisher, topic

Car: make, model, color, year, type (sports car, truck, sedan), engine size, number of cylinders, number of doors

Account: balance, transactions, type (checking, savings), overdraft limit, interest rate

Student: name, address, student id, classes taken, major, grades received, class, graduation year

Professor: name, address, employee id, department, title, year employed

13.3 In your own words, describe the difference between class and object.

Answers will vary. Look for whether they distinguish between classes as the blueprint for objects.

13.4 Imagine a Thermostat class that allows for a temperature to be set to specific values within minimum and maximum bounds. Discuss how encapsulation can be helpful for this class, as well as what problems might arise if data in the class were not encapsulated.

The temperature could be set internally to Celsius or Fahrenheit, and then conversions could expose the other type without problems.

If the thermostat is envisioned to allow a user to increase or decrease the temperature, methods could be provided for increase or decrease but leaving the temperature itself as read-only.

If the temperature property wasn't encapsulated, it would allow for it to be set to values outside the maximum and minimum limits, or to be set to a value that doesn't make sense according to the unit being used (e.g., setting temp to 22 has a very different meaning in Celsius versus Fahrenheit—one is very comfortable while the other is very cold).

Encapsulation can also help ensure that max is always higher than min.

13.5 Draw a UML class diagram for the Thermostat class in the previous question.

Thermostat
+ MaxValue: double +MinValue: double - temp: double + Temp: double --read-only
+ Increase(double) + Decrease(double)

13.6 What are the advantages and disadvantages of automatic properties?

Advantages:

- Less boilerplate code
- Faster to write code
- Code that is easier to read

Disadvantages:

- You have to know what the syntax means, as it isn't immediately intuitive.
- This only works for read/write properties.
- Changing a property to read-only means having to add both a backing instance variable and changing the property, as well as anywhere in the class the property is being written to.
- You cannot have any conversion or calculation code in an automatic property.

13.7   Why should fields always be marked as private?

This allows for a private implementation that allows for the implementation to be changed later. For instance, it may allow for conversion of a variable before being accessed. While any single instance variable may not need to be hidden, it is good practice to always make them private to allow future protection.

13.8   Why did the C# language designers choose to introduce properties instead of getters and setters like in Java?

In Java, getters and setters are a convention that is followed by programmers. The designers of C# chose to make it a language feature, as it made it clearer what the correct way to deal with properties is. This also allowed the compiler to use databinding with properties. Finally, it allowed for automatic properties that make the code a lot faster to write for many of the properties.

Chapter 14

14.1   Indicate whether the following concepts represent inheritance. If so, which is the super class and which is the subclass?
- Oven and Kitchen: No inheritance—A kitchen may contain an oven, but an oven is not a kitchen.
- Rectangle and Square: Inheritance—Rectangle is the super class. A square is a special form of a rectangle where all four sides have the same length.
- Vehicle and Truck: Inheritance—Vehicle is super class. Vehicle is a general description of a truck.
- Batman and Superhero: No inheritance—Batman is a specific instantiation of a superhero, so Batman might be an object of the Superhero class. There wouldn't likely be multiple different kinds of Batman objects.
- Student and Person: Inheritance—Person is the super class. Student is a Person. We can imagine other kinds of persons like Employee, Performer, Client and so on.
- Student and Employee: No inheritance—These are both specializations of a broader person concept. We can't say that all students are employees or that all employees are students.
- Order and Customer: It is common for business students to confuse these two concepts and try to set them up as inheritance, but this is an associative relationship where each customer would be associated with multiple order objects.
- Ferrari and Engine: No inheritance— A Ferrari has an engine, but it isn't an engine.

14.2   Identify at least two subclasses for each of the following classes:
- Car
    - Sedan, SportsCar, Compact
- Student
    - HighSchoolStudent, BusinessStudent, UpperClassStudent
- Order
    - (Trading): MarketOrder, LimitOrder, StopLossOrder
    - (Army): OPORD, Movement, ServiceSupport, Warning, Fragmentarty
    - (Oracle): InternalSales, Regular, Return
    - (SAP): vCash, Rush, SchedulingAgreement, Third-Party
- Account
    - CheckingAccount, SavingsAccount, InvestmentAccount
- Book
    - Hardcover, Ebook, Paperback, Novel, NonFiction
- Person
    - Student, Employee, Client, Patient, Instructor, Customer

14.3   If class A is a super class of class B, which is a super class of class C, which of the following are allowed?
- A q = new A();
- A x = new B();
- B y = new A();
- C z = new A();

The first and second are allowed. You can assign an object of a subclass to a variable declared as a super class, but not vice versa.

14.4   Given the classes (A, B, C) in the previous question, explain why the following is allowed:
```
List<A> letters = new List<A>();
letters.Add(new A());
letters.Add(new B());
letters.Add(new C());
```
The List here is declared to hold objects of Type A. Such a list can hold objects of the declared type and any subtypes, so we can add objects of type B and C since they are sub-types of A.

14.5   Given the classes (A, B, C) in the previous question, if class B implements a method P that is overridden in C, which version (in B or C) will be called in each of the following code snippets?
```
C c = new C();
c.P();

A a = new B();
a.P();

B b = new C();
b.P();
```

For this question, draw out a class diagram to see that the method lookup proceeds up the inheritance hierarchy.
- The version in C
- The version in B
- The version in C

14.6 Given the classes (A, B, C) in the previous question, if class A implements a method Q that is overridden in C (but not in B), which version (in A or C) will be called in each of the following code snippets?
```
C c = new C();
c.Q();

A a = new B();
a.Q();

B b = new C();
b.Q();
```

- The version in C
- The version in A
- The version in C

14.7 Given class Video, which is a subclass of class Media, which of the following are legal?
- `Video v1 = new Video();`
- `Video v2 = new Media();`
- `Media m1 = new Media();`
- `Media m2 = new Video();`

The first, third and fourth are all legal. The second statement is not, as you cannot assign a super class object to a subclass variable.

14.8 For each of the declarations in the previous question, indicate the static and dynamic type of the variable.
- `Video v1 = new Video ();`
  - Static: Video, Dynamic: Video
- `Video v2 = new Media();`
  - Not allowed.
- `Media m1 = new Media();`
  - Static: Media, Dynamic: Media
- `Media m2 = new Video();`
  - Static: Media, Dynamic: Video

14.9 Which of the following are NOT valid in C# if Drivable is an interface and Van is a class that implements it?
- `Drivable d;`

- `Drivable d = new Drivable();`
- `Drivable d = new Van();`
- `Drivable d = null;`

Only the second statement is illegal. You cannot create an object from an interface.

14.10 Describe what problems could occur if C# allowed multiple inheritance.

Multiple inheritance can lead to a problem where a method is defined in more than one parent but not overridden in a subclass. Which version of the method is invoked is then not clear. This is also known as "The Diamond Problem," where a class A defines a method that is overridden in subclasses B and C. But the method is not overridden in class D, which is a subclass of both B and C (the inheritance diagram takes on a diamond shape—hence the name of the problem). When trying to call the method on an object of D, it is unclear which version of the method should actually be called (the one in B or C?). While C# (and Java) doesn't allow multiple inheritance due to the problems seen in C++, some languages like Eiffel have done a better job of implementing Multiple Inheritance.

14.11 Why is it enough for the Sort method to know how to rank order two objects?

While we don't know how the Sort method works in detail, as long as it know how two objects are ranked, it can repeatedly rank objects in pairs until the entire list is sorted. A simple implementation of Sort might be to look at the first two elements, then swap them if they are out of order, then move on to the second and third elements and repeat the process. When the sort reaches the end of the list, it would go back to the beginning, unless the list was already sorted. This would go on until the list is sorted.

14.12 What are the benefits and disadvantages of using the Sort method compared to writing your own sort algorithm?
- Benefits: Much less code to write. Lower risk of making a mistake. Documentation for Sort method would list the speed that it sorts at.
- Disadvantages: Must trust that the implementation is good. Can't be customized to specific needs.

Chapter 15

15.1 Domain name: createspace.com
Host name: www

15.2 A static web page remains the same each time it is displayed, unless it is manually altered. Dynamic web pages are created on-the-fly using scripts when the page is requested. A dynamic page may look different each time it is displayed, depending on various factors like who the user is, the date/time, changes in the data read by the script and the parameter values passed through the script.

15.3 An Http request is a request sent by a browser to the server for a web page or related files by specifying the URL of the page.

15.4 Server side script is run on the server; so, it puts burden on the server. Client side script is run on the browser. Server side script cannot be viewed by the user in the browser; but client side script can be viewed.

Appendix C: Answers to Review Questions 581

15.5 In stateless mode, as soon as a client receives a response for its request, the connection between the client and server is terminated and the server keeps no information about the client. If the same client makes another request, the server has no way to know whether it is the same client or another client. So, stateless mode cannot meet the requirements of today's typical dynamic applications.

15.6 Web Forms development approach uses a design surface and drag-and-drop controls (rather than writing HTML code) to rapidly build rich user interfaces.

The MVC approach divides an application into three different components:
Model: the business entity, including the data, represented in the application
View: data as presented in the user interface
Controller: the component that maintains the model and the view

The MVC approach makes it easier to do *separation of concerns* (i.e., separation of different aspects of software functionality like business logic, user interface and data access) to make software easy to maintain. So, it is particularly suitable for large projects. Further, this approach gives full control over generated HTML markup.

15.7 In a Web Application Project, all class files, including code-behind files, are pre-compiled and deployed as a single assembly on the server. In a Web Site Project, typically the source code is stored on the server and compiled dynamically at run time, which takes extra time and makes the code less secure. However, Web Site Projects make it easier to update and test parts of the website.

15.8 Code-behind is code stored in a file separate from the HTML code. In-line code is embedded in HTML.

15.9 Start tag, attribute, content, end tag.

15.10 The browser doesn't understand ASP.NET code. It needs to be translated (rendered) by the server to produce HTML markup that the browser understands. The standard HTML markup is understood by the browser.

15.11 The head element contains information about the page that is not displayed in the page for the user. Contents of the head element include title element and meta elements.

15.12 The **runat** attribute specifies that the form is processed on the server so you can refer to the form and the controls on the page using server-side code.

15.13 If the AutoPostBack property of a TextBox is set to True, the web page will be posted back whenever the TextBox loses focus.

15.14 The runat="server" specifies that the ASP.NET code is to be processed on the server. The HTML markup is the code sent to the browser. Since the browser understands the HTML markup, there is no need to process it on the server.

15.15 Add the runat="server" attribute to the Input element.

15.16 The HTML controls are client side controls, which means they are not accessible by programs running on the server.
The data you enter into an HTML control is lost when the page is refreshed as a result of submitting a page.
Because HTML controls are part of the HTML language, there is no need to render them to HTML, as it is necessary for Web server controls.

15.17 Web server controls have better functionality than HTML server controls:
- Ability to detect the capabilities of the browser and render HTML markup accordingly. Thus, a control may look different in a mobile browser than in a laptop.
- A richer object model
- Ability to define custom layout for the controls using templates
- Ability to define a consistent look for controls using themes.

15.18 Client-side validation takes place in an input control when you make a change in the data and move focus away from the control. Server-side validation take place in every control that has a validator associated with it, whenever you click a button whose *CauseValidation* property is set to *True*.

15.19 An advantage of client-side validation is that it helps give immediate feedback to the user without a postback of the page that causes a roundtrip to the server. Client-side validation is subject to incompatibilities of the browser with the validation script. In general, client-side validation won't identify missing data.

15.20 Set the CauseValidation property of the button to False.

Chapter 16

16.1 DataTextField specifies the name of the field that is displayed in the DropDownList. DataValueField specifes the name of the field whose value is returned by the SelectedValue property.

16.2 Each row in a DataView is of type DataRowView.

16.3 A value of True for IsPostBack property indicates that the page is not loaded for the first time.

16.4 The data stored in class level variables is lost when the page is posted back.

16.5 View state stores information that is accessed from the same page by storing the information within the page itself in hidden fields.

16.6 The statement adds an item named BasePrice to the view state collection and assigns it the value of basePrice.

16.7 The Transfer method would display the source page's URL in the target page.

16.8 Transfer method is limited to transferring to target pages within the same website, whereas the Redirect method can display pages within the same website or other websites. The Transfer method also allows the target page to reference the source page using the **PreviousPage** property exposed by the Page class, which is not available with Redirect method.

16.9 Cross-page posting makes data available only from the previous page. So, the third page doesn't have direct access to the data on the first page.

16.10 Query string transfers data only between two pages. So, data from the first page needs to be transferred to the second page and then to the third page. Further Transfer method displays the URL of the source page on the target page.

16.11 Session state stores information that needs to be shared between multiple pages during a specific session of a user. Session state stores the data on the server, and the data is available till the user closes the browser or the session is timed out.

Appendix D

HTML Elements

Element	Description
`<!-- -->`	comment
`<!DOCTYPE>`	document type—the HTML version
`<a>`	hyperlink
`<audio>`	audio
`<b>`	bold
`<body>`	holds contents that are displayed on the page
` `	line break
`<button>`	clickable button
`<div>`	a non-semantic section that groups elements
`<em>`	emphasized text
`<form>`	HTML form
`<h1>` to `<h6>`	headings
`<head>`	contains information about the document
`<html>`	root element that contains all other elements
`<i>`	italics
`<img>`	image
`<input>`	control that accepts user input
`<label>`	label for an `<input>` element
`<li>`	list item
`<link>`	link to external resources like style sheets
`<menu>`	list of commands
`<menuitem>`	menu item
`<meta>`	metadata about the document like key words

`<nav>`	contains a set of navigation links
`<object>`	embedded objects like audio, PDF and java applets
`<ol>`	ordered list
`<option>`	option in a dropdown list
`<output>`	result of a calculation
`<p>`	paragraph
`<param>`	parameter for an `<object>`
`<pre>`	preformatted text
`<script>`	script
`<section>`	a group of related content.
`<select>`	dropdown list
`<small>`	makes the text one size smaller, like large to medium
`<source>`	media resources for media elements like video
`<span>`	in-line container for text or multiple elements
`<strong>`	important text
`<style>`	specifies CSS style
`<table>`	display data in tabular form
`<td>`	cell in a table
`<th>`	header cell in a table
`<time>`	date/time
`<title>`	title for the document
`<tr>`	row in a table
`<var>`	variable
`<video>`	video or movie

Index

@Page directive, 504
Abstract
   class, 463, 472, 473
   key word, 462
   method, 462
Access modifier, 142, 420, 436, 460, 486, 487
Adding C# Code to a Web Form, 509
Aggregation, 451
Argument, 145
   default values, 150
   passing by reference, 162
   passing by value, 149
Arrays, 205, 234
   2-Dimensional arrays, 263
   Array initializer, 209
   Array names, 217
   Array.Copy method, 218
   Array.IndexOf method, 221
   Jagged arrays, 269
   Passing arrays to methods, 259
ASP.NET, 495
ASP.NET Code, 513
aspx code, 500
Attributes of an element, 503
Autohide, 12
Automatic Property, 425
AutoPostBack property, 510
auto-styles, 510
base
   Class, 451
   constructor call, 453, 456
   key word, 453
   method call, 459
Behavior, 415, 417
Binary Files, 559
Binding a DropDownList, 529
Binding Source, 336, 362
Body element, 505, 511
Boolean, 26
Boolean expression, 83
Bootstrap, 506
break statement, 137
Button, 16
Camel Case, 14, 29

Casting, 56
CauseValidation property, 110
Character data, 27
CheckBoxes, 178
child form, 435, 440, 443
Class, 7, 22, 415, 417
   Partial class, 22
Class Library, 7
Clearing a ListBox, 46
Closing a Web Form, 510
Collections, 276
   Generic, 458
   sorting, 480
ComboBox, 186, 246, 354
   ComboBox styles, 189
Common Language Runtime (CLR), 6
Compare method, 482
CompareTo method, 480
Compiler, 6
Concatenating Strings, 66
connectionString, 367
Console Application, 120
   Console.ReadLine, 121
   Console.WriteLine, 121
constant, 43, 146
constructor, 420, 422
continue statement, 138
Control, 3, 16
Cross-page posting, 547
data bound control, 366
Data Flow Diagrams, 3
Data Source, 337, 528
Data tables, 336
Data Validation, 101
Database Systems, 335
   Creating a SQL Server database, 562
   Data tables, 336
   Dataset, 336
   Parameter query, 363
   Searching a key field, 351
DataBind() method, 533
Data-bound controls, 343
DataGridView, 340, 346
DataReader, 394

Index

DataRow, 370
DataRowView, 370
DateTime
    DateTime operations, 70
    Methods of DateTime, 75
    Today vs. Now, 73
DateTimePicker, 71
Debugging, 5
decimal type, 55
Decision Structure, 82
Default values of data types, 30
Dictionary, 277, 285
    Passing Dictionary as a parameter, 320
Directory, 252
Docking, 12
double type, 56
do-while Loop, 130
DropDownLists, 527
Dynamic dispatch, 467
Dynamic type, 467
Encapsulation, 417, 486
EndEdit method, 388
Enumerations, 168
ErrorProvider Control, 108
ExcecuteScalar method, 396
Exception handler, 49
ExecuteNonQuery method, 399
ExecuteReader method, 397
Fields, 47
Files, 226
    Append vs. Overwrite, 252
    Binary files, 226
    Directory class, 252
    SaveFileDialog, 255
    Sequential access files, 226
Filtering rows, 378
Floating point, 28, 55
Flowchart, 4
for loop, 131
for Loop, 213, 220
for vs. foreach, 216
foreach Loop, 138, 215, 283
Form
    Accessing from another form, 313
    Creating multiple instances, 316
    FormClosing event, 251
    Hiding vs. Closing, 316
    Load event, 250
    Login form, 325
    Sharing methods between forms, 318
    Show vs. ShowDialog, 317

getters, 424
Graphical User Interface, 3
has-a **relationship, 451**
Head element, 504
High-level languages, 6
host name, 492
HTML, 502
    controls, 514
    document, 504
    elements, 502
    server-controls, 515
    tags, 502
http, 492
HTTP
    request, 493
    response, 493
IComparable, 480
IComparer, 482
Identity, 416
if else if Statement, 92
if else Statement, 83
Implicit cast, 61
Infinite Loop, 128
Information Hiding, 417
Inheritance, 421, 450
Initializing a variable, 30
Input, 3
Input element, 515
instantiation, 416
int type, 27, 53
Integer, 26
Integer literals, 58
Integrated Development Environment (IDE), 6
Interface, 472, 474
Intermediate language, 6
internal (access modifier), 487
Interpreter, 6
is-a **relationship, 451**
IsPostBack Property, 532
Java language, 424
KeyValuePair collection, 290
Label, 16
    Autosize property, 16
Lifetime of a variable, 30
List<T> Collection, 278
ListBox, 196
    Items collection, 45
ListView, 298, 302
ListViewItem, 299
Literals, 56

Local variables, 46, 58
Locals Window, 58
Logic error, 5
Logical operators, 88
Low-level, 6
Machine language, 6
Machine Language, 6
Menus, 328
   Context Menu, 330
MessageBox, 49
Method, 22, 140, 259
   access modifier, 142
   call on object, 434
   cohesive methods, 151
   passing values to methods, 144
   public, 142
   static methods, 320
   void method, 142
Methods that Return a Value, 152
Microsoft Intermediate Language (IL) Assembly, 6
Multipage Websites, 543
Multiple inheritance, 473
Multiplication with large integers, 62
MVC, 498
Namespace, 21
Nested if Statements, 94
new (key word), 423
Non-integer (real) types, 27
Object, 415, 417
Object (class), 459, 468
Object composition, 451
Object diagram, 434
Object-oriented programming, 414
Output, 3
Overloading, 460
Overriding, 459
Parameter, 144, 320
   out parameter, 167
parameter query, 363
parent form, 325, 327, 435, 443
Parse, 33
Polymorphism, 467
PositionChanged event, 372
PostBack, 535
   Preserving data, 535
Precedence of Operators, 35
Prefixes for controls, 16, 416
Pre-test loops, 129
PreviousPage property, 548
private, 415, 423, 460, 486

Process, 3
Project, 8
Properties window, 11
Property, 418, 423, 424
protected, 436, 460, 487
Pseudo code, 4
public, 415, 436, 460, 486
RadioButtonList, 526
RadioButtons, 180
Real number, 26
Redirect method, 546
Reference types, 64
Relational operator, 83
Replace (string, string) method, 67
runat="server" attribute, 505
Running a web form, 510
Runtime error, 5
SaveFileDialog control, 255
Scope of Variables, 46
scripts, 494
   client-side, 494
   server-side, 494
Scroll Bars, 175
SelectCommand, 378
SelectedIndexChanged event, 192
Sequential Access Files, 226
Session State, 543
setters, 424
Solution, 8
Solution Explorer, 11
SortedDictionary, 277
SortedList, 277
SQL Server, 562
SqlCommand, 394
SqlCommandBuilder, 384
Startup form, 15
state, 416, 434
stateless protocol, 495
Static type, 467
StreamReader class, 227
StreamWriter class, 248
string Type, 28
   Concatenating strings, 66
   String operations, 65
   String.Format method, 44
   Substring method, 66
   ToLower method, 66
   Trim method, 65
String vs. Character, 68
Structure chart, 144
Subclass, 451, 458, 487

Super class, 451, 458, 487
switch Statement, 98
Syntax error, 4
Tab Pages, 308
Table adapter, 336, 362
TextBox, 16
this (key word), 421, 453
ToolStrip, 332
ToolTip, 13
Top-down design, 140, 154
ToString method, 421, 459, 468
Transfer method, 547
Trim() method, 65
Try-catch, 49
TryParse method, 102
typed dataset, 365
UML diagram, 3, 415, 434, 451, 474
unbound control, 366
Unobtrusive Validation, 519

untyped datase, 366
Using directive, 21
Validating Input, 182
   validating event, 106
Validation Controls, 517
Value types, 64
Value Types vs. Reference Types, 206
View state, 538
Web Application Project, 496
web form, 499
Web Forms template, 497
web page
   dynamic, 492
   static, 492
web server controls, 508
Web Site Project, 496
while Loop, 119
Widening conversion, 61
www, 492